Communications
in Computer and Information Science   2627

Series Editors

Gang Li ⓘ, *School of Information Technology, Deakin University, Burwood, VIC, Australia*

Joaquim Filipe ⓘ, *Polytechnic Institute of Setúbal, Setúbal, Portugal*

Zhiwei Xu, *Chinese Academy of Sciences, Beijing, China*

Rationale

The CCIS series is devoted to the publication of proceedings of computer science conferences. Its aim is to efficiently disseminate original research results in informatics in printed and electronic form. While the focus is on publication of peer-reviewed full papers presenting mature work, inclusion of reviewed short papers reporting on work in progress is welcome, too. Besides globally relevant meetings with internationally representative program committees guaranteeing a strict peer-reviewing and paper selection process, conferences run by societies or of high regional or national relevance are also considered for publication.

Topics

The topical scope of CCIS spans the entire spectrum of informatics ranging from foundational topics in the theory of computing to information and communications science and technology and a broad variety of interdisciplinary application fields.

Information for Volume Editors and Authors

Publication in CCIS is free of charge. No royalties are paid, however, we offer registered conference participants temporary free access to the online version of the conference proceedings on SpringerLink (http://link.springer.com) by means of an http referrer from the conference website and/or a number of complimentary printed copies, as specified in the official acceptance email of the event.

CCIS proceedings can be published in time for distribution at conferences or as post-proceedings, and delivered in the form of printed books and/or electronically as USBs and/or e-content licenses for accessing proceedings at SpringerLink. Furthermore, CCIS proceedings are included in the CCIS electronic book series hosted in the SpringerLink digital library at http://link.springer.com/bookseries/7899. Conferences publishing in CCIS are allowed to use our online conference service (Meteor) for managing the whole proceedings lifecycle (from submission and reviewing to preparing for publication) free of charge.

Publication process

The language of publication is exclusively English. Authors publishing in CCIS have to sign the Springer CCIS copyright transfer form, however, they are free to use their material published in CCIS for substantially changed, more elaborate subsequent publications elsewhere. For the preparation of the camera-ready papers/files, authors have to strictly adhere to the Springer CCIS Authors' Instructions and are strongly encouraged to use the CCIS LaTeX style files or templates.

Abstracting/Indexing

CCIS is abstracted/indexed in DBLP, Google Scholar, EI-Compendex, Mathematical Reviews, SCImago, Scopus. CCIS volumes are also submitted for the inclusion in ISI Proceedings.

How to start

To start the evaluation of your proposal for inclusion in the CCIS series, please send an e-mail to ccis@springer.com

Allel Hadjali · Emanuele Maiorana ·
Oleg Gusikhin · Carlo Sansone
Editors

Deep Learning Theory and Applications

6th International Conference, DeLTA 2025
Bilbao, Spain, June 12–13, 2025
Proceedings

Editors
Allel Hadjali [iD]
LIAS, ISAE-ENSMA
Poitiers, France

Emanuele Maiorana
Roma Tre University
Rome, Italy

Oleg Gusikhin
Ford Motor Company
Dearborn, MI, USA

Carlo Sansone
University of Naples Federico II
Naples, Italy

ISSN 1865-0929 ISSN 1865-0937 (electronic)
Communications in Computer and Information Science
ISBN 978-3-032-04338-2 ISBN 978-3-032-04339-9 (eBook)
https://doi.org/10.1007/978-3-032-04339-9

© The Editor(s) (if applicable) and The Author(s), under exclusive license
to Springer Nature Switzerland AG 2025

This work is subject to copyright. All rights are solely and exclusively licensed by the Publisher, whether the whole or part of the material is concerned, specifically the rights of translation, reprinting, reuse of illustrations, recitation, broadcasting, reproduction on microfilms or in any other physical way, and transmission or information storage and retrieval, electronic adaptation, computer software, or by similar or dissimilar methodology now known or hereafter developed.
The use of general descriptive names, registered names, trademarks, service marks, etc. in this publication does not imply, even in the absence of a specific statement, that such names are exempt from the relevant protective laws and regulations and therefore free for general use.
The publisher, the authors and the editors are safe to assume that the advice and information in this book are believed to be true and accurate at the date of publication. Neither the publisher nor the authors or the editors give a warranty, expressed or implied, with respect to the material contained herein or for any errors or omissions that may have been made. The publisher remains neutral with regard to jurisdictional claims in published maps and institutional affiliations.

This Springer imprint is published by the registered company Springer Nature Switzerland AG
The registered company address is: Gewerbestrasse 11, 6330 Cham, Switzerland

If disposing of this product, please recycle the paper.

Preface

This book contains the proceedings of the 6th International Conference on Deep Learning Theory and Applications. This year, DeLTA was held in collaboration with the University of Deusto, which hosted this event in Bilbao, Spain, on June 12–13, 2025. It was sponsored by the Institute for Systems and Technologies of Information, Control and Communication (INSTICC) and endorsed by the International Association for Pattern Recognition. DeLTA 2025 was also organized in cooperation with the ACM Special Interest Group on Artificial Intelligence, the Association for the Advancement of Artificial Intelligence, the Spanish Association of Artificial Intelligence and the International Neural Network Society.

Deep Learning and Big Data Analytics are currently two major areas of data science. As organizations increasingly collect massive volumes of data, the ability to extract meaningful insights for business decision-making has become essential, shaping both current practices and future technological developments. One of the key advantages of Deep Learning lies in its capacity to process large-scale, often unstructured data and to extract complex, high-level abstractions. This makes it an especially powerful tool for Big Data analytics, particularly in scenarios where the data is largely unlabeled. Machine learning and artificial intelligence are now integral to a wide range of real-world applications, including computer vision, information retrieval and summarization from structured and unstructured multimodal data sources, natural language understanding and translation, and many other domains. Deep Learning approaches, driven by the availability of Big Data, are surpassing traditional supervised and unsupervised methods by automatically learning relevant features and data representations. These models eliminate the need for extensive domain expertise or manual feature engineering, leading to more efficient and scalable solutions. Such capabilities are increasingly important in Internet of Things (IoT) applications, where data complexity and volume continue to grow.

DeLTA 2025 received 42 paper submissions from 23 countries, of which nine (21%) were accepted and published as full papers. A double-blind paper review was performed for each submission by at least 2, but usually 3 or more, members of the International Program Committee, which was composed of established researchers and domain experts.

The high quality of the DeLTA 2025 program was enhanced by the keynote lecture delivered by distinguished speakers who are renowned experts in their fields: Björn Schuller (University of Augsburg, Germany and Imperial College London, UK) and Wojciech Samek (TU Berlin, Germany).

All presented papers will be submitted for indexation by DBLP, Google Scholar, EI-Compendex, INSPEC, Japanese Science and Technology Agency (JST), Norwegian Register for Scientific Journals and Series, Mathematical Reviews, SCImago, Scopus, zbMATH and Web of Science/Conference Proceedings Citation Index.

Several awards, based on the combined marks of paper reviewing, as assessed by the Program Committee, and the quality of the presentation, as assessed by session chairs at the conference venue, were conferred at the conference's closing session as recognition for the best contributions.

The program for this conference required the dedicated effort of many people. Firstly, we must thank the authors, whose research efforts are reported here. Next, we thank the members of the Program Committee and the auxiliary reviewers for their diligent and professional reviewing. We would also like to deeply thank the invited speakers for their invaluable contribution and for taking the time to prepare their talks. Finally, a word of appreciation for the hard work of the INSTICC team; organizing a conference of this level is a task that can only be achieved by the collaborative effort of a dedicated and highly competent team.

We hope you all had an exciting and inspiring conference. We hope to have contributed to the development of our research community, and we look forward to having additional research results presented at the next edition of DeLTA, details of which are available at https://delta.scitevents.org.

June 2025

Allel Hadjali
Emanuele Maiorana
Oleg Gusikhin
Carlo Sansone

Organization

Conference Co-chairs

Oleg Gusikhin	Ford Motor Company, USA
Carlo Sansone	University of Naples Federico II, Italy

Program Co-chairs

Allel Hadjali	LIAS/ENSMA, University of Poitiers, France
Emanuele Maiorana	Roma Tre University, Italy

Program Committee

Marco Aceves-Fernández	Autonomous University of Queretaro, Mexico
Anish Acharya	University of Texas at Austin, USA
Abdel Belaid	LORIA, University of Lorraine, France
Guido Borghi	Università di Bologna, Italy
Geraldo Braz Júnior	Universidade Federal do Maranhão, Brazil
Marco Buzzelli	University of Milano-Bicocca, Italy
Dumitru-Clementin Cercel	National University of Science and Technology Politehnica Bucharest, Romania
Claudio Cusano	University of Pavia, Italy
Ashraf Elnagar	University of Sharjah, UAE
Enrique Fernández-Blanco	University of A Coruña, Spain
Raffaella Folgieri	Università degli Studi di Milano, Italy
Francesco Gargiulo	ICAR-CNR, Italy
Maki Habib	American University in Cairo, Egypt
Tarmizi Ahmad Izzuddin	Universiti Teknikal Malaysia Melaka, Malaysia
Ajay Jain	Independent Researcher, India
Mohit Jain	University of Delhi, India
Ido Kanter	Bar-Ilan University, Israel
Chutisant Kerdvibulvech	NIDA, Thailand
Constantine Kotropoulos	Aristotle University of Thessaloniki, Greece
Christos Kyrkou	University of Cyprus, Cyprus
Chih-Chin Lai	National University Kaohsiung, Taiwan
Chang-Hsing Lee	Ming Chi University of Technology, Taiwan

Wookey Lee	Inha University, South Korea
Fuhai Li	Washington University in St. Louis, USA
Tony Lindeberg	KTH Royal Institute of Technology, Sweden
Fadi Al Machot	University of Klagenfurt, Austria
Marc Masana	Graz University of Technology, Austria
Phayung Meesad	King Mongkut's University of Technology North Bangkok, Thailand
Anke Meyer-Baese	Florida State University, USA
George Papakostas	Democritus University of Thrace, Greece
Yash Patel	Capella University, USA
Gabriele Pieri	ISTI-CNR, Italy
Mircea-Bogdan Radac	Politehnica University of Timisoara, Romania
Sivaramakrishnan Rajaraman	National Library of Medicine, USA
Hatem Rashwan	Rovira i Virgili University, Spain
Tsang Ren	Universidade Federal de Pernambuco, Brazil
Erik Rodner	Friedrich Schiller University Jena, Germany
Samira Sadaoui	University of Regina, Canada
Klemens Schnattinger	Baden-Württemberg Cooperative State University (DHBW), Germany
Monika Sharma	Independent Researcher, India
Jitae Shin	Sungkyunkwan University, South Korea
Sunghwan Sohn	Mayo Clinic, USA
Minghe Sun	University of Texas San Antonio, USA
Ryszard Tadeusiewicz	AGH University of Science and Technology, Poland
Filippo Vella	National Research Council of Italy, Italy
Hai Wang	Saint Mary's University, Canada
Shengrui Wang	University of Sherbrooke, Canada
Theodore Willke	Intel Corporation, USA
Jianhua Xuan	Virginia Tech, USA

Additional Reviewer

Xingyi Liu	Mayo Clinic, USA

Invited Speakers

Wojciech Samek	TU Berlin, Germany
Björn Schuller	University of Augsburg, Germany and Imperial College London, UK

Invited Speakers

Inspecting AI Like Engineers: From Explanation to Validation with SemanticLens

Wojciech Samek

Chair for Machine Learning and Communications, TU Berlin, Berlin, Germany

Abstract. Human-designed systems are constructed step by step, with each component serving a clear and well-defined purpose. For instance, the functions of an airplane's wings and wheels are explicitly understood and independently verifiable. In contrast, modern AI systems are developed holistically through optimization, leaving their internal processes opaque and making verification and trust more difficult. This talk explores how explanation methods can uncover the inner workings of AI, revealing what knowledge models encode, how they use it to make predictions, and where this knowledge originates in the training data. It presents SemanticLens, a novel approach that maps hidden neural network knowledge into the semantically rich space of foundation models like CLIP. This mapping enables effective model debugging, comparison, validation, and alignment with reasoning expectations. The talk concludes by demonstrating how SemanticLens can help in identifying flaws in medical AI models, enhancing robustness and safety, and ultimately bridging the "trust gap" between AI systems and traditional engineering.

Brief Biography

Wojciech Samek is a Professor in the EECS Department at the Technical University of Berlin and the Head of the AI Department at the Fraunhofer Heinrich Hertz Institute (HHI) in Berlin, Germany. He earned an M.Sc. from Humboldt University of Berlin in 2010 and a Ph.D. (with honors) from the Technical University of Berlin in 2014. Following his doctorate, he founded the "Machine Learning" Group at Fraunhofer HHI, which became an independent department in 2021. He is a Fellow at BIFOLD – the Berlin Institute for the Foundation of Learning and Data and the ELLIS Unit Berlin. He also serves as a member of Germany's Platform for AI and sits on the boards of AGH University's AI Center, the Helmholtz Einstein School in Data Science (HEIBRiDS), and the DAAD Konrad Zuse School ELIZA. Dr. Samek's research in explainable AI (XAI) spans method development, theory, and applications, with pioneering contributions such as Layer-wise Relevance Propagation (LRP), advancements in concept-level explainability, evaluation of explanations, and XAI-driven model and data improvement. He has served as a senior editor for IEEE TNNLS, held associate editor roles for various other journals, and acted as an area chair at NeurIPS, ICML, and NAACL. He has received several best paper awards, including from Pattern Recognition (2020), Digital Signal Processing (2022), and the IEEE Signal Processing Society (2025). Overall, he has co-authored more than 250 peer-reviewed journal and conference papers, with several recognized as ESI Hot Papers (top 0.1%) or Highly Cited Papers (top 1%).

Deep Learning Deep Feelings: Large Models, Larger Emotions

Bjoern Schuller

University of Augsburg, Augsburg, Germany

Abstract. As AI systems permeate every corner of modern life, a crucial frontier emerges enabling machines not just to learn, but to feel— at least enough to understand our states and communicate with us in empathic manners. This keynote takes you on a journey from the current "Affective Intelligence" largely empowered by deep learning to the rising wave of large model exploitation in "Affective Intelligence 2.0". Blending advances in affective computing and the rapid progress in multimodal foundation models, emotionally aware AI is about to reshape human-machine interaction, digital health, multimedia, and will be a corner stone of AGI to come. Beyond algorithms, this talk explores the power—and responsibility—of creating machines that respond with empathy, adapt to individual emotional states, and navigate the complexities of real-world human affective experience. It will further highlight the potential of affective computing in "Friendly AI" and discuss potential of emotion as inspiration in deep learning. With a critical eye on ethical design and societal impact, the talk invites you to imagine an AI future that goes beyond data—into emotion, connection, and care.

Brief Biography

Björn W. Schuller is a distinguished academic and researcher with extensive expertise in Machine Intelligence and Signal Processing. He earned his diploma, doctoral degree, habilitation, and Adjunct Teaching Professor title in EE/IT from TUM in Munich, where he currently holds a Full Professorship as Chair of Health Informatics. Additionally, he is a Full Professor of Artificial Intelligence and Head of GLAM at Imperial College London. Schuller co-founded audEERING, an Audio Intelligence company, and has numerous affiliations, including roles at the Munich Data Science Institute and the Munich Center for Machine Learning. He has held multiple prestigious professorships globally and served as an independent research leader at the Alan Turing Institute. He is a Fellow of several prominent societies, including the ACM, IEEE, BCS, ELLIS, ISCA, and AAAC. With over 1,500 publications, more than 70,000 citations, and an h-index exceeding 110, he is highly influential in the field of Computer Science. He has held editorial positions, including Field Chief Editor of Frontiers in Digital Health, Editor in Chief of AI Open, and the IEEE Transactions on Affective Computing. Schuller has received over 50 awards, including being named one of 40 extraordinary scientists under 40 by the WEF in 2015. Currently, he is an ACM Distinguished Speaker and an IEEE Signal Processing Society Distinguished Lecturer. His work has been widely recognized in the media, with over 300 public press appearances and contributions to various international outlets including Newsweek, Scientific American, and Times.

Contents

End-to-End ASR Model with Iterative Attention Mechanism Enhanced
RNN Model for Phoneme Recognition 1
   Ke Fang and Yancong Deng

Diagnostic Trouble Codes Prediction with DTC-GOAT and Ensembles 14
   Abdul Basit Hafeez, Eduardo Alonso, and Atif Riaz

A Fast Fourier Transform-Aided Diffusion-Based U-Net Architecture
for Microscopic Medical Image Segmentation 28
   Saptarshi Pani, Gouranga Maity, Dmitrii Kaplun,
   Alexander Voznesensky, and Ram Sarkar

Non-cooperative Game Theory-Aided Learning of CNN Models for Skin
Lesion Classification 51
   Diptarka Mandal, Sujan Sarkar, Siddhant Majumder, Dmitrii Kaplun,
   Daria Sidorina, and Ram Sarkar

LoRA-Based Summarization of Data Privacy Clauses in Terms
and Conditions Documents Aligned with India's 2023 Digital Personal
Data Protection Act 70
   Preet Kanwal, Amish Gupta, J. Sai Mohananshu,
   and Prasad B. Honnavalli

Comparison of AI Speech-to-Text Systems and Their Application
in Artillery Command and Fire Control Systems 82
   Martin Blaha, Jaroslav Varecha, Jan Drábek, and Jiří Novák

Rhythm Fusion: Synchronizing Audio and Motion Features
for Music-Driven Dance Generation 96
   Nuha Aldausari, Gelareh Mohammadi, and David Cooper

Leveraging Synthetic Data for Deep-Learning-Based Road Crack
Segmentation from UAV Imagery 112
   Andriani Panagi and Christos Kyrkou

Trojan Vulnerabilities in Host-Based Intrusion Detection Systems 132
   Mark Cheung, Sridhar Venkatesan, and Rauf Izmailov

Identification of Key Feature Interactions via PDP Decomposition 151
   Selim Eren Eryilmaz and Ron Triepels

Variational Mode Decomposition (VMD) Parameter Selection Using
Sine-Cosine Algorithm (SCA): Application on Vibration Signals
for Rotating Machinery Monitoring 163
   *Ikram Bagri, Achraf Touil, Ahmed Mousrij, Aziz Hraiba,
and Karim Tahiry*

Forecasting Ethereum Prices with Machine Learning, Deep Learning,
and Explainable Artificial Intelligence Using Multi-source Market Articles
and Hybrid Sentiment Analysis 184
   *Naresh Kumar Satish, Mathieu Mercadier, Cristina Hava Muntean,
and Anderson Augusto Simiscuka*

Application of Neural Networks to Ultrasonic Data for Discrimination
of Fat Types in Muscle Tissue Models 204
   *Jegors Lukjanovs, Aleksandrs Sisojevs, Alexey Tatarinov,
and Tamara Laimiņa*

SwiNight: Class Imbalanced Night-Time Accident Detection with Swin
Transformer 217
   *Shrusti Porwal, Preety Singh, Anukriti Bansal, Saumilya Gupta,
Kartikay Goel, and Palakurthy Guneeth*

Enhancing Off-Policy Method SAC with KAN for Continuous
Reinforcement Learning 229
   Ali Bayeh, Malek Mouhoub, and Samira Sadaoui

Context-Aware Imputation for Parkinson's Disease Trajectories:
Systematic Benchmark of Cross-Sectional, Temporal, and Generative
Approaches 240
   *Moad Hani, Nacim Betrouni, Fatima Zahra Ouardirhi, Saïd Mahmoudi,
and Mohammed Benjelloun*

Investigating Zero-Shot Diagnostic Pathology in Vision-Language Models
with Efficient Prompt Design 263
   *Vasudev Sharma, Ahmed Alagha, Abdelhakim Khellaf,
Vincent Quoc-Huy Trinh, and Mahdi S. Hosseini*

Achieving Zero False Negatives: Optimizing Anomaly Detection
with Genetic Neural Architecture Search 280
   Rabie Najem and Mohammed Benjelloun

Whisper-Conformer: A Modified Automatic Speech Recognition for Thai
Speech Recognition 294
   Thanakron Noppanamas and Suronapee Phooomvuthisarn

RevCD: Reversed Conditional Diffusion for Generalized Zero-Shot
Learning 306
  *William Heyden, Habib Ullah, Muhammad Salman Siddiqui,
  and Fadi Al Machot*

Toward an Explainable Heatmap-Based Deep Neural Network for Product
Defect Classification and Machine Failure Prediction in Industry 4.0 325
  *Tojo Valisoa Andrianandrianina Johanesa, Lucas Equeter,
  Sidi Ahmed Mahmoudi, and Pierre Dehombreux*

Question Answering in a Low-Resource Language: Dataset and Deep
Learning Adaptations for Sinhala 336
  Janani Ranasinghe and Ruvan Weerasinghe

Author Index 353

End-to-End ASR Model with Iterative Attention Mechanism Enhanced RNN Model for Phoneme Recognition

Ke Fang[1(✉)] and Yancong Deng[2]

[1] University of Pennsylvania, Philadelphia, PA 19104, U.S.A.
anniefangke@163.com
[2] University of California San Diego, San Diego, CA 92126, U.S.A.

Abstract. Automatic Speech Recognition (ASR) systems have significantly advanced with the integration of deep learning techniques, particularly neural networks and attention mechanisms; however, challenges remain in accurately modeling variable-length sequences and capturing complex temporal dependencies inherent in speech data. In this paper, we propose an enhanced end-to-end ASR model that incorporates a novel iterative attention mechanism, which loops through the attention layer multiple times with independent Dense and Dropout layers in each iteration. This design enables the model to focus on different aspects of the input sequence, effectively enhancing its ability to capture nuanced temporal patterns. Evaluated on the TIMIT dataset—a benchmark known for its comprehensive phonetic coverage—our model achieves a Phoneme Error Rate (PER) of 15.9%, outperforming existing neural network-based models. The results demonstrate that our iterative attention mechanism offers a more flexible and accurate solution for speech recognition tasks, addressing challenges associated with variable-length sequences and noisy environments, and contributing to the development of more robust ASR systems.

Keywords: Automatic speech recognition · Attention mechanism · Recurrent neural networks · Artificial intelligence

1 Introduction

Automatic Speech Recognition (ASR) has recently found wide application in modern society, from voice-controlled assistants to automated transcription services. At present, most speech recognition approaches are based on deep learning, especially neural networks [1], which have demonstrated outstanding performance in processing continuous speech signals. Historically, the foundation of ASR systems has been based on classical statistical architectures that break down the entire recognition process into several quite distinct components: extraction of acoustic features, acoustic modeling, language modeling, and search algorithms based on Bayesian decision rules [2]. The development history of ASR is shown in Fig. 1. The journey of ASR started in the 1950s and 1960s [3], with the advent of systems designed for isolated word recognition. These early systems had a small scope and were highly dependent on template matching as well as simple statistical models. Due to limitations in hardware and computations, they were constrained to recognizing small vocabularies with no sophistication for continuous speech processing or handling large vocabulary tasks.

As technology improved, scientists began to explore more sophisticated methods, including using Hidden Markov Models (HMMs) [4], which became the standard for acoustic modelling in the 1980s and 1990s. HMM-based systems decomposed speech into smaller units, such as phonemes, and statistically modeled the transitions between them. These systems, however, needed unique acoustic and language simulation, which was more challenging and largely dependent on handcrafted characteristics.

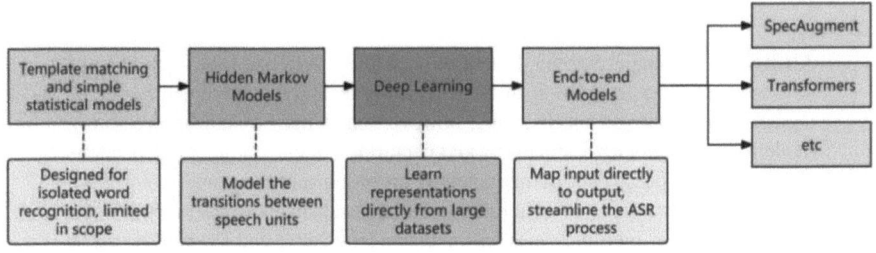

Fig. 1. The development history of ASR.

ASR has changed due to the growth of deep learning [5]. With deep learning methods, particularly neural networks, word error rates (WER) have decreased. A significant area of deep learning is ASR's ability to learn constructed information directly from massive data, eliminating the need for hand-made features and domain-specific experience. By removing individual models for acoustic, language, and pronunciation modeling, end-to-end models, which combine all components into a single structure and transfer speech signals directly into output text, have simplified the ASR process. ASR techniques perform better, specifically in exacting, real-world software, because of this simplicity and ease of training and use.

Numerous significant changes have contributed to the advancement of ASR. Techniques like SpecAugment [6], introduced in 2019, have helped improve ASR performance by augmenting training data, allowing designs to relate better to unseen information. Research into new architectures like Transformers has advanced the state-of-the-art for end-to-end ASR more [7]. Compared with conventional Recurrent Neural Networks (RNNs), these architectures are better suited to capture long-range dependencies of speech data.

Language models have recently contributed significantly to ASR by aiding in developing the most likely sentence hypotheses. Including a language model with an end-to-end structure is still a significant study area with methods like shallow fusion, deep fusion, and cold fusion being used to improve identification precision. Combining acoustic and language modeling into a single framework has produced more potent and specific audio techniques. As a result, the effectiveness of ASR methods has progressively increased over the years, with noticeable improvements noted in several indicators.

Despite significant developments in recent years, AI systems still struggle to be effective across various listeners, voices, and noisy environments. Although these improvements have been aided by developing novel end-to-end structures and data augmentation methods, more research is still needed. ASR will continue to advance as researchers look for ways to improve end-to-end models' solidity and develop ideas that can be successfully applied to products with few computational resources. Also, on-device speech recognition systems are being looked into, which would improve the effectiveness of real-time running without relying on cloud-based systems.

In this work, we focus on improving the attention mechanism in end-to-end ASR models, particularly for the TIMIT dataset [8]. Standard attention mechanisms typically calculate a weighted sum over all inputs to focus on particular parts of the input sequence. Our approach involves iterating the attention mechanism multiple times, with each iteration using its own distinct attention layer and dropout mechanism. This revision aims to improve the model's ability to focus on different aspects of the input sequence, leading to enhanced speech recognition accuracy.

2 Related Work

2.1 ASR System and Neural Network

Over time, Automatic Speech Recognition (ASR) systems have developed significantly. In order to model the temporal dynamics of speech, early systems relied on statistical models like Hidden Markov Models (HMMs) [4]. To improve acoustic modeling, Gaussian Mixture Models (GMMs) [8] were used with HMMs to represent continuous feature distributions. Mel-Frequency Cepstral Coefficients (MFCCs) [9] improved feature extraction by producing effective acoustic representations. In 2012, Hinton et al. [5] introduced Deep Neural Networks (DNNs) to ASR, leading to hybrid DNN-HMM systems that improved recognition accuracy. Then, Convolutional Neural Networks [10] were used to capture local dependencies in speech signals. Recurrent Neural Networks [11] and Long Short-Term Memory networks [12] were used to model temporal dependencies, with LSTMs succeeding on longer sequences and solving the vanishing gradient problem. Despite these improvements, dependence on the traditional HMM framework limited the potential of these models. This realization led to the development of end-to-end models that learn a direct mapping from speech to text.

2.2 End-to-End Model

End-to-end models were introduced to simplify the ASR pipeline by eliminating the need for separate components like GMMs and HMMs, instead training a model to map audio inputs directly to text outputs. One of the earliest and most influential approaches in this domain was Connectionist Temporal Classification (CTC) [13], which allowed for the alignment of input audio sequences with text sequences without the need for pre-segmented training data. CTC provided a way to model variable-length sequences and deal with timing uncertainty in speech recognition.

Based on CTC, sequence-to-sequence models [14] were introduced to improve the capabilities of end-to-end ASR. Encoder-decoder architectures that could solve varying input and output lengths were used in these models, giving them more adaptability. The encoder transfers the input sequence into a fixed-dimensional context vector, then the decoder generates the output sequence based on this context. This architecture facilitated the modeling of complex relationships between speech inputs and text outputs.

Despite their success, early end-to-end models faced challenges in accurately aligning speech signals with text outputs, particularly in long or noisy sequences. The fixed-length context vector in basic encoder-decoder models often struggled to obtain all the necessary information from the input, leading to degraded performance on longer utterances. This limitation led to the development of attention mechanisms that could dynamically concentrate on various parts of the input sequence during decoding, enhancing alignment and overall performance.

2.3 Attention Mechanism

Initially, neural networks struggled with capturing long-range dependencies in sequences because of issues like the vanishing gradient problem. While RNNs and LSTM networks addressed this to some extent, the introduction of attention mechanisms provided a more robust solution. The idea of attention was initially introduced by Bahdanau et al. [15] in the realm of neural machine translation, which allowed models to focus on specific parts of the input sequence while generating each word in the output sequence. The key idea was to compute a context vector as a weighted sum of all encoder hidden states, where the weights were determined by the alignment between the current decoder state and each encoder state. This approach enabled the model to selectively focus on relevant parts of the input sequence, improving the handling of long sentences and complex structures.

Following this, Luong et al. [16] refined the attention mechanism by introducing different alignment functions, such as dot-product attention, which calculated the similarity between the encoder's hidden states and the decoder's current hidden state more efficiently. In 2017, Vaswani et al. [7] proposed the Transformer model, where the attention mechanism was used not only to align encoder-decoder pairs but also within the encoder and decoder themselves through self-attention. The self-attention mechanism enables each location in the sequence to attend to all other locations, obtaining dependencies regardless of their distance within the sequence.

Introducing multi-head attention [17] was another development in the Transformer model. In multi-head attention, multiple heads operate in parallel, each focusing on different parts of the input. Then these heads are combined and passed through a linear transformation to generate the final output. This allows the model to capture various features and relationships within the data simultaneously, leading to more informative representations.

Despite being successful in various applications, attention mechanisms frequently require assistance handling highly long sequences, high computing efficiency, and various attention distributions. New research has focused on enhancing the effectiveness and flexibility of attention mechanisms in order to address these issues. For example, approaches like adaptive attention spans and sparse attention have been proposed to reduce computational complexity [18]. Our work builds upon these advancements by proposing a novel method that involves looping through the attention layer multiple times, treating each iteration as a distinct attention head. This method, coupled with independent dense and dropout layers for each iteration, enhances the model's capacity to focus on relevant information while reducing redundancy and overfitting. By iteratively refining the attention distribution, our approach aims to improve the model's capacity to capture complex temporal dependencies in speech data, leading to enhanced ASR performance.

3 Methodology

3.1 Encoder

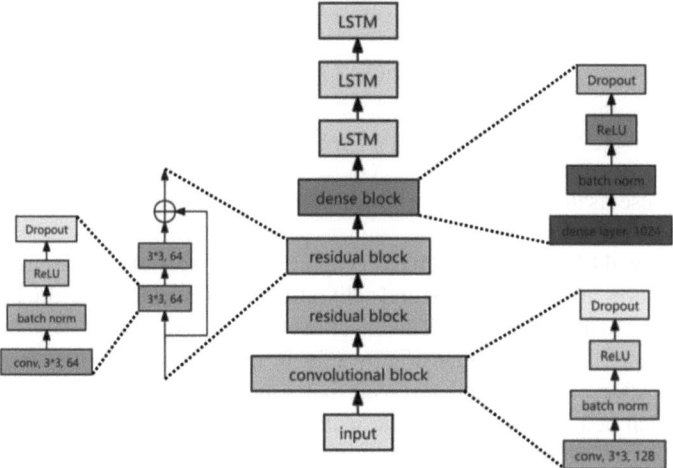

Fig. 2. The structure of the encoder in this model.

The encoder plays a crucial role in transforming the input speech signals into high-dimensional feature representations suitable for decoding. Figure 2 illustrates the architecture of the encoder used in our model. It comprises several key components: a Convolutional Block, two Residual Blocks, a Dense Block, and multiple Bidirectional Long Short-Term Memory (Bi-LSTM) layers. These components are sequentially arranged to capture various aspects of the input data, from local spatial features to long-range temporal dependencies. The structure ensures robustness through residual connections and dropout layers, which help prevent overfitting and improve generalization.

Convolutional Block. The Convolutional Block is responsible for capturing local spatial features from the input spectrograms. Initially, the input features are reshaped and split into multiple channels before being passed through a two-dimensional convolutional layer. The 2D convolution operation applies a set of filters to the input tensor, expressed as:

$$X^{(1)} = W * X + b \tag{1}$$

where X is the input tensor, W are the convolutional filters, ∗ denotes the convolution operation and b is the bias term.

After convolution, batch normalization is applied to normalize the output and mitigate internal covariate shifts, accelerating the training process. The normalized output is given by:

$$X^{(2)} = \gamma \cdot \frac{X^{(1)} - \mu}{\sqrt{\sigma^2 + \varepsilon}} + \beta \qquad (2)$$

where μ and σ^2 are the mean and variance of the input batch, ε is a small constant for numerical stability, γ and β are learned parameters for scaling and shifting.

Subsequently, the Rectified Linear Unit activation function is used to introduce non-linearity:

$$X^{(3)} = \text{ReLU}(X^{(2)}) = \max(0, X^{(2)}) \qquad (3)$$

$$X^{(4)} = X^{(3)} \odot D \qquad (4)$$

where $\odot$ represents the element-wise multiplication and D is a dropout mask with values 0 or 1.

Convolutional Block is responsible for capturing local spatial features from the input. The input features are first reshaped and split into multiple channels before being passed through a 2D convolutional layer. This layer uses 3×3 kernel and 1×3 stride, which effectively reduces the time dimension by a factor of three while preserving the spatial dimensions. After convolutional layer are batch normalization, ReLU activation and Dropout, which help in normalizing the feature distribution, introducing non-linearity and preventing over-fitting.

Residual Block. Following the Convolutional Block, the encoder incorporates two Residual Blocks. Each Residual Block is designed to capture higher-level features while maintaining the integrity of the original input through skip connections. The residual block follows the same computational principles as the convolutional block. Each residual block consists of two convolutional layers, each followed by batch normalization, ReLU activation and Dropout. The skip connection merges the original input with the output from the second convolutional layer, ensuring that the gradient can flow through the network without vanishing, even in deep architectures. This setup enhances the model's capability to capture complicated representations by combining the original and learned features.

Dense Block. After the Residual Blocks, the output is flattened and passed through a Dense Block. This block includes fully connected layer, batch normalization, ReLU activation and Dropout. The fully connected layer enables the model to learn global representations from the features extracted by the previous layers. This transformation acts as a bridge between the convolutional layers and the LSTM layers, converting spatial features into a format suitable for temporal processing.

LSTM Layers. The final stage of the encoder consists of multiple Bidirectional LSTM (Bi-LSTM) layers. Bi-LSTMs are effective in capturing temporal dependencies in sequence data by processing information in both directions. The operations within an LSTM cell at time step t are defined as in Eqs. (5)–(10):

$$f_t = \sigma(W_f \cdot [h_{t-1}, x_t] + b_f) \tag{5}$$

$$i_t = \sigma(W_i \cdot [h_{t-1}, x_t] + b_i) \tag{6}$$

$$o_t = \sigma(W_o \cdot [h_{t-1}, x_t] + b_o) \tag{7}$$

$$\tilde{C}_t = \tanh(W_C \cdot [h_{t-1}, x_t] + b_C) \tag{8}$$

$$C_t = f_t \odot C_{t-1} + i_t \odot \tilde{C}_t \tag{9}$$

$$h_t = o_t \odot \tanh(C_t) \tag{10}$$

Here f_t, i_t, o_t are the forget, input and output gates, C_t is the cell state and h_t is the hidden state at time step t.

Each LSTM layer in both directions is wrapped with a Dropout layer to prevent overfitting. The bidirectional structure enables the model to obtain dependencies from both past and future contexts, offering a thorough understanding of the temporal dynamics in the input.

3.2 Decoder

The decoder generates the output sequence based on the encoded representations and the attention mechanism. It is constructed using an LSTM cell, which is wrapped with a Dropout layer to enhance generalization. During training and evaluation, we utilize the TrainingHelper to feed the correct previous output token as the next input to the decoder, a technique known as teacher forcing. This approach stabilizes the learning process and accelerates convergence.

The decoder employs a BasicDecoder to process the input sequence step by step. The outputs from the decoder are passed through a fully connected layer to generate the final logits. For the inference phase, we use the BeamSearchDecoder to explore multiple possible output sequences by maintaining a beam of the most likely hypotheses. Beam search improves prediction accuracy by considering multiple potential outputs at each time step, balancing exploration and exploitation.

3.3 Attention Mechanism

The attention mechanism is a pivotal component that allows the decoder to focus on particular parts of the input sequence during decoding. We build upon the Luong Attention framework, which operates as follows:

$$c_t = \sum_{i=1}^{T} \alpha_{ti} h_i \tag{11}$$

$$\alpha_{ti} = \frac{\exp(e_{ti})}{\sum_{k=1}^{T} \exp(e_{tk})} \quad (12)$$

$$s_t = \tanh(W[s_{t-1}, y_{t-1}]) \quad (13)$$

$$e_{ti} = s_t^T W_a h_i \quad (14)$$

$$\tilde{s}_t = \tanh(W_c[s_t, c_t]) \quad (15)$$

$$o_t = \text{softmax}(V_{\tilde{s}_t}) \quad (16)$$

In Luong Attention, context vector c_t is computed as a weighted sum of the encoder hidden states h_i. The attention weights α_{ti} are calculated using a softmax function applied to the alignment scores e_{ti}. The alignment score is the dot product between the current decoder hidden state s_t and the encoder hidden state h_i, passed through a weight matrix W_a. After calculating the context vector, it is combined with the current decoder hidden state s_t to produce an updated hidden state $\tilde{s}_t$. Finally, the updated hidden state is passed through a fully connected layer, followed by a softmax function, to generate the output probabilities for the next word in the sequence.

The attention mechanism in this model builds upon the Luong Attention framework, incorporating a custom design that iterates through the attention layer multiple times. This modified approach deviates from the standard attention mechanism by assigning independent Dense and Dropout layers to each attention iteration, thereby enhancing the model's ability to focus on various parts of the input sequence during decoding. In each attention iteration, a distinct set of parameters is applied, resulting in multiple context vectors that capture different aspects of the input sequence. By looping through the attention mechanism multiple times, the model enhances its ability to capture diverse patterns within the input sequence, leading to more robust and context-aware decoding. This iterative process, combined with the individualized Dense and Dropout layers, enables the model to retain more control over the attention distribution, improving the accuracy and flexibility of the model.

4 Experiment and Analysis

4.1 Dataset

In this work, we use the TIMIT dataset for our experiments, following the data split procedure outlined in [18]. TIMIT is a widely used benchmark in speech recognition research, created by Texas Instruments and MIT. It consists of recordings from 630 speakers, including a 462-speaker training set and a 168-speaker testing set. The speakers represent eight major American English dialects, each reading ten phonetically diverse sentences. The TIMIT dataset is annotated with time-aligned phonetic transcriptions, making it valuable for training and evaluating ASR systems. Its detailed phonetic labels are instrumental for developing acoustic models. The dataset's comprehensive phonetic coverage and high-quality recordings make TIMIT a cornerstone in the field of speech processing and recognition. Due to its rich phonetic annotations, TIMIT enables precise evaluation of ASR systems at the phoneme level, aligning with our use of Phoneme Error Rate (PER) as the evaluation metric in this study.

4.2 Model Parameters

The training process was conducted using RMSprop Optimizer with a learning rate of 0.001, dropout of 0.5 to ensure stable convergence for the first 60 epochs. The batch size was 32 and beam width was 10. In the later phase, we reduced the learning rate to 0.0001 and applied a weight decay (L2 regularization) coefficient of 1×10^{-5}. The fine-tuning process was carried out over an additional 15 epochs. These hyperparameters were selected based on preliminary experiments to balance convergence speed and model generalization.

4.3 Evaluation Metric

We use the Phoneme Error Rate (PER) as the evaluation metric in this work. PER is commonly used to assess the performance of ASR systems, particularly in phoneme-level recognition tasks. It is computed similarly to the Word Error Rate (WER) but at the phoneme level. The PER is calculated as follows:

$$PER = \frac{S + D + I}{N} \times 100\% \tag{17}$$

where S is the number of substitutions, D is the number of deletions, I is the number of insertions, and N is the total number of phonemes in the reference transcription. The use of PER allows for a fine-grained evaluation of the model's phonetic transcription accuracy.

4.4 Experiment Results and Analysis

Table 1. Comparison of Experimental Data with Existing Data.

Network	PER
Augmented Conditional Random Fields[19]	23.0%
DBN-HMM Model[20]	22.8%
GMM-HMM Model[21]	21.7%
Deep RNN with LSTM[22]	17.7%
Attention-based Model[23]	17.6%
HMM over Time and Frequency Convolutional Net[24]	16.7%
Convolutional Model with Iterative Attention Mechanism	15.9%

Our model achieves a PER of 15.9%, outperforming other neural network-based models listed in Table 1. Traditional models like Augmented Conditional Random Fields (ACRF) and DBN-HMM exhibit higher PERs, primarily due to their reliance on manual feature extraction and heuristic methods. These models struggle to capture the nuances and complexities of speech data, particularly when handling variable-length inputs and long-range temporal dependencies. Hybrid models such as GMM-HMM also exhibit relatively high error rates, despite the introduction of probabilistic modeling techniques. While successful in early speech recognition tasks, their inability to capture complex temporal dependencies inherent in sequential data limits their performance.

Deep learning models, especially those based on Recurrent Neural Networks (RNNs) with Long Short-Term Memory (LSTM) units, as well as attention-based architectures, significantly improve upon these traditional methods. The deep RNN with LSTM achieves a PER of 17.7% [22], highlighting the advantages of neural networks in modeling sequential data. This improvement is largely due to LSTM's ability to address vanishing gradient issues and capture long-range dependencies in speech. The attention-based model with a PER of 17.6% [23] demonstrates that incorporating an attention mechanism further enhances the model's capacity to selectively focus on relevant parts of the input sequence, leading to more accurate predictions. This underscores the flexibility and robustness of attention mechanisms in dealing with variable-length inputs, making them particularly suited for tasks like speech recognition.

Our model further refines the focus on important features through multiple passes of the attention mechanism, enabling more accurate phoneme predictions. The iterative attention mechanism allows for dynamic weighting of different temporal parts of the input, making it highly suitable for variable-length speech data.

Table 2. Performance Comparison across Various Parameters.

	Optimizer	Number of Loops in Attention Mechanism	Number of Residual Blocks	PER
Experiment1	Adam	-	3	16.8%
Experiment2	Adam	8	3	16.5%
Experiment3	RMSProp	4	3	16.5%
Experiment4	RMSProp	8	3	16.4%
Experiment5	RMSProp	4	2	15.9%

Table 2 demonstrates the impact of different optimizers, attention loops, and residual blocks on model performance. One key insight from these experiments is the influence of the optimizer. The RMSProp optimizer outperforms Adam in our experiments, likely due to its ability to adapt learning rates based on the moving average of squared gradients, which can be beneficial for non-stationary objectives common in speech data.

Our experiments reveal that the number of loops in the attention mechanism plays a crucial role in the model's performance. As shown in Experiment 3 and Experiment

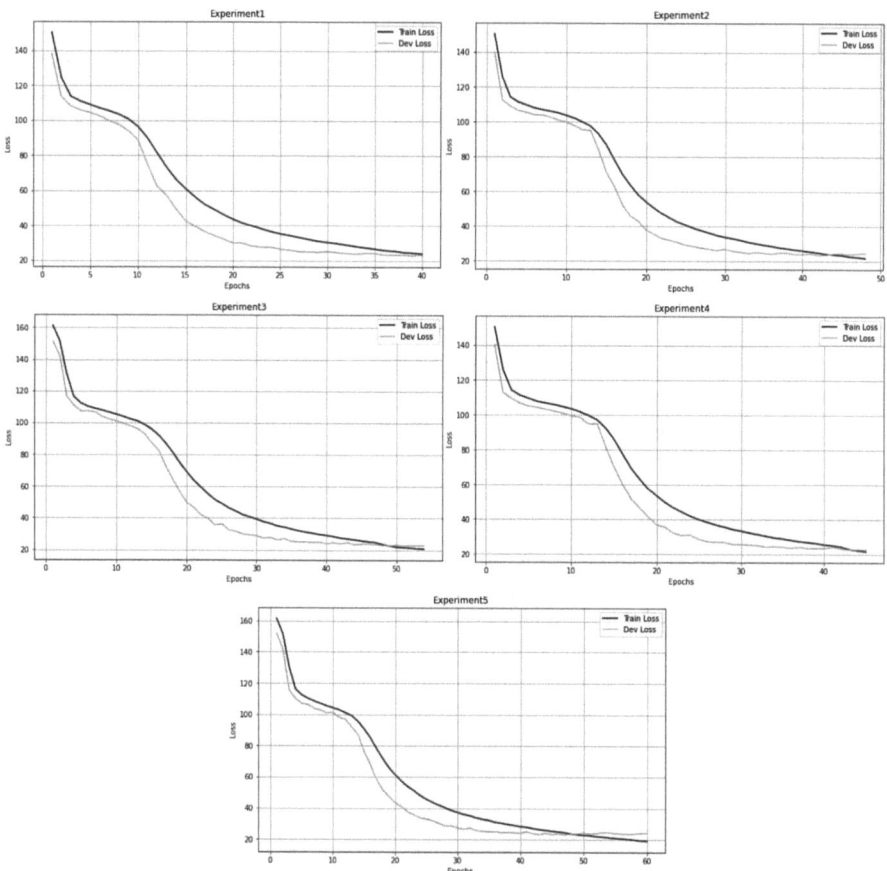

Fig. 3. The Loss Curves of Different Experiments.

5, a loop count of 4 strikes the best balance between attention complexity and performance, yielding a PER of 15.9%. This suggests that while iterative attention refines the focus on input features, an excessive number of iterations, such as 8 loops in Experiment 4, does not significantly improve performance and may introduce unnecessary complexity. Residual blocks also play a pivotal role in determining the model's accuracy. In Experiment 5, reducing the number of residual blocks from 3 to 2 led to a lower PER of 15.9%. This finding highlights the importance of controlling model complexity. While residual connections are instrumental in preventing vanishing gradient problems, an excessive number of residual blocks may lead to overfitting. By reducing the number of residual blocks, we effectively regularized the model and avoided overfitting while still maintaining the benefits of residual learning.

The loss curves in Fig. 3 highlight how different configurations of attention loops, residual blocks, and optimizers affect model performance. Experiment 1 shows a quick drop in loss but plateaus early, indicating limited ability to capture complex patterns. In Experiments 2 and 4, adding 8 attention loops helps reduce loss more effectively;

however, overfitting becomes apparent, as indicated by the gap between training and validation losses. Experiment 3, with 4 attention loops and RMSProp, strikes a better balance between loss reduction and overfitting, showing a more stable learning process. Experiment 5 achieves the best results by reducing the number of residual blocks, which prevents overfitting while still benefiting from the attention mechanism and RMSProp optimizer. This configuration achieves the lowest loss and the best generalization across the experiments.

5 Conclusion

In this paper, we presented an enhanced end-to-end ASR model that improves recognition accuracy by iterating through the attention mechanism multiple times with independent Dense and Dropout layers. This approach effectively enhances the model's focus on various parts of the input sequence, capturing complex temporal dependencies more efficiently. Our experimental results on the TIMIT dataset demonstrate that the proposed attention mechanism achieves a Phoneme Error Rate (PER) of **15.9%**, outperforming existing models. This highlights the efficacy of our method in addressing challenges such as variable-length sequences and noisy environments, offering a more flexible and accurate solution for speech recognition tasks.

References

1. McCulloch, W.S., Pitts, W.: A logical calculus of the ideas immanent in nervous activity. Neurocomputing Found. Res. **53**, 15–27 (1988)
2. Bayes, T.: An essay towards solving a problem in the doctrine of chances. Reprint of R. Soc. Lond. Philos. Trans., Rev. R. Acad. Sci. Exactas Fís. Nat. **53**(1–2), 11–60 (2001)
3. Davis, K.H.: Automatic recognition of spoken digits. J. Acoust. Soc. Am. **24**(6), 669 (1952)
4. Rabiner, L.R.: A tutorial on hidden Markov models and selected applications in speech recognition. In: Readings in Speech Recognition, pp. 267–296. Morgan Kaufmann Publishers Inc., San Francisco, CA, USA (1990)
5. Hinton, G., Deng, L., Yu, D., Dahl, G.E., Kingsbury, B.: Deep neural networks for acoustic modeling in speech recognition: the shared views of four research groups. IEEE Signal Process. Mag. **29**(6), 82–97 (2012)
6. Park, D.S., et al.: SpecAugment: a simple data augmentation method for automatic speech recognition (2019)
7. Vaswani, A., et al.: Attention is all you need. In: Proceedings of the 31st International Conference on Neural Information Processing Systems (NIPS 2017), pp. 6000–6010. Curran Associates Inc., Red Hook (2017)
8. Stauffer, C., Grimson, W.E.L.: Adaptive background mixture models for real-time tracking. In: Proceedings of the IEEE Computer Society Conference on Computer Vision and Pattern Recognition. IEEE (1999)
9. Abdul, Z.K., Al-Talabani, A.K.: Mel frequency cepstral coefficient and its applications: a review. IEEE Access **10**, 122136–122158 (2022)
10. Lecun, Y., Bottou, L.: Gradient-based learning applied to document recognition. Proc. IEEE **86**(11), 2278–2324 (1998)
11. Lipton, Z.C., Berkowitz, J., Elkan, C.: A critical review of recurrent neural networks for sequence learning. Comput. Sci. (2015)

12. Hochreiter, S., Schmidhuber, J.: Long short-term memory. Neural Comput. **9**(8), 1735–1780 (1997)
13. Graves, A., Fernández, S., Gomez, F., Schmidhuber, J.: Connectionist temporal classification: labelling unsegmented sequence data with recurrent neural networks. In: Proceedings of the 23rd International Conference on Machine Learning (ICML 2006), pp. 369–376. Association for Computing Machinery, New York (2006)
14. Sutskever, I., Vinyals, O., Le, Q.V.: Sequence to sequence learning with neural networks. In: Proceedings of the 27th International Conference on Neural Information Processing Systems (NIPS 2014), vol. 2, pp. 3104–3112. MIT Press, Cambridge (2014)
15. Bahdanau, D., Cho, K., Bengio, Y.: Neural machine translation by jointly learning to align and translate. Comput. Sci. (2014)
16. Luong, M.T., Pham, H., Manning, C.D.: Effective approaches to attention-based neural machine translation. Comput. Lang. (2015)
17. Cordonnier, J.B., Loukas, A., Jaggi, M.: Multi-head attention: collaborate instead of concatenate (2020)
18. Halberstadt, A.K., Glass, J.R.: Heterogeneous acoustic measurements and multiple classifiers for speech recognition. In: International Conference on Spoken Language Processing. DBLP (1999)
19. Hifny, Y., Renals, S.: Speech recognition using augmented conditional random fields. IEEE Trans. Audio Speech Lang. Process. **17**(2), 354–365 (2009)
20. Xie, Y., Zou, C.-R., Liang, R.-Y., Tao, H.-W.: Phoneme recognition based on deep belief network. In: Proceedings of the 2016 International Conference on Information System and Artificial Intelligence (ISAI), Hong Kong, China, pp. 352–355 (2016)
21. Sainath, T.N., Ramabhadran, B., Picheny, M., Nahamoo, D., Kanevsky, D.: Exemplar-based sparse representation features: from TIMIT to LVCSR. IEEE (8) (2011)
22. Graves, A., Mohamed, A.R., Hinton, G.: Speech recognition with deep recurrent neural networks. In: IEEE International Conference on Acoustics, vol. 38. IEEE (2013)
23. Chorowski, J., Bahdanau, D., Serdyuk, D., Cho, K., Bengio, Y.: Attention-based models for speech recognition. In: Proceedings of the 28th International Conference on Neural Information Processing Systems - (NIPS 2015), vol. 1, pp. 577–585. MIT Press, Cambridge (2015)
24. Toth, L.: Combining time- and frequency-domain convolution in convolutional neural network-based phone recognition. IEEE (2014)

Diagnostic Trouble Codes Prediction with DTC-GOAT and Ensembles

Abdul Basit Hafeez[1,2](✉), Eduardo Alonso[1,2], and Atif Riaz[1,2]

[1] City St George's, University of London, London, UK
{abdul.basit,e.alonso,atif.riaz.3}@citystgeorges.ac.uk
[2] CitAI Research Centre, City St George's, University of London, London, UK

Abstract. Diagnostic Trouble Codes (DTCs) produced by On-Board Diagnostic Systems (OBDs), and the research focused on their use for predictive maintenance have been around for a while now. In the last few years, we have witnessed advancement in terms of how these DTCs are utilised to perform self-supervised end-to-end prediction with the introduction of sequential prediction models, where the goal is to utilize past occurred fault events to predict the next DTC fault event. These models mainly use neural embeddings to encode the DTCs, along with their features, before applying neural networks capable, in turn, of processing sequential data. For instance, DTC-TranGru, which uses a GRU layer on top of a Transformer, has reported better results than LSTM and Attention-based models. In this paper, we first put forward an enhanced version of the DTC-TranGru model called DTC-GOAT (GRU's Optimized Alignment with Transformer), proposing optimizations including a better alignment of the Transformer with GRU's output, end-of-sequence EOS tokens, and strategically placed 1D spatial-dropout layers, to boost the accuracy of DTC prediction. Secondly, we also introduce an so-called Ensemble approach that uses multiple models for next-DTC prediction and show that it gives slightly higher top-5 accuracy results than the individual models.

Keywords: Diagnostic-trouble codes · Ensemble model · Transformer · GRU · Vehicles · Predictive maintenance

1 Introduction

On-Board Diagnostic systems (OBDs) have introduced a major shift in how predictive maintenance is performed in vehicles. They have replaced the traditional approach of using sensory data coming from sensors placed in the cars, with the diagnostic module-based fault events collected from OBDs and Electric Control Units (ECUs).

In contrast to sensory data, which has a numeric format and allows researchers to use several anomaly detection methods for predictive maintenance [7,8,14], data coming from these ECUs is in textual events format, and has multiple features with a high number of unique classes in each feature. This makes it difficult to apply machine learning algorithms out of the box and hence forces researchers to focus on a small group of DTCs or simple algorithms [3,16,22]. In the last few years, a new approach named next-DTC prediction has emerged, which uses sequential algorithms like Long Short-Term Memory (LSTM) [4], attention mechanisms [5], Transformers [17] and hybrid models [6] to predict the next DTC in the sequences, depending on the last N events in the sequence. These methods use neural embeddings to represent the DTC events, and the majority of them employ separate embedding layers for each feature associated with the DTC event.

One of these models, named DTC-TranGru [6], introduced a hybrid model combining transformer and Gated Recurrent Unit (GRU) architectures to predict the next DTC in a sequence. This model demonstrated superior performance compared to existing approaches, achieving a top-5 accuracy of 81.4% in predicting the next DTC event, including its associated features. However, the complex nature of vehicle diagnostics and the potential for further improvements motivated us to explore additional enhancements to this model. In this paper, we have made the following two main contributions:

1. **Architectural Refinements.** We introduce several modifications to the DTC-TranGru [6] architecture, including better alignment and optimized combination of transformer output with GRU output, introduction of an EOS token, and strategic placement of 1D spatial-dropout layers. These changes enhanced the model's ability to capture complex dependencies in DTC sequences. Our architecture was able to reduce the validation loss to 4.28 and achieve the top-5 accuracy of 82.13%.
2. **Ensemble Approach.** We develop an ensemble approach that combines our enhanced DTC-GOAT model with two complementary architectures. This combination of models leverages the strengths of different approaches, leading to better top-5 accuracy of 83.15%, which is greater than individual architectures.

The rest of this paper is organized as follows: Sect. 2 provides background on DTC prediction and related work. Section 3 details our methodology, including the DTC-GOAT architecture and the ensemble approach. Section 4 presents our experimental setup and results. Section 5 discusses the implications of our findings, and Sect. 6 concludes the paper with suggestions for future work.

2 Background and Related Work

This section first presents the background on established predictive maintenance approaches used in the industry before the shift towards On-Board Diagnostic Systems. We will afterwards discuss the work related to the use of diagnostic trouble codes coming from OBDs and ECUs. We will wrap up by sharing recent research which has focused on a self-supervised next-DTC prediction approach.

Predictive maintenance has historically been concentrated on using outlier detection methods on sensory data coming from the sensors installed in machines, especially vehicles. Examples include using non-parametric models [20], multi-level hybrid models [23], ensemble approaches [9], deep learning models [1] and autoencoder models [10] for outlier selection and deletion of outliers in sensory data.

The availability of OBDs and Electric Control Units (ECUs) in vehicles has replaced the type of data collected from automobiles with textual fault events often called Diagnostic Trouble Codes (DTCs), and has resulted in increased usage of this data in research papers to come up with predictive maintenance algorithms. Since these DTCs lack numeric representation, and often have multiple high cardinality features corresponding to different granularity, researchers have focused on predicting a few DTCs or casting the problem to some classification task. Examples of these include starter motor component's fault prediction using Random Forest [19] and the classification of DTC sequences into faulty and fault-free sequences [16].

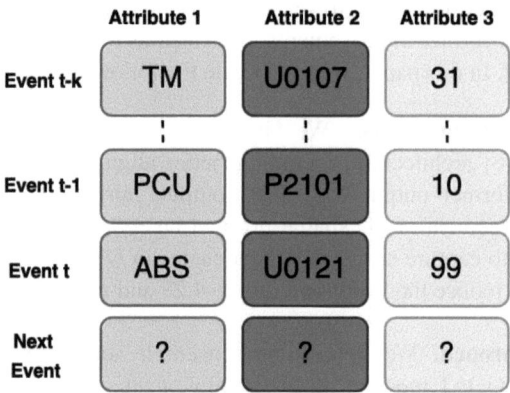

Fig. 1. The next-DTC prediction problem is explained using a simplified DTC fault event sequence with N DTC events. Each DTC fault event has three features. Given the last N events, the goal is to predict the next DTC event at time $t + 1$ with all three features.

Recent research has introduced a novel approach to vehicle fault prediction using a self-supervised next-DTC (Diagnostic Trouble Code) prediction task [4]. This method removes constraints to model the prediction problem into a classification task and to focus on a few DTCs at a time. Instead, it analyzes the entire sequence of DTC events associated with a vehicle. The model uses the most recent N events as input to predict the $(N + 1)^{th}$ event. Each DTC event is characterized by multiple features (attributes), offering diverse insights into the fault's characteristics, specificity, and origin. For example, the ECU (Electronic Control Unit) feature identifies which specific control unit reported the issue. The base-DTC represents the actual diagnostic code, while the fault-byte provides the most detailed information, potentially down to the chip level.

Figure 1 illustrates a simplified DTC sequence example for the next-DTC prediction task. The sequence begins with an event from the Telemetry ECU module (TM), with "U0107" as the base-DTC indicating a dead battery or a faulty Throttle Actuator Control (TAC) module and "31" as the fault-byte. This pattern of recording events using these three features continues up to the current timestep t. The objective is to predict all three features of the subsequent DTC event.

Given that these features are categorical and often have numerous categories (for instance, 419 categories for the base-DTC features), traditional methods like One-Hot Encoding (OHE) prove inefficient for event representation. To address this, recent self-supervised models employ neural embedding layers to create compact, dense representations. These representations are significantly smaller than the total number of distinct classes in each feature, thereby enhancing the efficiency of next-DTC prediction. The process involves developing separate embedding layers for each feature. These embeddings are then combined and refined concurrently with the prediction task. This approach has been utilized in several recent studies [4–6, 17].

All proposed models combine these neural embeddings with some sort of sequential processing algorithm to learn complex dependencies, for instance, Long Short-Term Memory (LSTM) [4], Luong attention followed by a Gated Recurrent Unit (GRU) layer [5], and a transformer decoder [17]. Some models use a hybrid of multiple models, like the combination of Transformer and GRU [6].

3 Methodology

This section provides details of our proposed architecture, which we call DTC-GOAT. We share how our model enhances the DTC-TranGru model [6] by incorporating different architectural changes. Furthermore, we provide details about our ensemble approach for predicting the next DTC event in the sequence. In both contributions, the core idea remains to predict the next Diagnostic Trouble Code (DTC) event in a sequence but with improved accuracy and robustness.

3.1 DTC-GOAT Building Blocks

Our architecture has some common layers with DTC-TranGru, but differs in how these layers are connected, and introduces some new layers along with changes in parameters, enabling increased capacity of the model with better alignments and connections across the layers. Figure 2 presents a high-level overview of the new architecture and Algorithm 1 provides implementation details for it.

The subsections ahead describe important blocks in our architecture.

Pre-Transformer Layers. We employ three separate embedding layers, one for each feature of the DTC event. Every layer has a different embedding size depending on the number of unique classes of that particular feature. We combine these embedding layers across time dimension and pass these to a 1D spatial-dropout layer, which applies the dropout across a time dimension.

Since the transformer layer does not cater for the positional information of the tokens, we apply the positional encoding technique described in the original transformer model [21] to concatenated embeddings, before passing it to the transformer block.

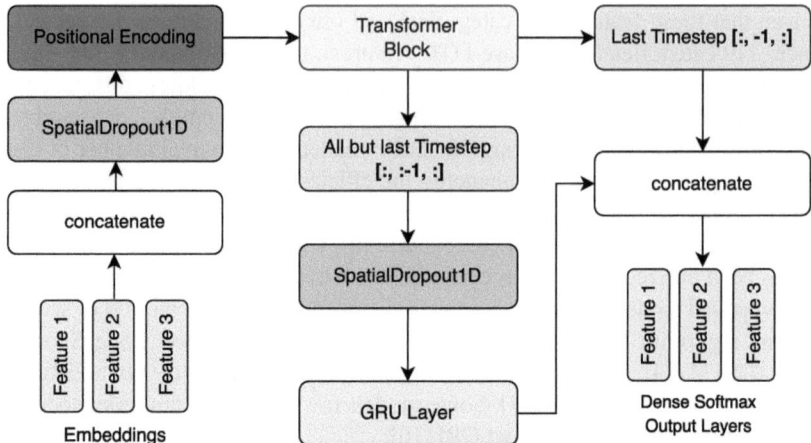

Fig. 2. Overall architecture of DTC-GOAT. Pre-transformer layers shown on the left side of the figure are common with DTC-TranGru [6]. This includes individual embedding layers for each feature (attribute), followed by concatenation along the time axis, 1D spatial-dropout, and positional encoding. The middle and right part of the figure depicts the true difference between proposed model and DTC-TranGru, where we pass all but the last timestep (EOS token) from the Transformer output to the GRU layer and concatenate the last transformer timestep with the hidden state of the last timestep from GRU, before passing it to all 3 dense layers. The figure shows that we introduce a 1D spatial-dropout before the GRU layer.

Transformer Block. We use two transformer encoder layers, each starting with a multi-head attention layer consisting of 3 heads. It is followed by residual connection combining positional encoding (or the previous encoder layer's output) with multi-head attention's output. Afterwards, the output of the residual connection is passed to the first layer-normalization operation, which is followed by a Feed-Forward Network (FFN) block consisting of two dense layers, the first one with 256 dimensions and the second one with 52 dimensions.

After the FFN block, we have a residual connection between the output of the first layer normalization operation and the second FFN dense layer, followed by the second and final normalization layer. Figure 3 presents a detailed view of the single encoder layer of the transformer block.

Transformer to GRU Layer. As seen in Fig. 6 as a part of preprocessing, we introduce an end-of-sequence (EOS) token at the end of each feature's sequence at $(N + 1)^{th}$ index. In many transformer architectures, the representation associated with the EOS token is argued to entail enough information about all timesteps to be used solely to perform downstream tasks, e.g., classification in BERT [2].

We pass the output of the transformer excluding the last timestep (EOS) to the 1D spatial-dropout first and eventually to the GRU layer.

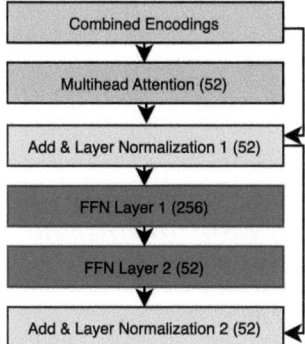

Fig. 3. Detailed view of transformer encoder layer. The encoder layer receives either a 52-dimensional combined positional encoding vector or the output of the previous encoder layer as an input. This input is passed to the multi-head attention layer and added to the output of the multi-head attention layer to act as a residual connection. This added residual connection is then passed to the first layer normalization operation. After layer normalization, we employ a Feed-Forward Network (FFN) block, which first increases the dimensionality of the input to 256 dimensions before bringing it back to the original size of 52 dimensions.

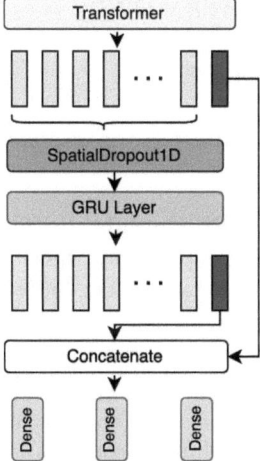

Fig. 4. This figure shows the detailed view of how the Transformer and the GRU outputs are combined using an EOS token. The last timestep in transformer output corresponds to the EOS token, and we pass all but this timestep of transformer output to the 1D spatial-dropout layer followed by the GRU layer. The last timestep from the transformer is then concatenated with the hidden state of the last timestep from the GRU layer and is then passed to each dense layer for individual features.

Algorithm 1. DTC-GOAT.

Require: $seq_1 \ldots seq_N$
Ensure: DTC event $(feat^1, feat^2, feat^3)$ for $seq_1 \ldots seq_N$
  for $epoch \leftarrow 1$ to N **do**
    $f_{EMB}^1 \leftarrow EMB(f_{OHE}^1)$
    $f_{EMB}^2 \leftarrow EMB(f_{OHE}^2)$
    $f_{EMB}^3 \leftarrow EMB(f_{OHE}^3)$
    $f_{EMB} \leftarrow CONCAT(feat_{EMB}^1, feat_{EMB}^2, feat_{EMB}^3)$
    $pos\_enc \leftarrow POSITIONAL\_ENCODING(f_{EMB})$
    $pos\_enc \leftarrow 1DSpatialDropout(pos\_enc)$
    $enc \leftarrow pos\_enc$
    for $encoder\_layer \leftarrow 1$ to N **do**
      $multihead \leftarrow multihead_n(enc)$
      $l\_norm\_1 \leftarrow Layer\_Norm(pos\_enc + multihead)$
      $ffn\_l1 \leftarrow Dense(l\_norm\_1)$
      $ffn\_l2 \leftarrow Dense(ffn\_l1)$
      $l\_norm\_2 \leftarrow Layer\_Norm(ffn\_l2 + l\_norm\_1)$
      $enc \leftarrow l\_norm\_2$
    end for
    $LAST\_TS \leftarrow enc[:, -1, :]$
    $ALL\_BUT\_LAST\_TS \leftarrow enc[:, : -1, :]$
    $spatial\_dropout \leftarrow 1DSpatialDropout(ALL\_BUT\_LAST\_TS)$
    $GRU\_OUTPUT \leftarrow GRU(spatial\_dropout)$
    $CONCATENATED \leftarrow CONCAT([GRU\_OUTPUT, LAST\_TS])$
    $feat\_out^1 \leftarrow Dense(CONCATENATED)$
    $feat\_out^2 \leftarrow Dense(CONCATENATED)$
    $feat\_out^3 \leftarrow Dense(CONCATENATED)$
    $loss \leftarrow Loss(feat\_out^1, feat\_out^2, feat\_out^3)$
    OptimizeParameters(loss)
  end for

Combining Transformer and GRU Outputs. For the GRU, we used one layer with 256 units. With the input of N timesteps coming from the transformer, ignoring the EOS token, the GRU produces N hidden state vectors. Since the last hidden dimension of the GRU is believed to contain all the essential information from the complete sequence, we use this hidden state and ignore other timesteps from the GRU.

As seen in Fig. 4, we concatenate this last hidden dimension of GRU (256-dimension) with the representation of EOS (52-dimension), which is the last timestep of transformer output, to get a 308-dimensional vector.

Dense Output Layers. This concatenated vector is then passed to the output block. In the output block, we have 3 output dense layers, each corresponding to an individual feature for the DTC event. We use softmax as an activation function in each layer, and the categorical cross-entropy function to calculate the loss. Total loss is calculated by summing the individual feature losses with same weights

$$L(\hat{y}, y) = \sum_{F^i}^{F} (-\sum_{k}^{K_{f^i}} y^{(k)} \log(y^{(\hat{k})})), \quad (1)$$

where K_{f^i} represents the number of unique classes in feature (attribute) f_i, F denotes the number of total features, while $\hat{y}$ is the predicted class, and y represents the actual class.

3.2 Ensemble of Models

As a second contribution to our work, we propose an ensemble approach for predicting the top-5 accuracy of the next DTC. As shown in Fig. 5, instead of taking the class with the highest probability for each feature from one individual model, we first take the average of the probabilities from all 3 models, across features. After the average prediction is computed for each feature, we calculate top-5 accuracy, which counts how many times the actual class for each feature has simultaneously existed in the top 5 predicted classes from the model.

Algorithm 2 shows how the feature level average is calculated using predictions with three different models.

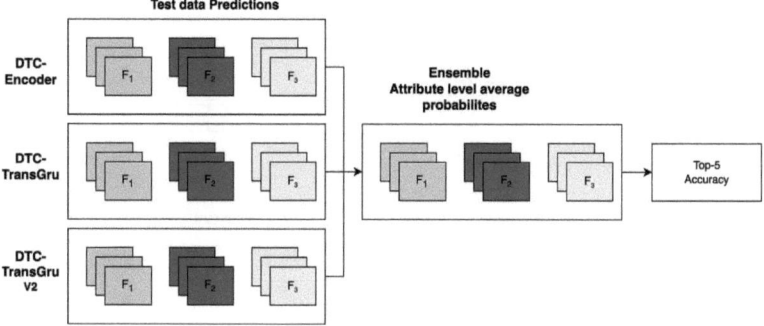

Fig. 5. We first pass the test dataset to three different models (DTCEncoder, DTC-TranGru, and DTC-GOAT) to get three sets of predictions, and then for each feature, we take the average of probabilities for all class predictions across three models. These average class probabilities are then used to calculate top-5 accuracy.

Algorithm 2. Ensemble Approach.

Require: $test\_pred_1, test\_pred_2, test\_pred_3$
Ensure: Ensemble predictions ($average\_pred$)
  for $f \leftarrow 1$ to F do
    for $row \leftarrow 1$ to N do
      $average\_pred[f][row] \leftarrow (test\_pred_1[f][row] + test\_pred_2[f][row] + test\_pred_3[f][row])/3$
    end for
  end for

---

4 Experiments and Results

In this section, we will provide details about the data, data preprocessing, and experimental setup for the DTC-GOAT, and share the results of the experiments at the end of the section.

4.1 Dataset and Data Preprocessing

We used a dataset provided by *Name hidden for peer review*, which consisted of 250k sequences, each corresponding to a unique vehicle and containing a varying number of multivariate DTC fault events.

As a part of preprocessing, the events within each sequence were sorted as per the time when they occurred. As seen in Fig. 6, we form a separate vector of each feature for all sequences to facilitate the application of a separate embedding layer per feature. We also append a special EOS token at the end of each feature sequence.

We trim longer sequences to the last N timesteps, which in our case was 10 timesteps, and pad the shorter sequences with a special token ('0'). For the train-test split, we kept 4,750 sequences for validation purposes, 12,500 as testing data and the rest of 232,750 sequences were used as training data.

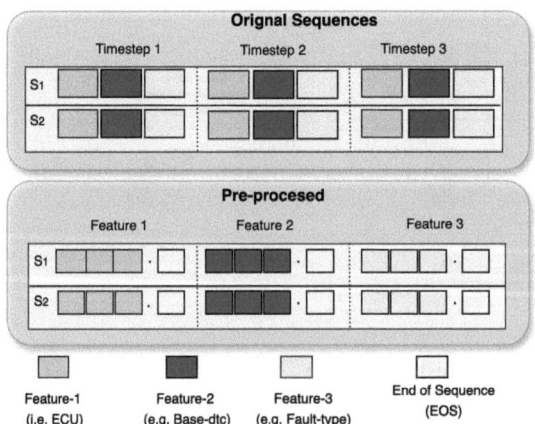

Fig. 6. The figure shows 2 unique DTC sequences before and after preprocessing. In the preprocessing step, we first separate each feature into separate vectors, keeping the original time order. We restrict the number of DTCs to 3 for the example, but in the original dataset each sequence has at least 5 DTC events. Each feature vector is appended with an end-of-sequence (EOS) token, to reinforce the completion of the sequence and to concatenate the representation of this token with GRU's output.

4.2 Experimental Setup and Hyperparameter Tuning

We tried multiple hyperparameters and parameters mentioned in Table 1 for our model, and the final parameters were selected with Hyperband [11] hyperparameter tuning method, executed with keras-tuner [15] python library. For the ablation study, apart from comparing with related models, we also compared with modified versions of these models, which were enhanced by adding the 1D spatial-dropout layer before recurrent layers and changing the parameters to the ones obtained in our experiments. The goal was to perform a true comparison and show that the results we achieved are not influenced by these optimizations only.

On the model level, the main choice was the learning rate; we used a min value of $1e^{-4}$ and a max value of 0.1 as the learning parameter, and we got the best performance with the value of 0.0005. Another important choice was the value for dropout in 1D spatial-dropout layers in the model, for which we tried the values up to 0.5, but found the best value to be 0.1.

In the transformer block, we experimented with 2 to 5 encoder layers and found two encoder layers to be best performing. We saw that 3 heads performed most optimally in the multi-head attention layer, and 256 dimensions were the best performance for the first layer of the FFN block. Unlike DTC-TranGru, in our case, the GRU layer gave the best result with 256 units.

Table 1. Min and max values of hyperparameters and parameters tried with Hyperband. Some of the important choices include learning rate, number of encoder layers, number of heads in multi-head attention, embedding size and the number of GRU layer units.

Choice	min	max	final
Learning rate	1e−4	0.1	0.005
Number of heads	1	6	3
Number of encoder layers	1	5	2
FFN Dimension	128	256	256
Feature-1 (module) embedding	4	24	8
Feature-2 (base-dtc) embedding	12	56	32
Feature-3 (fault-byte) embedding	4	32	12
Spatial dropout after embeddings	0.0	0.5	0.1
GRU layer units	96	256	256

We used the RELU [13] activation function in the GRU layer and softmax in the output dense layers.

4.3 Results

We compare the proposed model with several others, including one that introduces the next-DTC prediction problem and implements an LSTM-based architecture utilizing neural embeddings [4]. Another comparison involves a model that encodes DTC [5], applying long-attention [12] to GRU outputs and utilizing a dense bottleneck for compact representations. Additionally, we compare our proposed model with a modified small-GPT-2 [18] transformer decoder architecture [17], before comparing it to DTC-TranGru [6], which employs a GRU immediately after the transformer [21] and uses the GRU output solely in the final layers.

The results of the experiments and comparisons with other architectures can be seen in Table 2. It shows that our model DTC-GOAT was able to achieve the best test top accuracy of 82.13% and the lowest validation loss of 4.28. As a part of the ablation study, we also show that the parameter changes and optimization borrowed from our architecture did improve other architectures slightly, but did not surpass the metrics of our architecture. We show that allowing the direct influence of transformers representation on the output via concatenation of EOS token with GRU, instead of passing transformers representation indirectly via GRU, results in an increment of model capacity. The concatenation of the last hidden state of the GRU and the transformer's output for the EOS tokens index acts similarly to a skip connection and provides a better flow of information through the network.

For the ensemble approach, it can be seen in the Table 3 that the top-5 accuracy achieved by the ensemble model is 83.15%, which is higher than the accuracies achieved by individual models. This reaffirms our assumption that in the situation where the dataset is limited and having a large model isn't possible, we can use multiple small models to get better results. Since each model learns different features and focuses on different latent patterns, taking the average probability for each feature has the potential to reduce uncertainty and hence boost the overall top-5 accuracy.

Table 2. Experiment results for DTC-GOAT compared with other models, like SMFP [4], DTCEncoder [5], Transformer Decoder model [17], DTC-TranGru [6]. For the ablation study and to analyse whether changes in parameters and the spatial-dropout layer improved the performance alone, we tried enhanced versions of all the compared models by introducing minor optimization changes used in our model. As highlighted by the bold text, DTC-GOAT achieved the best results among all other models

Architecture	Validation Loss	Top-5 Test Accuracy
SMFP	4.50	76.15%
SMFP (Opt)	4.49	76.23%
DTCEncoder	4.36	79.21%
DTCEncoder (Opt)	4.35	79.32%
Transformer-Decoder	7.6	40%
Transformer-Decoder (Opt)	7.54	41.4%
DTC-TranGru	4.33	81.4%
DTC-TranGru (Opt)	4.32	81.47%
DTC-GOAT	**4.28**	**82.13%**

Table 3. Top-5 accuracy results from an ensemble of multiple models compared with individual models. Combining predictions from three models boosts the top accuracy, and turns out to be higher than the top-5 accuracy achieved by individual models

Architecture	Top-5 Test Accuracy
DTCEncoder	79.32%
DTC-TranGru	81.47%
DTC-GOAT	82.13%
Ensemble of Models	**83.15%**

5 Discussion and Conclusion

This paper introduces DTC-GOAT, which incorporates several optimizations and changes to the architecture of DTC-TranGru, to improve the top-5 accuracy of the next-DTC prediction task. The optimizations we propose include combining the output of the transformer model and the GRU model in a better way using the EOS token. We argue that since the output of the transformer is now reaching separately to the output layer with the help of the EOS token's representation concatenation with the GRU output, instead of being shadowed by GRU, it increases the model capacity and doesn't overfit early as opposed to other models.

We also introduce an ensemble approach of combining multiple models to attain the class probabilities of each feature in DTC predictions, which gives a better top-5 accuracy of 83.15% as compared to the individual participating models in the ensemble approach. We believe that making use of multiple small models, which might all have learned different patterns, is useful to increase the accuracy of the task where data is limited and it is not possible to build a very large model.

Acknowledgements. This research is supported by Diagnostic Software engineering Solutions Design at Robert Bosch GmbH (Now ETAS).

References

1. Davari, N., Veloso, B., Ribeiro, R.P., Pereira, P.M., Gama, J.: Predictive maintenance based on anomaly detection using deep learning for air production unit in the railway industry. In: 2021 IEEE 8th International Conference on Data Science and Advanced Analytics (DSAA), pp. 1–10 (2021). https://doi.org/10.1109/DSAA53316.2021.9564181
2. Devlin, J., Chang, M.W., Lee, K., Toutanova, K.: BERT: pre-training of deep bidirectional transformers for language understanding. In: North American Chapter of the Association for Computational Linguistics (2019). https://api.semanticscholar.org/CorpusID:52967399
3. Fransson, M., Fåhraeus, L.: Finding patterns in vehicle diagnostic trouble codes: a data mining study applying associative classification (2015). http://uu.diva-portal.org/smash/get/diva2:828052/FULLTEXT01.pdf
4. Hafeez, A., Alonso, E., Ter-Sarkisov, A.: Towards sequential multivariate fault prediction for vehicular predictive maintenance. In: 2021 20th IEEE International Conference on Machine Learning and Applications (ICMLA). IEEE, Pasadena (2021)

5. Hafeez, A.B., Alonso, E., Riaz, A.: DTCEncoder: a swiss army knife architecture for DTC exploration, prediction, search and model interpretation. In: 2022 21st IEEE International Conference on Machine Learning and Applications (ICMLA), pp. 519–524. IEEE, Nassau (2022). https://doi.org/10.1109/ICMLA55696.2022.00085
6. Hafeez, A.B., Alonso, E., Riaz, A.: DTC-TranGru: improving the performance of the next-DTC prediction model with transformer and GRU. In: Proceedings of the 39th ACM/SIGAPP Symposium on Applied Computing, SAC 2024, pp. 927–934. Association for Computing Machinery, New York (2024). https://doi.org/10.1145/3605098.3635962
7. Hermine, A., et al.: Data improving in time series using ARX and ANN models. IEEE Trans. Power Syst. 1 (2017). https://doi.org/10.1109/TPWRS.2017.2656939
8. Hui-xin, T., et al.: An outliers detection method of time series data for soft sensor modeling, pp. 3918–3922 (2016). https://doi.org/10.1109/CCDC.2016.7531669
9. Iftikhar, N., Baattrup-Andersen, T., Nordbjerg, F.E., Jeppesen, K.: Outlier detection in sensor data using ensemble learning. Procedia Comput. Sci. **176**, 1160–1169 (2020). https://doi.org/10.1016/j.procs.2020.09.112, https://www.sciencedirect.com/science/article/pii/S1877050920320123. Knowledge-Based and Intelligent Information & Engineering Systems: Proceedings of the 24th International Conference KES2020
10. Kuzin, T., Borovicka, T.: Early failure detection for predictive maintenance of sensor parts. In: Conference on Theory and Practice of Information Technologies (2016). https://api.semanticscholar.org/CorpusID:5844744
11. Li, L., Jamieson, K., DeSalvo, G., Rostamizadeh, A., Talwalkar, A.: Hyperband: a novel bandit-based approach to hyperparameter optimization. J. Mach. Learn. Res. **18**(185), 1–52 (2018). http://jmlr.org/papers/v18/16-558.html
12. Luong, M.T., Pham, H., Manning, C.D.: Effective approaches to attention-based neural machine translation (2015). https://doi.org/10.48550/ARXIV.1508.04025, https://arxiv.org/abs/1508.04025
13. Nair, V., Hinton, G.: Rectified linear units improve restricted Boltzmann machines Vinod Nair, vol. 27, pp. 807–814 (2010)
14. Nanduri, A., Sherry, L.: Anomaly detection in aircraft data using recurrent neural networks (RNN). In: 2016 Integrated Communications Navigation and Surveillance (ICNS), pp. 5C2-1–5C2-8 (2016). https://doi.org/10.1109/ICNSURV.2016.7486356
15. O'Malley, T., et al.: Kerastuner (2019). https://github.com/keras-team/keras-tuner
16. Pirasteh, P., et al.: Interactive feature extraction for diagnostic trouble codes in predictive maintenance: a case study from automotive domain. In: Proceedings of the Workshop on Interactive Data Mining, WIDM 2019. Association for Computing Machinery, New York (2019)
17. Poljo, H.: Transformer decoder as a method to predict diagnostic trouble codes in heavy commercial vehicles. Master's thesis (2021)
18. Radford, A., et al.: Language models are unsupervised multitask learners. OpenAI blog **1**(8), 9 (2019)
19. Schapire, R.E.: Explaining adaboost. In: Empirical Inference (2013). https://api.semanticscholar.org/CorpusID:7122892
20. Subramaniam, S., Palpanas, T., Papadopoulos, D., Kalogeraki, V., Gunopulos, D.: Online outlier detection in sensor data using non-parametric models. In: Proceedings of the 32nd International Conference on Very Large Data Bases, VLDB 2006, pp. 187–198. VLDB Endowment (2006)
21. Vaswani, A., et al.: Attention is all you need (2023)

22. Virkkala, L., Haglund, J.: Modelling of patterns between operational data, diagnostic trouble codes and workshop history using big data and machine learning. Ph.D. thesis, Uppsala universitet (2016). https://www.diva-portal.org/smash/get/diva2:909003/FULLTEXT01.pdf
23. Zhang, M., Li, X., Wang, L.: An adaptive outlier detection and processing approach towards time series sensor data. IEEE Access **7**, 175192–175212 (2019). https://doi.org/10.1109/ACCESS.2019.2957602

A Fast Fourier Transform-Aided Diffusion-Based U-Net Architecture for Microscopic Medical Image Segmentation

Saptarshi Pani[1], Gouranga Maity[2], Dmitrii Kaplun[3,4(✉)], Alexander Voznesensky[4], and Ram Sarkar[2]

[1] Department of Electrical Engineering, Jadavpur University, Kolkata, India
[2] Department of Computer Science and Engineering, Jadavpur University, Kolkata, India
[3] Artificial Intelligence Research Institute, China University of Mining and Technology, Xuzhou, China
dikaplun@etu.ru
[4] Department of Automation and Control Processes, Saint Petersburg Electrotechnical University "LETI", Saint Petersburg, Russia

Abstract. The rapid development of deep learning techniques has led to major advancements in medical image segmentation. The majority of segmentation models now in use are discriminative, i.e., they are mostly aimed at developing a mapping between segmentation masks and the input image. These discriminative techniques, however, suffer from an unstable feature space and ignore the underlying data distribution of input samples. This issue is highly pertinent to the segmentation of microscopic medical images, which often have low contrasts and intricate patterns. This paper suggests at using a generative models understanding of the underlying data distribution to supplement discriminative segmentation techniques. Hence, a diffusion based segmentation model is proposed in this study in combination with the Fast Fourier Transform (FFT). The proposed model integrates diffusion principles in the frequency domain. After that, a U-Net architecture with an FFT-based feature extraction aided by an attention mechanism is designed to enhance the segmentation performance. By combining frequency-domain processing, attention module and iterative noise reduction, the model effectively captures both global and local features, enabling precise segmentation of complex structures in microscopic medical images. The effectiveness of the model has been evaluated on three publicly available standard and complex microscopic medical image datasets. The proposed model has obtained Dice scores of 88.13%, 88.52% and 98.57% on TNBC, CPM17 and GlaS datasets, respectively, which are better than many recently proposed models found in the literature. The code implementation of the methodology is available at: FCAM-Diffusion.

Keywords: Image Segmentation · Diffusion Model · Fast Fourier Transform · Attention Mechanism · U-Net · Microscopic Medical Image

---

S. Pani and G. Maity—Authors contributed equally to this research.

1 Introduction

Medical imaging technologies have been instrumental in advancing diagnosis and treatment within the healthcare industry. Among the critical tasks in medical imaging, segmentation stands out as a key component, forming the foundation for computer-aided diagnostic tools and systems used in image-guided treatments. Microscopic medical image segmentation is a pivotal task in modern healthcare and biomedical research. It involves the precise identification and delineation of specific regions within microscopic medical images, such as cells, tissues, or pathological features. Microscopic medical image segmentation differs from conventional imaging tasks in that it requires models to precisely detect fine structures in detailed images, which often feature low contrast and complex patterns. Past researcher had utilized traditional segmentation techniques, such as color-based thresholding [35], watershed methods [26], level sets [15], super-pixels [30], edge detection [3], morphological [] operations and graph cut [24] approaches. Subsequently, machine learning-based segmentation methods, including Support Vector Machines [9] and clustering techniques [8], were developed to improve segmentation accuracy.

Recently, deep learning techniques have been widely explored across various image segmentation fields, including computer-aided diagnosis of pathological images. Convolutional neural networks (CNNs), such as fully convolutional networks (FCNs) [19], U-Net [23], U-Net++ [40], ResUNet++ [16] and Pyramid Scene Parsing Network [38] have demonstrated outstanding performance in image segmentation tasks, laying a strong foundation for their application in medical image segmentation. Specifically, U-Net and its variants have become the preferred approaches due to their capacity to capture multi-scale features. Nevertheless, improving these models to enhance the segmentation accuracy on diverse microscopic medical image datasets remains a key research area, particularly for tasks that demand both fine spatial precision and broader contextual understanding.

In modern times, diffusion probabilistic models [13] have gained significant attention as a leading category of generative models, with their applications extending to diverse fields like inpainting, super-resolution, and semantic editing, including medical imaging. Several studies [1,34] have explored integrating diffusion models into segmentation tasks, leveraging their ability to effectively model complex data distributions and enhance the performance of segmentation networks. These models operate in two key phases: noise addition and denoising. During the noise addition phase, the original image is progressively perturbed with noise over multiple time steps until it approximates noise, mimicking a physical diffusion process. Conversely, in the denoising phase, the process is reversed, where the model gradually removes the noise to reconstruct a high-quality image by leveraging a network trained to learn this reversal.

Microscopic medical image segmentation plays a vital role in various healthcare applications, such as automated histopathology, cellular analysis, and cancer diagnosis. For instance, precise segmentation of triple-negative breast cancer (TNBC) [22] images helps assess tumor heterogeneity, a crucial factor in predicting patient outcomes. Additionally, in datasets like CPM17 [31] and the GlaS (Gland Segmentation in Colon histology images) [29], identifying individual nuclei or glandular structures helps researchers and medical professionals understand cellular organization, detect abnormalities, and assess tissue health. By enhancing segmentation precision, advanced models support more consistent diagnoses and offer valuable insights for personalized treatment strategies.

Microscopic medical image segmentation poses several challenges despite its potential. The similarity between different classes and the variation within a single class makes it difficult to distinguish cellular structures, especially in areas with overlapping or densely packed cells. Low contrast and noise further obscure boundaries, while scale variations in microscopic medical images require models that can capture both detailed and broad features. Moreover, issues like data imbalance and the computational limitations of high-resolution imaging restrict the scalability of current approaches. To address these challenges, we propose a Fast Fourier Transform-aided diffusion-based U-Net model based model whose architecture is represented in Fig. 1. The **key contributions** of this paper are listed below.

- The proposed model incorporates a diffusion-based framework wherein Gaussian noise is initially introduced into the microscopic medical image. The noisy image is subsequently transformed into the Fourier domain, where it undergoes an iterative denoising process. At each iteration, the processed image is reverted to the spatial domain through the application of inverse Fourier transform, enabling the reconstruction of spatial features while mitigating noise artifacts.
- A U-Net architecture is employed to predict the segmentation mask for each denoised microscopic medical image during each iteration of the diffusion process. The architectural design of the U-Net facilitates the integration of both the spatial and frequency-domain features, thereby enhancing the accuracy and robustness of segmentation predictions.
- The proposed model demonstrates state-of-the-art performance on three complex microscopic medical image segmentation datasets, namely TNBC, GlaS, and CPM17.

2 Related Study

TNBC: The TNBC dataset [22], characterized by its diverse and detailed cellular structures, is frequently utilized to test and evaluate segmentation techniques in the analysis of histopathological images. A novel architecture, called 2MSPK-Net [37], is proposed that utilizes a multi-scale, multi-dimension attention module to capture both local and global dependencies while fusing features across scales. Additionally, the integration of prior knowledge from the Segment Anything Model (SAM) enhances the networks ability to differentiate ambiguous features, making this the first approach to leverage SAM in nuclei segmentation. A reverse erasing strategy and cross-layer information flow further refine feature detail recovery, improving segmentation accuracy. The authors of [27] introduced a modified MultiResU-Net for efficient segmentation, which incorporates a Gaussian distribution-based Attention Module (GdAM) to highlight important spatial regions using relevant text information. The model further improves feature transfer through Controlled Dense Residual Blocks (CDRB), ensuring better information flow across encoder and decoder layers. Testing on two breast cancer datasets, TNBC and MonuSeg, shows that this approach outperforms existing methods.

DRPVit model [17] was designed to tackle issues such as noise, blurring, and unclear boundaries in nucleus segmentation for pathology images. The framework incorporates IE-IFANet to enhance nucleus clarity and the RegProxy model, which combines a transformer encoder for semantic modeling with a classifier for segmentation tasks. A hybrid loss function, integrating boundary and Tversky losses, ensures improved edge and region segmentation. Additionally, post-processing removes noise and fills gaps, while paramedical tools analyze nucleus size and morphology for better diagnostic accuracy. The authors of AWGU-NET [28] proposed a segmentation model combining U-Net with a DenseNet-121 backbone to capture both spatial and contextual information. The model includes a Wavelet-guided channel attention module for better boundary detection and a global attention module for channel-specific focus. Experiments on Monuseg and TNBC datasets demonstrate the model's effectiveness in improving histopathological image analysis and cancer diagnosis. The TSCA-net [11] is an encoder-decoder based architecture for medical image segmentation that integrates CNN and Transformer models. It uses spatial and channel attention modules to capture global information at various layers, enhancing feature extraction at different scales. A spatial and channel feature fusion block is designed in the decoder to improve the fusion of Transformer features.

The authors of [32] proposed two modules to improve medical image segmentation by enhancing feature discrimination and multi-scale context. The Label-Aware Attention (LAA) module improves foreground and background feature distinction by fusing same-label local features using label-aware affinity learning. The Multi-scale Feature Boosting (MFB) module captures scale context through parallel convolutions with different receptive fields and a cross-encoder reference. A Tree-like Branch Encoder Network TBE-Net [36] was proposed, featuring a tree-like encoder for better feature extraction and preservation. The Depth and Width Expansion (DWE) module is introduced to enhance the networks depth and width with minimal parameters, improving performance. Additionally, a Deep Aggregation Module (DAM) aggregates encoder features for more effective decoding and segmentation.

Collectively, these techniques emphasize the increasing ability to achieve precise and interpretable segmentation in complex histopathology datasets such as TNBC.

GlaS: The GlaS dataset [29] is particularly valuable because it provides high-resolution images with complex features, such as varying tissue structures, which are common in histological analysis. This makes it an excellent resource for testing the performance of segmentation models in a medical context, especially when it comes to fine-grained tasks like gland segmentation. Meng et al. [21] proposed the AFC-UNet, which combines multi-scale feature fusion, attention mechanisms, and a CNN-Transformer hybrid module to overcome these limitations. The model includes the Multi-feature Fusion Attention Gates (MFAG) to focus on relevant features, and a Convolutional Hybrid Attention Transformer (CHAT) to capture long-range dependencies. Zhu et al. [41] proposed MFCNet, a network for medical image segmentation that addresses the underuse of multi-scale convolutional layers and channel fusion. The model employs dual-scale convolutional kernels at each encoder level and integrates a Dual-scale Channel-wise Cross-fusion Transformer (DCCT) in the skip connections for better feature fusion. A spatial attention mechanism further amplifies important areas in the features.

Chen et al. [4] proposed the attention-guided and noise-resistant (AGNR) framework for medical image segmentation, which integrates attention mechanisms and noise resistance to handle artifacts and occlusions. The model combines spatial and channel attention to prioritize key regions and accentuate relevant features. A noise-resistant semantic distillation (NRSD) module enhances robustness against interference, while an information bottleneck (IB) technique reduces ambiguity in feature representation. The AGNR framework improves segmentation accuracy and reliability in noisy medical images. Huang et al. [14], proposed DRA-Net, a dynamic regional attention network for medical image segmentation, addressing the limitations of CNNs and Transformers. DRA-Net adapts its feature extraction scope by measuring feature similarity and focusing on dynamic regions, reducing information loss. It also enhances local edge detail learning through regional feature interaction and ordered shift multilayer perceptron (MLP) blocks. Zhou et al. [39] proposed DAUNet, a novel model for colonic crypt segmentation in confocal laser endomicroscopy (CLE) images, addressing challenges like blurred boundaries and morphological diversity. DAUNet employs a dual-branch feature extraction module and feature fusion guided module with attention mechanisms to improve feature representation. It also incorporates a local multi-layer perceptron for refining features and enhancing segmentation accuracy.

Wang et al. [33] proposed CRMEFNet, a novel medical image segmentation network designed to overcome limitations in modeling multi-level features and identifying complex boundaries. The network features three key components: a Coupled Refinement Module (CRM) for optimizing low and high-frequency features, a Multiscale Exploration and Fusion Module (MEFM) for adaptive feature fusion, and a Cascaded Progressive Decoder (CPD) for fine-grained pixel recognition. Liu et al. [18] proposed MSGAT, a model for medical image segmentation that addresses challenges like complex structures and blurred boundaries. It uses Pyramid Vision Transformer (PVT) as the backbone and introduces the Channel Received Field Block (CRFB) to filter high-level features. The Multiscale Feature Enhancement Module (MSFEM) and Multiscale Parallel Partial Decoder (MSPPD) enhance and aggregate features, while the gated axial reverse attention (GARA) refines segmentation boundaries. Bougourzi et al. [2] proposed PAG-PVTUnet, a hybrid CNN-Transformer model for medical image segmentation. The model utilizes Pyramid input to emphasize features at different scales, incorporates a PVT transformer for long-range dependency capture, and employs a dual-attention gate mechanism to fuse CNN and Transformer features.

CPM17: Chen et al. [5] proposed HistoNeXt, which is a hybrid architecture that combines CNNs with dual-mechanism feature pyramid fusion, dense connections, and attention mechanisms to improve cell nuclei segmentation and classification. It aims to tackle challenges like class imbalance, resolution sensitivity, and computational efficiency. Zhu et al. [42] proposed Clustering Vision Mamba (CCViM) model, which enhances medical image segmentation by addressing the challenge of capturing both global and local feature interactions. While existing Vision Mamba (ViM) models excel at long-range feature interaction with linear complexity, they struggle to preserve short-range local dependencies due to fixed scanning patterns. CCViM improves upon this by incorporating a context clustering module, which dynamically segments image tokens into adaptable windows.

Luna et al. [20] proposed a dense feature pyramid network with a feature mixing module to enhance the field of view for nuclei segmentation while preserving pixel-level details. It introduces a multi-scale self-attention guided refinement module for improved output quality and enables better instance clustering by separating clustering objectives from pixel-related tasks. Chen et al. [6] proposed BRP-Net, a Boundary-assisted Region Proposal Network for accurate nucleus segmentation. It introduces a Task-aware Feature Encoding (TAFE) network that efficiently extracts features for both semantic segmentation and instance boundary detection by considering their inter-dependencies. Coarse nucleus proposals are generated from these tasks and refined through an instance segmentation network. Graham et al. [12], proposed Hover-Net for joint nuclear segmentation and classification in H-E stained histology images. By utilizing vertical and horizontal distances of nuclear pixels to their centers of mass, the network effectively separates clustered nuclei, ensuring accurate segmentation in overlapping regions. Additionally, an up-sampling branch is used to predict the type of each segmented nucleus.

3 Methodology

The methodology integrates a Diffusion model, in which Gaussian noise is first added to a microscopic medical image and then the noisy image is denoised iteratively in T timesteps. A robust U-Net based architecture with additional components, including Fast Fourier Transform (FFT) [7] blocks and Attention mechanisms is used to predict the segmentation mask of the noisy image in each iteration of denoising, to enhance medical image segmentation. FFT blocks inside the U-Net model convert spatial domain information into frequency domain, allowing the model to effectively isolate and process noise through frequency noise reduction. Attention blocks further refine these features by highlighting relevant regions, which enhances precise segmentation. Segmentation performance is assessed on diverse datasets such as TNBC, GlaS, and CPM17, over which the framework demonstrates notable segmentation results, outperforming state-of-the-art methods.

3.1 Diffusion Model

Diffusion models are probabilistic generative models widely recognized for their effectiveness in tasks like image synthesis and segmentation. These models simulate a diffusion process, adding noise to data and reversing it to produce realistic outputs. Similar to particle diffusion, this approach evolves data toward equilibrium states.

The implementation introduces a noise injection mechanism, governed by time steps. Gaussian noise is progressively added to input images via a smooth scaling function defined by α and β, ensuring controlled degradation. The noisy image is transformed into the frequency domain using FFT, where iterative denoising is performed at each timestep. The reverse process, critical for denoising, uses a mathematical procedure within the frequency domain itself to predict and correct noise. After denoising, the image is converted back to the spatial domain using inverse FFT, yielding clean outputs.

A key innovation is the integration of frequency noise reduction in the Fourier domain, targeting specific bands to minimize irrelevant noise. Leveraging Fourier analysis improves feature delineation and reduces artifacts in microscopic images. The model incorporates conditional guidance using target masks, aligning the generative process with domain-specific needs. This conditioning ensures precise denoising while balancing generalization and specificity.

Through iterative refinement, the model generates denoised images optimized for segmentation. The interplay of forward and reverse diffusion, FFT, and inverse FFT establishes a probabilistic framework addressing noise and uncertainties in real-world data. By merging generative modeling with Fourier analysis and conditioning, this diffusion model provides a robust solution for medical image segmentation, advancing the field in noisy, complex scenarios.

Forward Process. The forward process of the diffusion model simulates the progressive addition of noise to an image over a series of timesteps, transforming the original image into a highly noisy version. Let the clean input image be x_0, and the process is parameterized by a predefined schedule of variance terms $\beta_t \in (0,1)$, where $t = 1, 2, \ldots, T$ represents the timesteps, I is the identity matrix, which is used to specify the covariance structure of the Gaussian distribution. At each step, noise is incrementally added to the image according to the Markovian process defined as:

$$(f(a_t|a_{t-1}) = \mathcal{N}(a_t; \sqrt{1-\beta_t}a_{t-1}, \beta_t I)) \tag{1}$$

This process iteratively corrupts the image such that the distribution at any timestep t can be expressed as

$$(f(a_t|a_0) = \mathcal{N}(a_t; \sqrt{\bar{\alpha}_t}a_0, (1-\bar{\alpha}_t)I)) \tag{2}$$

where $\alpha_t = 1 - \beta_t$ and $\bar{\alpha}_t = \prod_{i=1}^{t} \alpha_i$ accumulates the scaling factors. In practice, to directly sample a noisy image at timestep t, the equation

$$(a_t = \sqrt{\bar{\alpha}_t}a_0 + \sqrt{1-\bar{\alpha}_t}\epsilon) \tag{3}$$

is used, where $\epsilon \sim \mathcal{N}(0, I)$ represents Gaussian noise. This progressive degradation ensures that a_T approximates pure noise, enabling the subsequent reverse process to focus on recovering a_0. This mathematical framework underpins the generative capabilities of the diffusion model, making it particularly robust for tasks involving complex data distributions.

Reverse Process. The reverse process in the diffusion model corresponds to a generative denoising procedure, which sequentially reconstructs the pure image a_0 from a noisy observation a_T by iterating through the backward timesteps $t = T, T-1, \ldots, 1$. At each timestep t, the model learns a conditional distribution

$$(d_\theta(a_{t-1}|a_t) = \mathcal{N}(a_{t-1}; \mu_\theta(a_t, t), \Sigma_\theta(a_t, t))) \tag{4}$$

where μ_θ and Σ_θ are parameterized by a U-Net architecture. The segmentation mask M serves as an auxiliary conditioning input, guiding the reconstruction process to ensure that the generated image adheres to the spatial structure of the target regions. Specifically, the U-Net leverages the concatenation of noisy images a_t and the segmentation mask M as inputs to predict the noise component $\epsilon_\theta(a_t, t, M)$, enabling the estimation of the denoised image as

$$a_{t-1} = \frac{1}{\sqrt{\alpha_t}} \left(a_t - \frac{\beta_t}{\sqrt{1-\bar{\alpha}_t}} \epsilon_\theta(a_t, t, M) \right) + \sigma_t z \tag{5}$$

where $z \sim \mathcal{N}(0, I)$ introduces stochasticity. This iterative refinement incorporates spatial attention mechanisms and frequency-aware processing via FFT blocks within the U-Net to enhance the recovery of fine details critical for accurate segmentation. By aligning the denoising process with the underlying anatomical structures represented in M, the reverse process achieves precise restoration, ensuring high-fidelity segmentation outputs even in challenging noisy scenarios.

Fast Fourier Transform. The FFT [7] plays a pivotal role in the reverse process of Diffusion model and the U-Net based architecture implemented in the study, specifically through the integration of FFT blocks at each encoder level and in the bottleneck of the model. This integration leverages the frequency domain's ability to represent spatial information more effectively, enabling enhanced feature extraction and noise reduction. By transforming spatial domain data into the frequency domain using a 2D FFT, the model extracts critical spectral features, which significantly improve the segmentation mask prediction. Mathematically, the FFT operation is expressed as,

$$(A(f_x, f_y) = \sum_{x=0}^{W-1} \sum_{y=0}^{H-1} a(x, y) e^{-j2\pi(f_x x/W + f_y y/H)}) \tag{6}$$

where $A(f_x, f_y)$ represents the transformed frequency domain representation of the spatial input $a(x, y)$, and W and H are the width and height of the image, respectively. This transformation decomposes the image into its magnitude $|A(f_x, f_y)|$ and phase $\angle A(f_x, f_y)$, capturing essential patterns and relationships in the frequency space. The FFT block utilizes these two components by concatenating them along the channel dimension to form a rich feature representation.

Within the U-Net model, these FFT features are processed through a specialized convolutional block consisting of a series of 1×1 convolutions, batch normalization layers, and activation functions to extract frequency-specific patterns. The processed features are then passed through an attention mechanism, further refining the frequency-domain information by focusing on salient regions that contribute most to accurate segmentation. This design ensures that both global and local frequency features are effectively captured and integrated into the network's learning pipeline.

The advantage of operating in the frequency domain stems from its ability to delineate high-frequency noise from low-frequency structural patterns, a distinction that is particularly valuable for medical image segmentation. By reducing irrelevant high-frequency components, the FFT block enhances the signal-to-noise ratio, ensuring that critical anatomical structures are preserved and emphasized. This capability is particularly important when dealing with noisy medical imaging data, such as histopathological slides, where subtle morphological features are crucial for accurate segmentation.

3.2 U-Net Architecture

The U-Net architecture in the proposed model is structured with an encoder-decoder framework aided with FFT-based feature extraction and attention mechanisms. The encoder consists of four convolutional blocks, each followed by an FFT block to transform spatial feature maps into the frequency domain, where an attention block is applied to enhance critical frequency features. The spatial feature maps and their FFT-transformed counterparts are concatenated at every layer to preserve both spatial and frequency information. At the bottleneck layer, the most compressed representation of the image undergoes a similar transformation, with FFT and attention mechanisms refining the feature maps further before passing them to the decoder. The decoder mirrors the encoder with four upsampling blocks, to generate the segmentation mask of the corresponding noisy medical image. This integration enables the U-Net to effectively combine spatial and frequency features, facilitating precise segmentation mask predictions even for noisy inputs.

Frequency Channel Attention Module (FCAM). The FCAM [25] is a key enhancement within the proposed model, operating in conjunction with the FFT block to improve segmentation mask predictions. This mechanism refines the representation of frequency-domain features by selectively emphasizing the most relevant spectral information while suppressing less critical components. Such a targeted focus accelerates the convergence of the overall diffusion model, reducing the number of epochs or timesteps required for accurate segmentation mask predictions as is demonstrated in tables Table 1, Table 2, Table 3, where the number of training epochs is reduced substantially upon application of the FCAM.

At its core, the attention mechanism is implemented as a sequence of operations that compute relationships between the feature map's channels. Let $(F \in \mathbb{R}^{B \times C \times H \times W})$ represent the frequency-domain feature map, where B, C, H, and W are the batch size, number of channels, height, and width, respectively. The attention mechanism begins by transforming F into a query, key, and value space using 1×1 convolutions. Mathematically, these transformations are expressed as:

$$(Q = W_q F, \quad K = W_k F, \quad V = W_v F) \tag{7}$$

where W_q, W_k, W_v are learnable parameters of the 1×1 convolutional layers, and $Q, K, V \in \mathbb{R}^{B \times C' \times H \times W}$, with $C' = C/8$.

The attention map is computed by measuring the similarity between the query and key representations. This is achieved via the scaled dot-product operation, normalized by the softmax function:

$$(\text{Attention}(Q, K) = \text{softmax}\left(\frac{Q \cdot K^T}{\sqrt{C'}}\right)) \tag{8}$$

where $(Q \cdot K^T \in \mathbb{R}^{B \times (H \times W) \times (H \times W)})$ represents the affinity between spatial positions in the frequency domain. The resulting attention map encodes global relationships among spectral features.

The value matrix (V) is weighted by the attention map, enabling the model to focus on the most salient frequency components:

$$(F_{\text{att}} = \text{Attention}(Q, K) \cdot V) \tag{9}$$

where $F_{\text{att}} \in \mathbb{R}^{B \times C' \times H \times W}$ represents the refined feature map. This output is reshaped and added back to the original feature map F, incorporating residual learning:

$$(F_{\text{out}} = \gamma F_{\text{att}} + F) \tag{10}$$

where γ is a learnable parameter that balances the influence of the attention-enhanced features. The residual connection ensures gradient flow and stability during training.

The FCAM enhances the model's capacity to discern critical spectral patterns, particularly in complex medical imaging datasets. By efficiently capturing relationships in the frequency domain, the mechanism reduces the reliance on prolonged training or iterative refinement through numerous timesteps. This improvement directly impacts the training efficiency and predictive performance of the overall diffusion model, achieving high-quality segmentation masks with fewer computational resources.

3.3 Overall Model Architecture

The overall model architecture shown in Fig. 1 introduces a unified framework that integrates the diffusion process and a U-Net architecture aided by FFT-based feature extraction and attention mechanisms to achieve efficient and accurate segmentation masks. This design leverages the complementary strengths of these components, making it particularly effective for complex domains like microscopic medical imaging.

The model begins with a diffusion process where Gaussian noise is systematically added to a microscopic medical image over multiple timesteps, creating progressively noisier inputs. These noisy images are transformed to the frequency domain using the FFT block, enabling the model to capture global patterns, such as texture and structure, that are critical for robust segmentation in noisy environments. Within the frequency domain, the diffusion process iteratively denoises the image at each timestep by leveraging learned noise prediction and correction mechanisms. Once the denoising for a specific timestep is complete, the image is transformed back to the spatial domain using inverse FFT. This iterative noise addition and removal framework, guided by the forward and reverse diffusion steps, ensures that the model learns to reconstruct clean images effectively.

Every denoised image in the spatial domain in each timestep is subsequently passed through a U-Net architecture for segmentation mask prediction. The U-Net is augmented with FFT blocks and attention mechanisms at multiple levels, which enhance its ability to process both spatial and frequency-domain features. By incorporating magnitude and phase components from the FFT representation, the model ensures a comprehensive spectral encoding of the image. The attention mechanism further refines these features by selectively focusing on spatial and channel-wise dependencies, dynamically prioritizing critical aspects of the input while filtering irrelevant information. This selective refinement improves generalization across datasets and reduces computational overhead, making the model more efficient.

During training, the iterative nature of the diffusion model allows the U-Net to progressively refine its segmentation predictions over multiple timesteps. The enriched features from the FFT and attention mechanisms enable the network to make accurate and contextually informed predictions. This integrated approach not only accelerates convergence by reducing the number of epochs required but also maintains superior performance across diverse datasets. By combining spectral analysis, attention blocks, and the diffusion process, our model achieves high precision and efficiency in generating segmentation masks for challenging microscopic medical images.

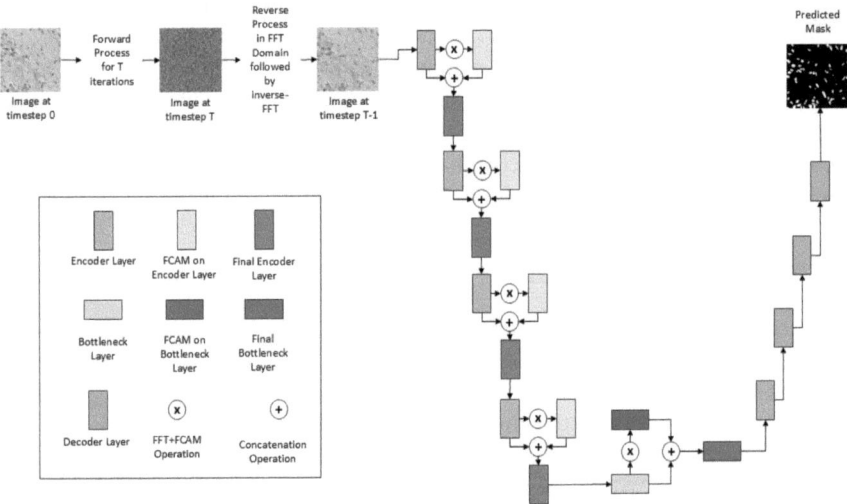

Fig. 1. Overall workflow of the proposed FFT-based diffusion process-aided U-Net model used for microscopic medical image segmentation.

3.4 Loss Function

The model introduced in this work relies solely on Binary Cross Entropy (BCE) loss, which is vital for training in image segmentation tasks. The BCE loss quantifies the difference between the predicted and actual ground truth, steering the optimization process toward producing accurate segmentation masks. By minimizing this loss, the model improves its ability to differentiate between foreground and background pixels, leading to more precise segmentation results. It is given by the following expression:

$$(\text{BCE Loss} = -\frac{1}{N}\sum_{i=1}^{N}[y_i \log(p_i) + (1-y_i)\log(1-p_i)]) \tag{11}$$

where (y_i) represents the true label for the i-th pixel, (p_i) is the predicted probability of the pixel being part of the object (foreground), and (N) is the total number of pixels. This loss function penalizes large deviations between the predicted probabilities and the actual labels, making it ideal for segmentation tasks where pixel-wise accuracy is crucial. BCE loss is particularly effective in scenarios where there is an imbalance between the foreground and background, as it assigns a higher weight to incorrect predictions.

4 Results And Discussion

4.1 Experimental Setup

The experimental setup for this study involved evaluating the proposed model on TNBC, GlaS, and CPM17 datasets. Each image was resized to a resolution of (128 × 128) pixels to standardize the input format. Model training was done for 100 epochs using a batch size of 2 and a learning rate of 0.001, optimized using the Adam optimizer. The experiments were executed using PyTorch, and all computations were performed on an NVIDIA TESLA P100 GPU, which ensured the efficient handling of deep learning workloads. This configuration facilitated the systematic evaluation of the proposed methodology and ensured reproducibility in performance metrics.

4.2 Dataset Description

TNBC: The TNBC (Triple-Negative Breast Cancer) [22] dataset comprises images associated with a highly metastatic subtype of breast cancer, known for its lack of approved therapeutic targets and significant contribution to cancer mortality in women. The dataset is critical for research aimed at understanding the molecular pathways and designing effective treatments to inhibit cancer stem cell (CSC) growth, suppress metastatic metabolic signatures, and counteract tumor immunosuppression. For this study, the dataset was split into 35 images for training, 10 for validation, and 5 for testing, ensuring a balanced and robust evaluation of the proposed model.

GlaS: The GlaS (Gland Segmentation in Colon Histology) [29] dataset comprises 165 H&E-stained histological images derived from 16 colorectal adenocarcinoma sections, representing stages T3 and T42. These sections, each from a different patient, were processed independently, leading to significant inter-subject variability in stain distribution and tissue architecture. The digitized images, captured at a pixel resolution of 0.465 m using a Zeiss MIRAX MIDI Slide Scanner, provide a rich resource for gland segmentation tasks. In this study, the dataset was divided into 68 images for training, 17 for validation, and two distinct test sets: Test Dataset A with 60 images and Test Dataset B with 20 images, enabling comprehensive evaluation of the proposed model.

CPM17: The CPM17 [31] dataset was featured in the 2017 Digital Pathology Challenge, hosted by the Medical Image Computing and Computer-Assisted Intervention Society (MICCAI). This dataset comprises histopathological tissue images from patients diagnosed with various cancers, including non-small cell lung cancer (NSCLC), head and neck squamous cell carcinoma (HNSCC), glioblastoma multiforme (GBM), and low-grade glioma (LGG). The dataset emphasizes the critical task of nuclei segmentation in histology images, targeting challenges such as incorporating nuclei contour information, segmenting small or irregularly shaped nuclei in dense regions, and accurately locating scattered nuclei. This dataset supports the development of contour-based dual-path instance segmentation networks, enabling independent extraction and fusion of nuclei and contour features at multiple scales during encoding and decoding stages. In this study, 25 images were used for training, 7 for validation, and 32 for testing, facilitating the evaluation of the proposed segmentation approach.

4.3 Evaluation Metrics

IoU: Intersection over Union (IoU) is a commonly used metric to evaluate the overlap between the predicted segmentation mask and the ground truth mask. It is defined as the ratio of the intersection between the predicted and ground truth masks to their union. Mathematically, it is expressed as:

$$IoU = \frac{\sum (P \cap T)}{\sum (P \cup T)} \tag{12}$$

where P represents the predicted mask and T represents the ground truth mask. IoU is a robust measure for segmentation tasks as it penalizes both false positives and false negatives, providing a balanced assessment of segmentation accuracy. An IoU of 1 indicates perfect overlap, whereas an IoU of 0 indicates no overlap.

Dice Score: The Dice coefficient, or Dice score, is another evaluation metric that quantifies the similarity between two sets, commonly used for medical image segmentation. It is computed as the harmonic mean of precision and recall and is defined as:

$$DiceScore = \frac{2 \times \sum (P \cap T)}{\sum P + \sum T} \tag{13}$$

This metric emphasizes the overlap between the predicted and ground truth masks while being less sensitive to small discrepancies. It ranges from 0 (no overlap) to 1 (perfect overlap) and is particularly effective for datasets with imbalanced foreground and background regions.

Accuracy: Accuracy is a straightforward metric that measures the proportion of correctly classified pixels in the predicted mask relative to the total number of pixels. It is computed as:

$$Accuracy = \frac{\sum (P == T)}{N} \tag{14}$$

$P == T$ denotes pixels where the prediction matches the ground truth, and N is the total number of pixels. While accuracy provides a general measure of performance, it may not be sufficient in scenarios where the data is imbalanced, as it could be dominated by the background class.

Precision: Precision measures the proportion of correctly predicted positive pixels out of all pixels predicted as positive. It is defined as:

$$Precision = \frac{TruePositives}{TruePositives + FalsePositives} \tag{15}$$

Recall: Recall, also known as sensitivity, measures the proportion of correctly predicted positive pixels out of all actual positive pixels in the ground truth. It is given by:

$$Recall = \frac{TruePositives}{TruePositives + FalseNegatives} \quad (16)$$

Number of Epochs: The number of epochs used during training is a critical evaluation metric in assessing model efficiency and optimization. While maintaining high performance on metrics such as Dice Score, IoU, Accuracy, Precision, and Recall is essential for ensuring the quality of predictions, reducing the number of epochs required to achieve these results is equally important. A lower number of training epochs signifies an optimized model capable of converging faster, thereby reducing computational costs and training time. This consideration is particularly relevant in scenarios with limited resources or time constraints, where achieving high accuracy and segmentation quality must be balanced with training efficiency.

4.4 Ablation Study

To identify the optimal architecture and parameters for the proposed model, an extensive ablation study has been conducted. The experiments performed on the dataset are as follows:

(i) U-Net with Diffusion
(ii) U-Net with Diffusion + FFT Block
(iii) U-Net with Diffusion + FFT Block + FCAM

Table 1, Table 2, and Table 3 present the significant impact of integrating frequency domain components and attention mechanisms on the segmentation performance on TNBC, GlaS and CPM17 datasets, respectively. The Bar chart, as shown in Fig. 2, compares the Dice and IoU scores of three different models across TNBC, GlaS, and CPM17 datasets. Each dataset is represented in a separate section, with bars grouped by an individual metric (Dice Score and IoU), illustrating the segmentation performance of the models. Figures 3, 4, and 5 shows the dice score and loss curves during training and validation process. Additionally, Figs. 6, 7, and 8 highlight qualitative comparisons, showcasing the segmentation results and the incremental contribution of each module to the overall model's performance. The results underscore the importance of combining

Table 1. Results of the ablation study considering the components of the segmentation model conducted over the TNBC dataset. All scores are represented in %. Bold values indicate best scores

Model	Dice Score	IoU	Accuracy	Precision	Recall	No. of epochs
(i)	87.34	77.37	84.05	86.88	87.43	300
(ii)	87.51	77.85	84.45	87.17	87.86	300
(iii)	**88.13**	**78.82**	84.37	**87.44**	**88.44**	**100**

Table 2. Results of the ablation study considering the components of the segmentation model conducted over the GLAS dataset. All scores are represented in %. Bold values indicate best scores

Model	Dice Score	IoU	Accuracy	Precision	Recall	No. of epochs
(i)	97.96	96.10	88.58	97.90	97.80	300
(ii)	98.06	96.21	88.68	98.04	98.10	300
(iii)	**98.57**	**97.18**	**90.31**	**98.51**	**98.62**	**100**

Table 3. Results of the ablation study considering the components of the segmentation model conducted over the CPM17 dataset. All scores are represented in %. Bold values indicate best scores

Model	Dice Score	IoU	Accuracy	Precision	Recall	No. of epochs
(i)	88.35	79.29	69.11	87.58	87.54	300
(ii)	88.82	79.87	69.67	89.11	88.36	300
(iii)	**89.07**	**80.32**	**74.53**	**89.32**	**88.82**	**100**

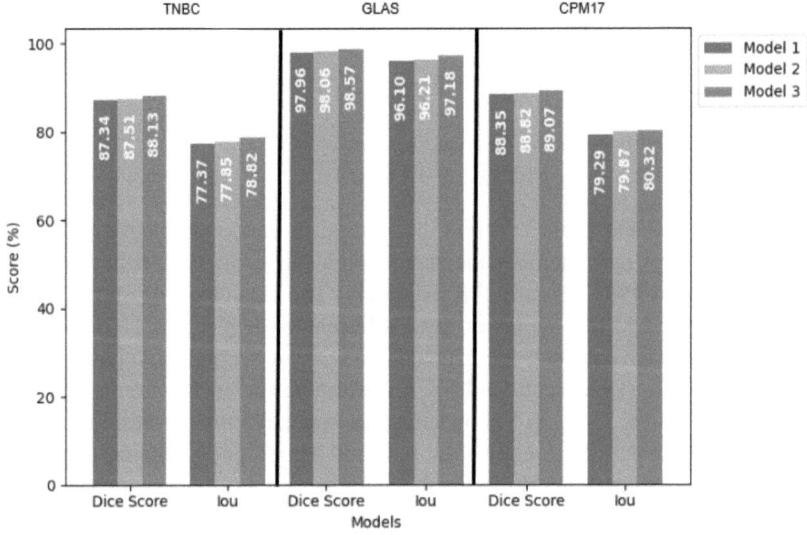

Fig. 2. The Bar chart comparison of segmentation performances across TNBC, GlAS, and CPM17 datasets.

diffusion and frequency domain processing, attention mechanism to achieve superior outcomes. Specifically, the FFT block enhances the model's capacity to capture global spatial features, while the FCAM further refines the feature representation by emphasizing critical frequency channels. The inclusion of diffusion in all configurations ensures

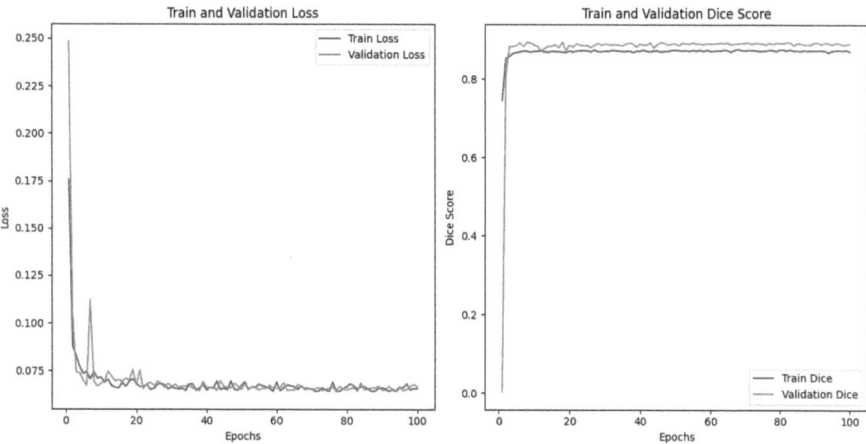

Fig. 3. Training and Loss curves representing the training and validation losses and dice scores respectively on the TNBC dataset.

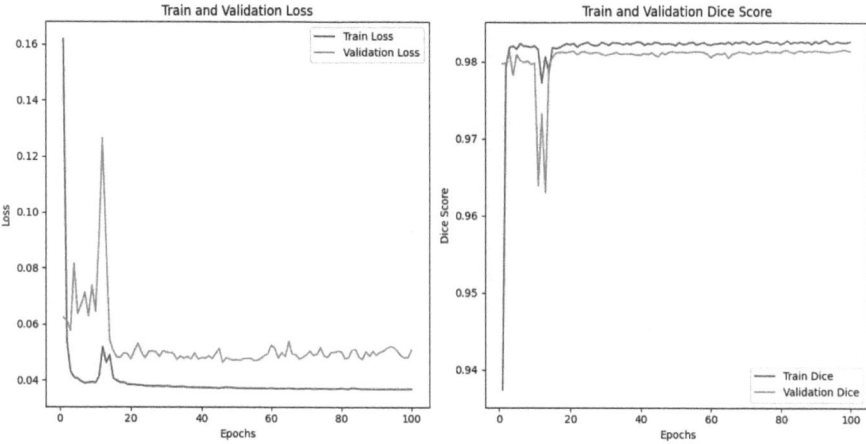

Fig. 4. Training and Loss curves representing the training and validation losses and dice scores respectively on the GlaS dataset.

the effective propagation of feature information throughout the network, contributing to improved segmentation accuracy.

Figures 6, 7, 8 show the visual representation of the heatmap images for an image from the testing set of each dataset. For each heatmap image, the following sub-plots are provided sequentially - Original Image, Ground Truth Mask, Predicted Mask, Image from Encoder Layer 1, Image from FFT Layer 1, Image from Encoder Layer 2, Image from FFT Layer 2, Image from Encoder Layer 3, Image from FFT Layer 3, Image from Encoder Layer 4, Image from FFT Layer 4, Image from Bottleneck Layer, Image

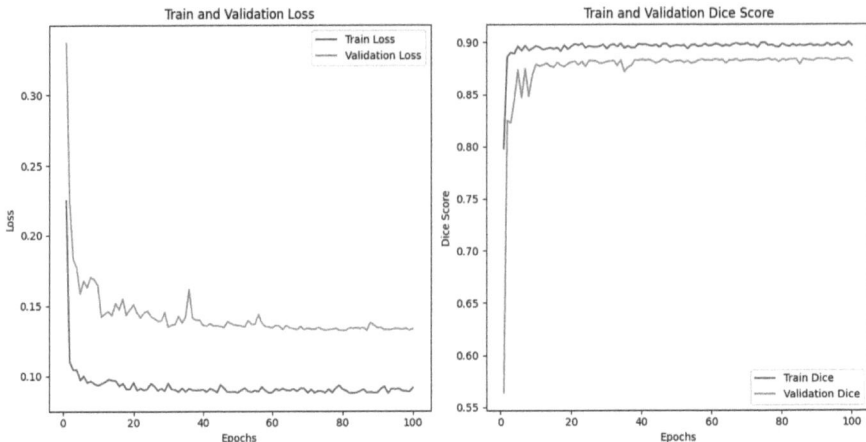

Fig. 5. Training and Loss curves representing the training and validation losses and dice scores respectively on the CPM17 dataset.

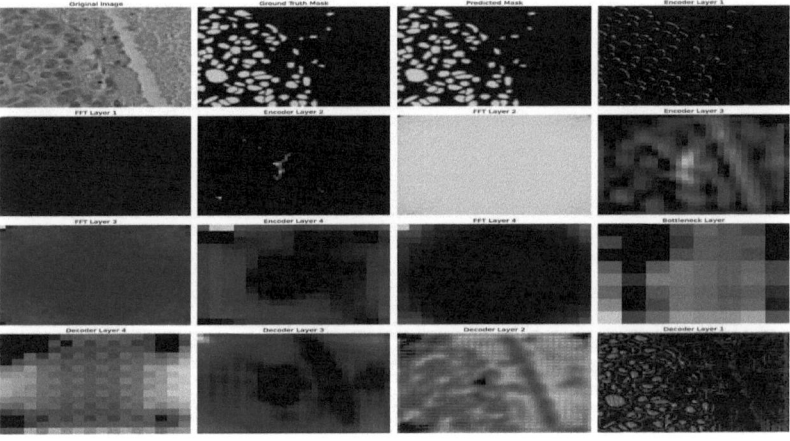

Fig. 6. Heatmap images representing segmentation results on some sample images taken from the TNBC dataset.

from Decoder Layer 1, Image from Decoder Layer 2, Image from Decoder Layer 3 and Image from Decoder Layer 4.

4.5 SOTA Comparison

Table 4 presents a comprehensive performance comparison of the proposed model against state-of-the-art (SOTA) methods on the TNBC, GlaS, and CPM17 datasets. The results highlight that the proposed model achieves the highest dice scores across all three datasets, significantly surpassing previous benchmarks. On the TNBC dataset, the

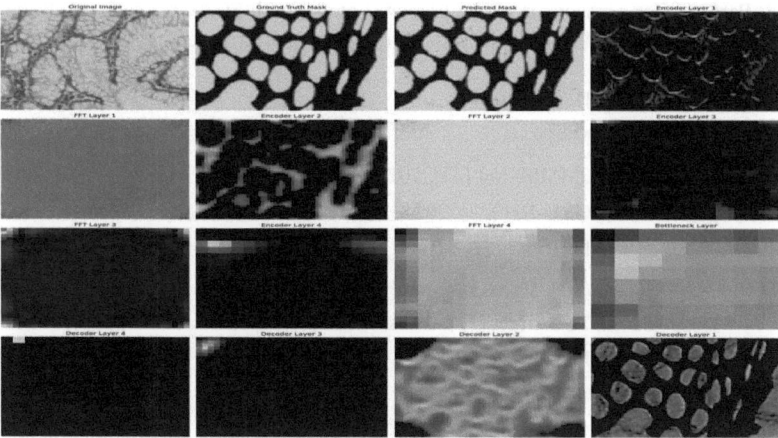

Fig. 7. Heatmap images representing segmentation results on some sample images taken from the GlaS dataset.

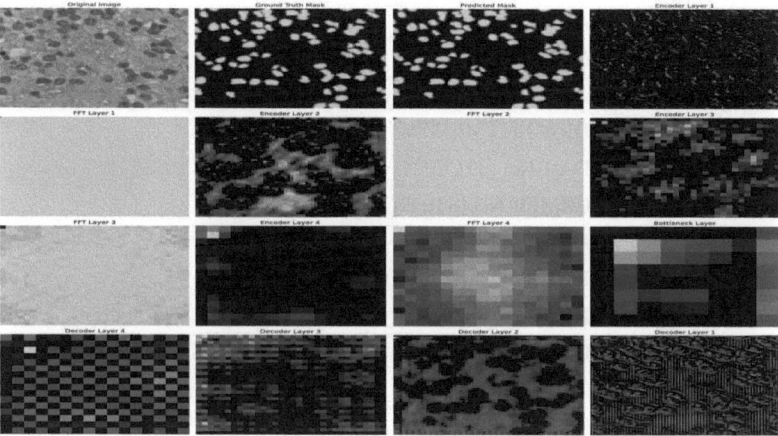

Fig. 8. Heatmap images representing segmentation results on some sample images taken from the CPM17 dataset.

model achieves a dice score of 88.13%, outperforming recent methods such as TBE-Net [36] and LAA [32]. For the GlaS dataset, the model demonstrates exceptional performance with a Dice score of 98.14%, surpassing advanced approaches like PAG-PVTUnet [2] and MSGAT [18]. Similarly, on the CPM17 dataset, the model attains a Dice score of 89.07%, exceeding the performance of methods such as CCViM [42] and HistoNeXt [5]. These results demonstrate the effectiveness of the proposed model in accurately segmenting diverse microscopic medical image datasets, establishing it as a superior choice for this domain.

Table 4. Performance comparison of the proposed model with SOTA methods on all the three datasets. All scores are in %. Bold values indicate superior performance.

Dataset: TNBC		
Model	Dice Score	IoU
2MSPK-Net [37], 2025	84.01	72.41
GRU-Net [27], 2024	80.24	66.25
DRPVit [17], 2024	81.20	69.60
AWGU-Net [28], 2024	81.65	69.18
TSCA-Net [11], 2024	81.90	69.75
LAA [32], 2024	82.96	71.13
TBE-NET [36], 2024	84.14	72.67
LEECAU-Net [10], 2024	84.20	72.70
Proposed, 2025	**88.13**	**78.82**
Dataset: GlaS		
Model	Dice Score	IoU
AFC-Unet [21], 2025	90.74	83.67
MFCNet [41], 2025	92.12	86.09
AGNR-Net [4], 2024	89.9	82.9
DRA-Net [14], 2024	90.23	82.45
DAU-Net [39], 2024	91.36	84.83
CRMEF-Net [33], 2024	92.29	86.20
MSGAT [18], 2024	93.50	90.82
PAG-PVTUnet [2], 2024	94.20	89.29
Proposed, 2025	**98.57**	**97.18**
Dataset: CPM17		
Model	Dice Score	IoU
HistoNeXt [5], 2025	87.40	77.62
CCViM [42], 2025	88.35	79.13
DenseFPN [20], 2024	81.30	68.50
BRP-Net [6], 2020	87.70	78.10
Hover-Net [12], 2019	86.9	76.83
Proposed, 2025	**89.07**	**80.32**

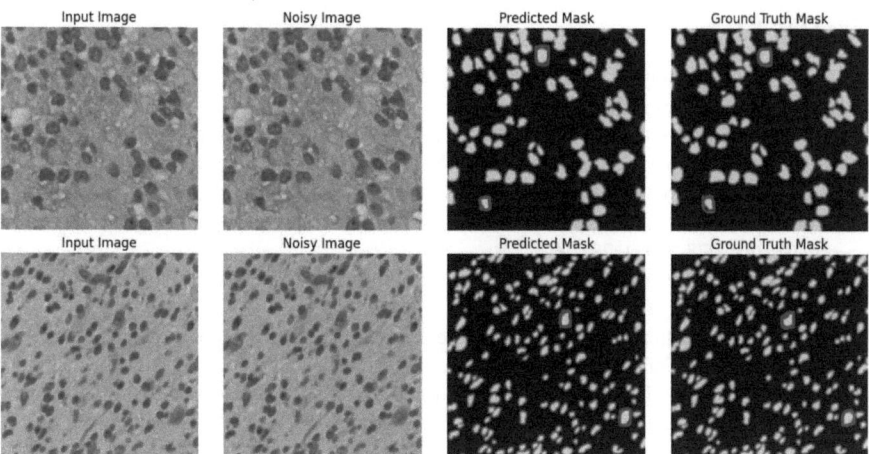

Fig. 9. Some examples of segmentation masks showing complex boundary patterns being hard to predict using the proposed model.

5 Conclusion and Future Scope

Microscopic medical image segmentation demands models that can accurately identify minute components in intricate images, which frequently have complex patterns and low-contrast. Recently, diffusion-based probabilistic models have become more popular in computer vision tasks because of their remarkable image generating capabilities. In this paper, a diffusion based U-Net model in combination with the FFT has been proposed. The proposed model employs a diffusion process to iteratively denoise images by transforming input images to the Fourier domain, where noise reduction is performed, and subsequently reverting them to the spatial domain. The U-Net architecture integrates spatial and frequency features, enhancing segmentation accuracy by leveraging feature extraction and reconstruction capabilities. The proposed model has tested on challenging microscopic medical images, and it has produced Dice scores of 88.13%, 89.07% and 98.57% on TNBC, CPM17 and GlaS datasets, respectively. Although the results are better than SOTA models considered here for comparison, it has some limitations. For example, it produces low Dice score on TNBC and CMP17 datasets, where images have very complex boundary patterns (Fig. 9).

Also the proposed model requires comparatively a high number of epochs to converge compared to non-diffusion based models. In the future, we aim to resolve the issues. We also plan to use some pre-processing techniques to enhance the input image qualities. In order to verify the robustness of our approach, we will also apply it to additional picture modalities, such as ultrasound, CT scans, and MRIs.

References

1. Amit, T., Shaharbany, T., Nachmani, E., Wolf, L.: SegDiff: image segmentation with diffusion probabilistic models (2022). https://arxiv.org/abs/2112.00390
2. Bougourzi, F., Dornaika, F., Taleb-Ahmed, A., Truong Hoang, V.: Rethinking attention gated with hybrid dual pyramid transformer-CNN for generalized segmentation in medical imaging. In: Antonacopoulos, A., Chaudhuri, S., Chellappa, R., Liu, C.L., Bhattacharya, S., Pal, U. (eds.) ICPR 2024. LNCS, vol. 15304, pp. 243–258. Springer, Cham (2025). https://doi.org/10.1007/978-3-031-78128-5_16
3. Canny, J.: A computational approach to edge detection. IEEE Trans. Pattern Anal. Mach. Intell. **PAMI-8**(6), 679–698 (1986)
4. Chen, B., et al.: Attention-guided and noise-resistant learning for robust medical image segmentation. IEEE Trans. Instrum. Meas. **73**, 1–13 (2024)
5. Chen, J., Wang, R., Dong, W., He, H., Wang, S.: HistoNeXt: dual-mechanism feature pyramid network for cell nuclear segmentation and classification. BMC Med. Imaging **25**(1), 9 (2025)
6. Chen, S., Ding, C., Tao, D.: Boundary-assisted region proposal networks for nucleus segmentation. In: Martel, A.L., et al. (eds.) MICCAI 2020. LNCS, vol. 12265, pp. 279–288. Springer, Cham (2020). https://doi.org/10.1007/978-3-030-59722-1_27
7. Cochran, W., et al.: What is the fast Fourier transform? Proc. IEEE **55**(10), 1664–1674 (1967)
8. Coleman, G., Andrews, H.: Image segmentation by clustering. Proc. IEEE **67**(5), 773–785 (1979)
9. Cortes, C.: Support-vector networks. Mach. Learn. (1995)
10. Das, S., Maity, G., Mandal, D., Kaplun, D., Voznesensky, A., Sarkar, R.: LEECAU-net: a latent entropy-quantized and efficient channel attention-aided U-net model for microscopic medical image segmentation. In: 2024 IEEE International Conference on Progress in Informatics and Computing (PIC), pp. 259–267. IEEE (2024)
11. Fu, Y., Liu, J., Shi, J.: TSCA-net: transformer based spatial-channel attention segmentation network for medical images. Comput. Biol. Med. **170**, 107938 (2024)
12. Graham, S., et al.: Hover-net: simultaneous segmentation and classification of nuclei in multi-tissue histology images. Med. Image Anal. **58**, 101563 (2019)
13. Ho, J., Jain, A., Abbeel, P.: Denoising diffusion probabilistic models. In: Larochelle, H., Ranzato, M., Hadsell, R., Balcan, M., Lin, H. (eds.) Advances in Neural Information Processing Systems, vol. 33, pp. 6840–6851. Curran Associates, Inc. (2020). https://proceedings.neurips.cc/paper_files/paper/2020/file/4c5bcfec8584af0d967f1ab10179ca4b-Paper.pdf
14. Huang, Z., Wang, L., Xu, L.: DRA-net: medical image segmentation based on adaptive feature extraction and region-level information fusion. Sci. Rep. **14**(1), 9714 (2024)
15. Husham, A., Hazim Alkawaz, M., Saba, T., Rehman, A., Saleh Alghamdi, J.: Automated nuclei segmentation of malignant using level sets. Microsc. Res. Tech. **79**(10), 993–997 (2016)
16. Jha, D., et al.: ResUNet++: an advanced architecture for medical image segmentation. In: 2019 IEEE International Symposium on Multimedia (ISM), pp. 225–2255 (2019)
17. Li, L., He, K., Zhu, X., Gou, F., Wu, J.: A pathology image segmentation framework based on deblurring and region proxy in medical decision-making system. Biomed. Signal Process. Control **95**, 106439 (2024)
18. Liu, Y., Yun, H., Xia, Y., Luan, J., Li, M.: MSGAT: multi-scale gated axial reverse attention transformer network for medical image segmentation. Biomed. Signal Process. Control **95**, 106341 (2024)
19. Long, J., Shelhamer, E., Darrell, T.: Fully convolutional networks for semantic segmentation. In: Proceedings of the IEEE Conference on Computer Vision and Pattern Recognition, pp. 3431–3440 (2015)

20. Luna, M., Chikontwe, P., Nam, S., Park, S.H.: Attention guided multi-scale cluster refinement with extended field of view for amodal nuclei segmentation. Comput. Biol. Med. **170**, 108015 (2024)
21. Meng, W., Liu, S., Wang, H.: AFC-UNet: attention-fused full-scale CNN-transformer UNet for medical image segmentation. Biomed. Signal Process. Control **99**, 106839 (2025)
22. Naylor, P., Laé, M., Reyal, F., Walter, T.: Segmentation of nuclei in histopathology images by deep regression of the distance map. IEEE Trans. Med. Imaging **38**(2), 448–459 (2018)
23. Olaf Ronneberger, P.F., Brox, T.: U-net: convolutional networks for biomedical image segmentation. In: Proceedings of the IEEE/CVF Conference on Computer Vision and Pattern Recognition, pp. 234–241 (2015)
24. Qi, J.: Dense nuclei segmentation based on graph cut and convexity-concavity analysis. J. Microsc. **253**(1), 42–53 (2014)
25. Qin, Z., Zhang, P., Wu, F., Li, X.: FcaNet: frequency channel attention networks. In: Proceedings of the IEEE/CVF International Conference on Computer Vision (ICCV), pp. 783–792 (2021)
26. Ranefall, P., Egevad, L., Nordin, B., Bengtsson, E.: A new method for segmentation of colour images applied to immunohistochemically stained cell nuclei. Anal. Cell. Pathol. **15**(3), 145–156 (1997)
27. Roy, A., Pramanik, P., Ghosal, S., Valenkova, D., Kaplun, D., Sarkar, R.: GRU-net: gaussian attention aided dense skip connection based multiresunet for breast histopathology image segmentation. In: Yap, M.H., Kendrick, C., Behera, A., Cootes, T., Zwiggelaar, R. (eds.) MIUA 2024. LNCS, vol. 14859, pp. 300–313. Springer, Cham (2024). https://doi.org/10.1007/978-3-031-66955-2_21
28. Roy, A., Pramanik, P., Kaplun, D., Antonov, S., Sarkar, R.: AWGUNET: attention-aided wavelet guided U-net for nuclei segmentation in histopathology images. arXiv preprint arXiv:2406.08425 (2024)
29. Sirinukunwattana, K., et al.: Gland segmentation in colon histology images: the GlaS challenge contest. Med. Image Anal. **35**, 489–502 (2017)
30. Sornapudi, S., et al.: Deep learning nuclei detection in digitized histology images by superpixels. J. Pathol. Inform. **9**(1), 5 (2018)
31. Vu, Q.D., et al.: Methods for segmentation and classification of digital microscopy tissue images. Front. Bioeng. Biotechnol. **7** (2019)
32. Wang, L., Xu, P., Cao, X., Nappi, M., Wan, S.: Label-aware attention network with multi-scale boosting for medical image segmentation. Expert Syst. Appl. **255**, 124698 (2024)
33. Wang, Z., Yu, L., Tian, S., Huo, X.: CRMEFNet: a coupled refinement, multiscale exploration and fusion network for medical image segmentation. Comput. Biol. Med. **171**, 108202 (2024)
34. Wu, J., et al.: MedSegDiff: medical image segmentation with diffusion probabilistic model (2023). https://arxiv.org/abs/2211.00611
35. Xu, H., Lu, C., Mandal, M.: An efficient technique for nuclei segmentation based on ellipse descriptor analysis and improved seed detection algorithm. IEEE J. Biomed. Health Inform. **18**(5), 1729–1741 (2014)
36. Yang, S., Zhang, X., He, Y., Chen, Y., Zhou, Y.: TBE-net: a deep network based on tree-like branch encoder for medical image segmentation. IEEE J. Biomed. Health Inform. **29**(1), 521–534 (2025)
37. Yue, G., Ma, X., Li, W., An, Z., Yang, C.: 2MSPK-net: a nuclei segmentation network based on multi-scale, multi-dimensional attention, and SAM prior knowledge. Biomed. Signal Process. Control **100**, 107140 (2025)
38. Zhao, H., Shi, J., Qi, X., Wang, X., Jia, J.: Pyramid scene parsing network. In: Proceedings of the IEEE Conference on Computer Vision and Pattern Recognition (CVPR) (2017)

39. Zhou, J., Xiong, H., Liu, Q.: A novel dual-branch asymmetric encoder–decoder segmentation network for accurate colonic crypt segmentation. Comput. Biol. Med. **173**, 108354 (2024)
40. Zhou, Z., Rahman Siddiquee, M.M., Tajbakhsh, N., Liang, J.: UNet++: a nested U-net architecture for medical image segmentation. In: Stoyanov, D., et al. (eds.) DLMIA/ML-CDS -2018. LNCS, vol. 11045, pp. 3–11. Springer, Cham (2018). https://doi.org/10.1007/978-3-030-00889-5_1
41. Zhu, C., et al.: Multi-perspective feature compensation enhanced network for medical image segmentation. Biomed. Signal Process. Control **100**, 107099 (2025)
42. Zhu, Y., Zhang, D., Lin, Y., Feng, Y., Tang, J.: Merging context clustering with visual state space models for medical image segmentation. IEEE Trans. Med. Imaging 1 (2025)

Non-cooperative Game Theory-Aided Learning of CNN Models for Skin Lesion Classification

Diptarka Mandal[1], Sujan Sarkar[2], Siddhant Majumder[2], Dmitrii Kaplun[3,4(✉)], Daria Sidorina[4], and Ram Sarkar[2]

[1] Department of Applied Mathematics, Defence Institute of Advanced Technology, Pune, India
[2] Department of Computer Science and Engineering, Jadavpur University, Kolkata, India
[3] Artificial Intelligence Research Institute, China University of Mining and Technology, Xuzhou, China
[4] Department of Automation and Control Processes, Saint Petersburg Electrotechnical University "LETI", Saint Petersburg, Russia
dikaplun@etu.ru

Abstract. Millions of people worldwide are afflicted with skin cancer every year. It happens when DNA damage from UV radiation from the sun or tanning beds causes skin cells to proliferate out of control. Skin lesion analysis is one of the many medical imaging jobs where computer vision models are widely used. Convolutional Neural Networks (CNNs), in particular, are deep learning models that have demonstrated proficiency in extracting pertinent information from images and offer high accuracy in classification tasks. In order to train several CNN models simultaneously on a dataset, this work suggests a novel non-cooperative game theory-based method that involves forcing the models to compete with one another. The interaction between the models not only improves their overall performance, but also underscores the efficiency of our approach in optimizing training time. This approach is evaluated on two publicly available skin lesion datasets, namely HAM10000 dataset for multi-class classification yielding 92.93% accuracy, and PH2 for binary class classification achieving 99.88% accuracy. The code of the proposed methodology can be found at: GitHub repository.

Keywords: Medical image · Skin cancer · Game theory · Collaborative learning · Deep learning · Ensemble approach

1 Introduction

Early detection as well as accurate diagnosis is crucial for successful treatment of diseases. With the progress in image processing, machine learning is steadily being used to perform early detection of diseases by analyzing various medical images. Deep learning models, especially Convolutional Neural Networks (CNNs), have proved specially adept at extracting useful features from images and are being used successfully in different medical imaging tasks including segmentation, detection and classification.

D. Mandal, S. Sarkar and D. Kaplun—Authors contributed equally to this research.

© The Author(s), under exclusive license to Springer Nature Switzerland AG 2025
A. Hadjali et al. (Eds.): DeLTA 2025, CCIS 2627, pp. 51–69, 2025.
https://doi.org/10.1007/978-3-032-04339-9_4

Skin cancer is the most common type of cancer in the United States, with more than 5 million cases diagnosed each year. Non-melanoma skin cancers, including basal cell carcinoma and squamous cell carcinoma, are the most frequently occurring [25]. The 5-year survival rate for melanoma, the deadliest form of skin cancer, is 99% when detected early but drops significantly if it has spread to other parts of the body. Early detection is crucial for successful treatment. Skin lesions vary widely in appearance due to factors like skin type, lesion type, and stage of development. This diversity makes it challenging to create a universal classification system that accurately categorizes all possible lesions. Some sample images taken from two different skin lesion datasets show the challenges in skin lesion classification. Figure 1 shows the sample images of the HAM10000 dataset and the sample images are shown in Fig. 2 taken fron the PH2 dataset.

Game theory is the study of strategic interactions, where players make decisions to maximize their payoffs. Non-cooperative game theory studies strategic interactions, where each player acts independently to maximize their own payoff, often without cooperation. In deep learning, it is applied in areas such as adversarial training, where models are trained to improve by competing against each other, such as in Generative Adversarial Networks (GANs) [12]. In the recent past, this approach has been successfully used in different image processing and computer vision applications. For example, game theory concept has been in the domain of feature selection for text classifications by [11], robotics by [30], document image binarization by [5] etc. [29] proposed a model that combined the output of three separately trained attention-based CNNs: ResNet50, InceptionResNetV2, and Dense-Net201, using the Dempster-Shafer theory which were assisted by a soft-attention module. [4] applied the Choquet integral for ensembles and proposed a novel method for evaluating fuzzy measures: coalition game theory, information theory, and the Lambda fuzzy approximation. Three different sets of fuzzy measures were calculated using three different weighting systems, in addition to coalition game theory and information theory. These three fuzzy measure sets were used to calculate three Choquet integrals, and the results were then combined.

Motivation and Contributions. Our work proposes a novel skin lesion classification method by training multiple CNN models using a game theory based approach by making the models play a non-cooperative game. In tackling the challenges of medical image classification, particularly the optimization of deep learning models' efficiency, we introduce a novel non-cooperative game theory-based framework where two base models compete to minimize their validation loss, thereby accelerating convergence and reducing the number of training epochs needed, conserving computational resources. To enhance the effectiveness of this approach, we employ Shapley values to ensemble the results of the competing models, ensuring that the final prediction fairly reflects the contribution of each model, leading to more accurate outcomes. This approach is validated using HAM10000 (multi-class) and PH2 (binary-class) datasets, demonstrating its effectiveness in improving the accuracy of medical image classification, specifically for skin lesion analysis.

- We propose a non-cooperative game theory-based approach, where two CNN models compete to minimize their respective validation losses.

- Each model operates with an adaptive learning rate, adjusting dynamically to outperform the another.
- Each model's contribution is taken care of and the final prediction is made using a Shapely value-based ensemble of the models.
- Competition between models accelerates the training process, leading to a significant reduction in validation losses.
- The interaction between models not only improves overall performance, but also underscores the efficiency of the approach in optimizing training outcomes.

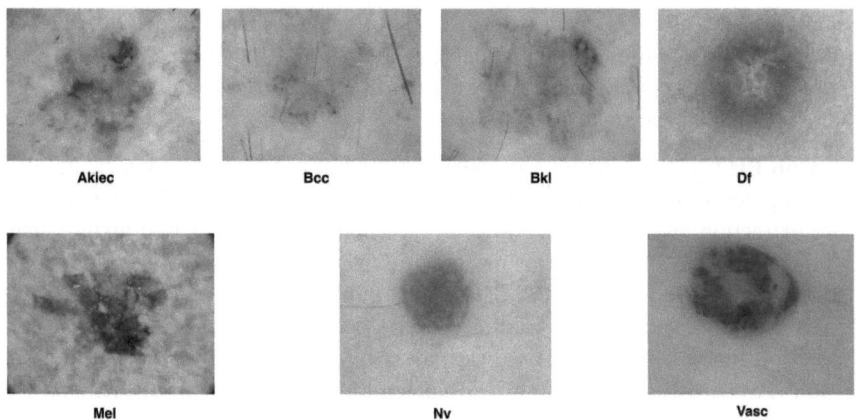

Fig. 1. Sample images from different classes of the HAM10000 dataset.

2 Literature Review

In this section, we discuss some recent methods developed for skin cancer classification.

Survey on HAM10000: [19] developed deep learning frameworks to address class imbalance in skin lesion classification and improve the interpretability of the decision-making process. They also proposed an end-to-end healthcare system via an Android application. [13] developed a CNN-based approach for skin cancer detection, utilizing data pre-processing techniques such as sampling, dull razor, and segmentation with an autoencoder-decoder on the HAM10000 dataset. [26] proposed the Margin-Aware Adaptive-Weighted (MAAW) loss function to address class imbalance and reduce overfitting by separating overlapping classes. [31] proposed a two-stage, low-cost data augmentation strategy for skin lesion classification using the EfficientNet model, validated on the HAM10000 dataset. The approach achieved a balanced accuracy of 0.853, outperforming many ensemble models, and is suitable for deployment in low-resource settings. [14] proposed a Levy Stable-based ensemble method using MobileNet-V2 and EfficientNet-B0 for skin cancer classification on the HAM10000 dataset, achieving an accuracy of 83%. The model was optimized using the OpenVINO toolkit, resulting in real-time inference speeds of up to 110.3 FPS on an Intel i7 CPU. [28] introduced a classification method where multiple clients utilized the same model architecture and refined weights iteratively based on their local data. This approach enhanced model learning, maintained data privacy, and boosted accuracy by incorporating diverse datasets. Using a VGG11 model with a self-attention layer and four clients, the method achieved 92.84% accuracy, 93.02% precision, 92.84% recall, and a 92.89% F1-score on the HAM10000 dataset. [8] proposed a system including a multi-class CNN and six binary CNNs that were assimilated to players. The players' approach was to first group the pigmented lesions (melanoma, nevus, and benign keratosis) into confidence levels (confident, medium, unsure) using the recently proposed method of evaluating the predictions' level of confidence. A team of players could then improve the diagnosis for difficult lesions with medium and unclear prediction. EfficientNetB5 served as the foundation for our networks, and they used the ISIC dataset to evaluate their methods by achieving the balanced accuracy of 86%.

Survey on PH2: [36] proposed a machine learning model for early skin cancer diagnosis using dermoscopic images. The model employed the HSV [23] color space for feature extraction and gray level co-occurrence matrix (GLCM) based texture analysis, followed by the support vector machine (SVM) based classification. Trained and tested on 200 images from the PH2 dataset, the model achieved a 96% accuracy. [32]. proposed an automated machine learning-based system for skin cancer diagnosis using dermoscopic images. The model used various texture features and Particle swarm optimization (PSO) for feature selection. Evaluated on 1500 images from PH2 and Dermofit, it achieved classification accuracies of 97.79% with SVM and 97.54%, respectively with the K-Nearest Neighbor (KNN) classifier. [15] used a pre-trained AlexNet for feature extraction in their classification model, freezing the initial layers and replacing the final layers with a softmax classifier. The model was tested on the PH2 dataset to classify melanoma, common nevus, and atypical nevus lesions.

[24] proposed a new approach for skin lesion recognition using a CNN model with 69 layers. The proposed method was tested on three benchmark datasets of skin lesions including PH2. Their method obtained 97.2% of accuracy on the PH2 dataset. [9] proposed an automated framework for melanoma detection that uses a pre-trained CNN

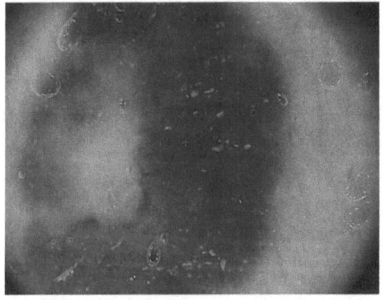
Melanoma

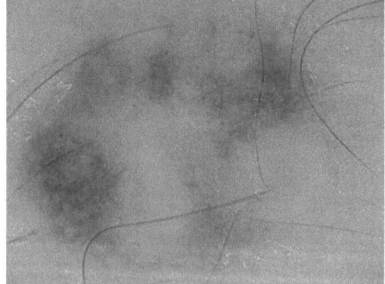
Others

Fig. 2. Sample images from different classes of the PH2 dataset.

model to extract visual features from dermoscopy images, followed by classifiers for diagnosis. They evaluated deep features from eight CNN models and studied the effects of boundary localization and normalization techniques.

3 Methodology

We have proposed a non-cooperative game theory-based resource efficient deep learning framework for accurate classification of skin lesions. In the subsequent sections, we have described the workflow of our proposed methodology in depth along with a brief discussion about the base models that act as individual players in our framework. The framework is demonstrated in Fig. 3.

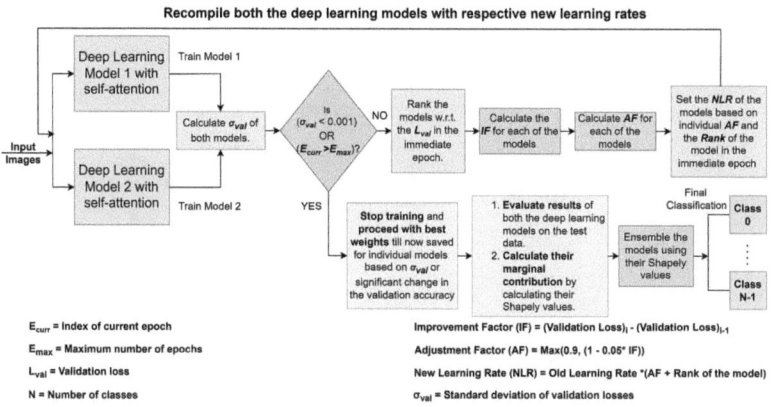

Fig. 3. Workflow of the proposed approach used for skin lesion classification

3.1 DenseNet201

DenseNet201 leverages dense connectivity, where each layer is connected to every other layer, enhancing feature propagation and gradient flow. This deep architecture, with 201 layers, efficiently captures intricate patterns necessary for medical image classification, especially in tasks requiring fine-grained analysis. Its parameter efficiency and strong generalization make it ideal for applications with limited computational resources, ensuring reliable performance in diverse medical imaging scenarios.

3.2 EfficientNet-B0

EfficientNet-B0 utilizes a compound scaling approach that balances network depth, width, and resolution, achieving high performance with minimal computational cost. The inclusion of Squeeze-and-Excitation blocks enhances its focus on relevant features, crucial for medical imaging tasks. Its efficiency and scalability make it well-suited for deployment in resource-constrained environments, providing robust and accurate classification in various medical imaging applications.

3.3 Self-attention

Incorporating self-attention blocks [35] with DenseNet201 and EfficientNet-B0 enhances their ability to focus on the most critical regions of medical images, thereby improving classification accuracy. Self-attention mechanism shown in Fig. 4 allow the models to dynamically weigh the importance of different features across the entire image, enabling them to capture contextually relevant details that might be subtle or dispersed. This is particularly valuable in medical imaging, where small but significant features can indicate the presence of a disease. By integrating self-attention, DenseNet201 and EfficientNet-B0 not only retain their strengths—such as efficient feature reuse and balanced resource utilization—but also gain a heightened capacity for discerning complex patterns, leading to obtain more precise and reliable diagnostic outcomes.

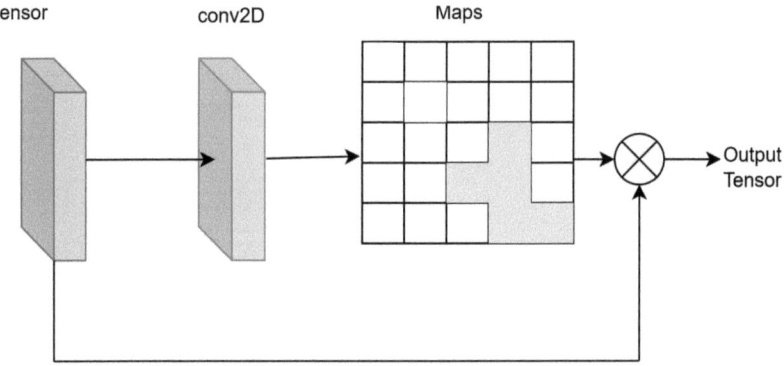

Fig. 4. A block diagram of the self-attention module.

3.4 Improvement Factor and Adjustment Factor

We define the difference between the validation loss of a model in the current epoch from the validation loss of the same model in the previous epoch as the **Improvement Factor**. This gives us a measure of how much a model has improved because of the current epoch. The **Improvement Factor** of a model is normalized to a value in (0,1). This value is termed the **Adjustment Factor** since its used for dynamic learning rate adjustment according to the performances of the models.

3.5 Shapley Value

Shapely value [37] is a concept in the game theory used to find the average marginal contribution of each players. In our classification problem, there are two players (i.e., models). Shapely value provides us a measure of the contribution made by a single model prediction in the combined prediction. These shapely values are then used as weights to perform a weighted fusion for an improved prediction.

3.6 Competitive Learning

In this work, the two models are made to take part in a game during their training, which dynamically adjusts the learning rate of the models depending on their performance. Standard deviation of the validation loss of the two models are obtained after completion of an epoch. If the standard deviation of both the base model's validation losses is less than the models' previous lowest standard deviation of the validation losses or the validation accuracies obtained is better than the previous best validation accuracies, then

---

Algorithm 1. Proposed Approach.

Input: Model set, $\mathbf{M} = \{M_1, M_2, \ldots, M_n\}$; Maximum number of epochs, $\mathbf{E_{max}}$; a threshold, $\mathbf{T}$; set of learning rates of M, $\mathbf{LR\_list}$; a training set, $\mathbf{Train}$; a validation set, $\mathbf{Valid}$; a test image, $\mathbf{Test\_img}$;
Output: Output class, $\mathbf{C_{out}}$;
$OLR \leftarrow LR\_list$
$min\_std\_dev \leftarrow \infty$
for i ← 1 **to** n **do**
  $max\_val\_acc \leftarrow -\infty$
end for
for e ← 1 **to** E_{max} **do**
  for i ← 1 **to** n **do**
    $Train\_model(M[i], Train, Valid, e)$
    $V[i] \leftarrow CalculateValLoss(M[i])$
    $val\_acc[i] \leftarrow CalculateValAccuracy(M[i])$
  end for
  $std\_dev \leftarrow CalculateStandardDeviation(V)$
  if $(std\_dev >= T)$ **then**
    $Rank\_list \leftarrow CalculateRank(M, V)$
    for i ← 1 **to** n **do**
      $IF[i] \leftarrow CalculateIF(M[i])$
      $AF[i] \leftarrow CalculateAF(M[i])$
      $NLR[i] \leftarrow FindNLR(AF[i], Rank\_list[i], OLR[i])$
    end for
    for i ← 1 **to** n **do**
      if $(std\_dev < min\_std\_dev)$ **then**
        SaveWeights(M[i])
        $min\_std\_dev \leftarrow std\_dev$
      else if $(val\_acc[i] > max\_val\_acc[i])$ **then**
        SaveWeights(M[i])
        $max\_val\_acc[i] \leftarrow val\_acc[i]$
      end if
    end for
  end if
end for
$P \leftarrow PredictConfidenceScore(M, Test\_img)$
$MC \leftarrow CalculateShapleyValue(M)$
$C_{out} \leftarrow FinalPrediction(P, MC)$

the model weights are saved. If the standard deviation is less than the threshold then the game reaches the equilibrium, and the training of models is stopped. Otherwise, learning rates are adjusted and the training continues. For the adjustment of learning rates, the **Adjustment Factor** (see Subsect. 3.4) is employed for each model after completion of an epoch. To get the new learning rate, the **Adjustment Factor** is multiplied with the current learning rate and the maximum of the product and 10^{-6} is taken as the new learning rate to ensure that the new learning rate is not negligible.

The in-depth algorithm behind this game-theory based training of CNN models is described in Algorithm 1.

3.7 Shapley Value-Based Classification

To ensemble two models M_i and M_j using Shapley values, we calculate the contribution of each model by considering all possible permutations and combinations of the models. The Shapley value for each model, denoted as ϕ_{M_i} and ϕ_{M_j}, represents the average marginal contribution of that model to the prediction. In Eq. 1, $f(\{M_i\})$ represents the prediction made by model M_i, $f(\emptyset)$ is the baseline prediction, and $f(\{M_i, M_j\})$ is the prediction using both models together. The final ensemble prediction P is computed as a weighted sum of the individual model predictions P_{M_i} and P_{M_j}, with the Shapley values as weights using Eq. 2.

$$\phi_{M_i} = \frac{1}{2}[f(\{M_i\}) - f(\emptyset)] + \frac{1}{2}[f(\{M_i, M_j\}) - f(\{M_j\})] \qquad (1)$$

$$P = \phi_{M_i} \times P_{M_i} + \phi_{M_j} \times P_{M_j} \qquad (2)$$

4 Experimental Results

In this section, we have reported the experimental setup, dataset details, comparison with state-of-the-art models, ablation studies along with some visualizations like loss curves, and confusion matrices. We have also provided the results of Wilcoxon ranksum test to ensure that the proposed framework provides a statistically significant result. Our model is evaluated on the said datasets in terms of some standard evaluation metrics that include classification accuracy, precision, recall and F1-score.

4.1 Experimental Setup

An initial learning rate of 0.001 and the ADAM optimizer are used in our experiments. The batch size has been set to 8 and the models are trained for a maximum of 60 epochs. All methods are implemented using PyTorch on an NVIDIA TESLA P100 GPU.

4.2 Dataset Description and Preprocessing

HAM10000: The Human Against Machine dataset [33], also known as HAM10000, is a publicly available skin lesions dataset comprising dermoscopic images sized at 450×600 pixels. These images primarily originated from Cliff Rosendahl's skin cancer practice in Queensland, Australia, and the Dermatology Department of the Medical University of Vienna, Austria. Collected from diverse populations, various acquisition and cleaning methods were employed, along with the development of semi-automatic workflows, to address diversity issues. The dataset contains 10,015 images categorized into seven skin disease groups: 327 actinic keratosis and intraepithelial carcinoma (Akiec) images, 514 basal cell carcinoma (Bcc) images, 1099 benign keratosis (Bkl) images, 115 dermatofibroma (Df) images, 1113 melanoma (Mel) images, 6705 melanocytic nevus (Nv) images, and 142 vascular malformation (Vasc) images.

PH2: The PH2 dataset [18] comprises a total of 200 dermoscopic images, consisting of 80 common nevi, 80 atypical nevi, and 40 melanomas. The images are sized at 768×560 pixels. Each image includes ground truth diagnosis, age, sex, and lesion location. As the number of samples in the dataset is pretty low, we have augmented the original dataset using different augmentation techniques like HorizontalFlip, VerticalFlip, Cropping, Zooming in, Zooming Out and have splitted it into 70% for training, 10% for validation and 20% for testing. We have converted this dataset into a 2-class dataset to check the robustness of our model.

MSID: To ensure the robustness of the proposed model, it has additionally been tested on another medical image classification dataset, namely the Monkeypox dataset [2]. The MSID dataset comprises a limited number of images. Following data augmentation, the total number of images per class in the training and test sets includes Chickenpox (1118 images), Measles (939 images), Monkeypox (2651 images), and Normal (2736 images). The dataset is split into an 80/20 ratio for training and testing, with 10% of the training data reserved for validation.

We have also tested our model's performance on the CIFAR-10 [17], a benchmarking dataset to check the domain adaption capability of our proposed methodology.

4.3 Comparison with Past Models

Different approaches have been developed over the time for accurate classification of different skin lesions. Among them, some of the models have been proven to be highly effective in early diagnosis of skin cancer. Here we have compared different performance aspects of our proposed methodology with some state-of-the-art models' performance on HAM10000 and PH2 datasets and it has outperformed existing models on HAM10000 (shown in Table 1) and PH2 (shown in Table 2). [10] developed an ensemble of deep learning models, primarily EfficientNets, to classify skin lesions by leveraging multi-resolution inputs, extensive data augmentation, loss balancing, and integration of patient metadata through a dense neural network. The approach included different cropping strategies and a data-driven method to handle an unknown class in the test set. According to the accuracy achieved on HAM10000 dataset, they are the second highest one after our proposed methodology. [13] proposed a method utilizing CNNs for the classification of skin cancer lesions. The process includes data preprocessing steps like sampling, noise removal using the Dull Razor technique, and image segmentation using encoder-decoder architectures, followed by training the model using transfer learning techniques such as DenseNet169 and ResNet50. This technique gave an prominent result on the HAM10000 dataset keeping them in the third best position. [3] used DenseNet201 for deep feature extraction and ANN for classification purposes. They achieved an accuracy of 99% on the PH2 dataset, making them the second best on the list, after our proposed methodology. [9] proposed an automated framework for melanoma detection in dermoscopy images using deep features from eight CNN models and various classifiers. They found DenseNet-121 with a multi-layer perceptron (MLP) achieved the accuracy of 98.33% on the PH2 dataset, giving them the third best on the list.

Table 1. Performance comparison between proposed model and some existing methods on the HAM10000 dataset. All values are in %.

Model	Accuracy	Precision	Recall	F1-Score
[10]	92.60	–	–	–
[38]	86.40	–	–	–
[16]	88.80	90.50	88.80	89.10
[31]	85.30	–	–	–
[19]	82.00	–	–	–
[13]	91.20	–	–	91.70
[14]	83.00	83.00	83.00	83.00
[26]	–	72.30	70.60	70.20
[20]	88.00	88.00	88.00	88.00
[6]	87.75	–	–	–
[1]	91.43	–	–	–
Proposed	**92.93**	**93.02**	**92.84**	**92.89**

4.4 Additional Results

Our model outperforms most of the existing methods as listed in Table 4, when tested on the MSID dataset.

The confusion matrix of our proposed model on the Monkeypox dataset is shown in Fig. 5.

From Table 3, we can see that a threshold of **0.001** results in the best performance for our model.

Table 2. Performance comparison between proposed model and some past models on the PH2 dataset. All values are in %.

Model	Accuracy	Precision	Recall	F1-Score
[36]	96.00	–	–	–
[22]	92.50	–	–	–
[32]	97.80	–	–	–
[7]	93.30	–	92.30	–
[3]	99.00	–	–	–
[9]	98.30	–	–	–
[24]	97.20	–	–	–
[27]	98.00	–	100.0	–
Proposed	**99.88**	**99.60**	**99.67**	**99.63**

Table 3. Comparison of the model performance across various threshold values. Scores are in %.

Threshold	Dataset	Accuracy	Precision	Recall	F1 Score
0.0001	HAM10000	90.28	89.45	91.34	90.32
	PH2	99.14	98.86	99.20	98.81
	MSID	94.28	93.10	94.51	95.25
	CIFAR-10	95.45	97.12	94.20	95.46
0.0005	HAM10000	91.52	91.73	90.23	91.65
	PH2	99.35	99.12	98.82	99.21
	MSID	95.17	94.25	95.66	96.58
	CIFAR-10	96.51	96.82	96.21	96.50
0.001	HAM10000	**92.93**	93.02	92.84	92.89
	PH2	**99.88**	99.60	99.67	99.63
	MSID	**96.59**	96.22	95.84	97.11
	CIFAR-10	**98.97**	99.10	98.80	98.95

Table 4. Performance comparison between the proposed model and some past models on the Monkeypox (MSID) dataset. Scores are in %.

Model	Methodologies used	Accuracy
[34]	Two deep learning models with the LSTM network.	85.00
[21]	Transfer learning models like residual networks and SqueezeNet.	91.00
[2]	A newly modified DenseNet-201-based deep CNN model	93.19
Proposed	Self-attention aided EfficientNetB0 and DenseNet201 models with a game-theory based framework	**96.59**

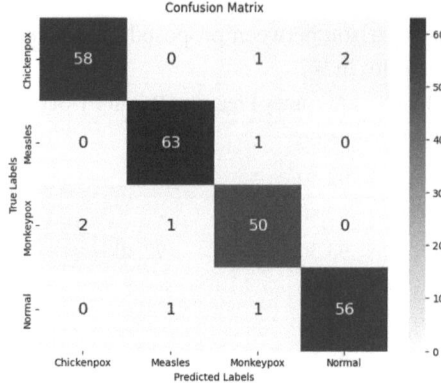

Fig. 5. Confusion matrix of the proposed model on the Monkeypox dataset.

4.5 Ablation Studies

In this section, we have shown the results of ablation studies on HAM10000 and PH2 datasets to make sure that our proposed framework works well on both the datasets. Additionally we have tested our proposed model on the CIFAR-10 dataset to check whether our model is generalizable to all kinds of datasets. In doing so, we have also recorded performances of the two base CNN models with and without the self-attention module.

Table 5. Ablation study on the HAM10000 dataset.

Model	Accuracy	Convergence epoch	Validation loss
1. EfficientNetB0	0.8967	60	0.0010
2. DenseNet201	0.9082	60	0.0017
3. EfficientNetB0 + self-attention	0.9024	60	0.0011
4. DenseNet201 + self-attention	0.9125	60	0.0019
5. (1)+(2)+ game-theory framework	0.9293	18	0.0020

4.6 Statistical Analysis

For statistical analysis of the results, we have performed the non-parametric Wilcoxon rank-sum test. We have compared p-values of the base models with our proposed model, as shown in Fig. 6, to see the if p-values are lower than the standard 0.05. It should be noted that we have considered the loss values of those models. From this figure, we can clearly see our proposed model's loss values are significantly different from the individual base models with self-attention mechanism. Hence, it disproves the null hypothesis, which assumes our proposed model is statistically similar to other models.

Table 6. Ablation study on the PH2 dataset.

Model	Accuracy	Convergence epoch	Validation loss
1. EfficientNetB0	0.9883	60	0.0019
2. DenseNet201	0.9897	60	0.0016
3. EfficientNetB0 + self-attention	0.9907	60	0.0018
4. DenseNet201 + self-attention	0.9942	60	0.0015
5. (1)+(2)+ game-theory framework	0.9988	22	0.0017

Table 7. Ablation study on the CIFAR-10 dataset.

Model	Accuracy	Convergence epoch	Validation loss
1. EfficientNetB0	0.9436	60	0.0021
2. DenseNet201	0.9680	60	0.0015
3. EfficientNetB0 + self-attention	0.9572	60	0.0020
4. DenseNet201 + self-attention	0.9725	60	0.0012
5. (1)+(2)+ game-theory framework	0.9897	27	0.0015

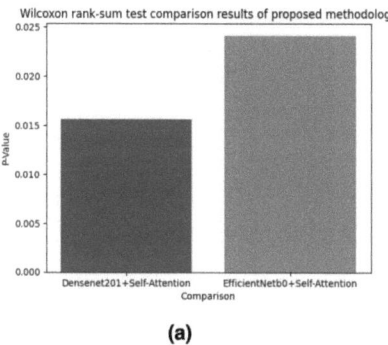
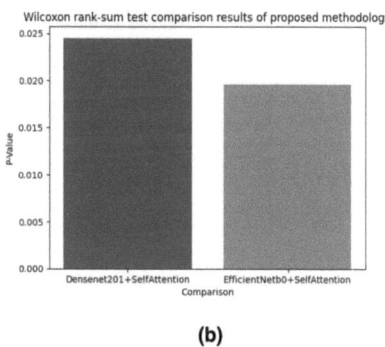

(a) (b)

Fig. 6. Comparison of p-values of the validation losses on (a) HAM10000 and (b) PH2 datasets

5 Discussion

Classifying skin lesions accurately involves a great deal of intricacy. HAM10000 and PH2 datasets present unique difficulties in this regard. Using the suggested non-cooperative game theory-based paradigm, we introduce a unique method to solve issue that shows great promise for optimizing the training of deep learning models. The method efficiently reduces the number of training epochs needed by accelerating convergence to reduced validation losses by generating a competitive dynamic between two CNN models. The gradual convergence of the validation losses is demonstrated in

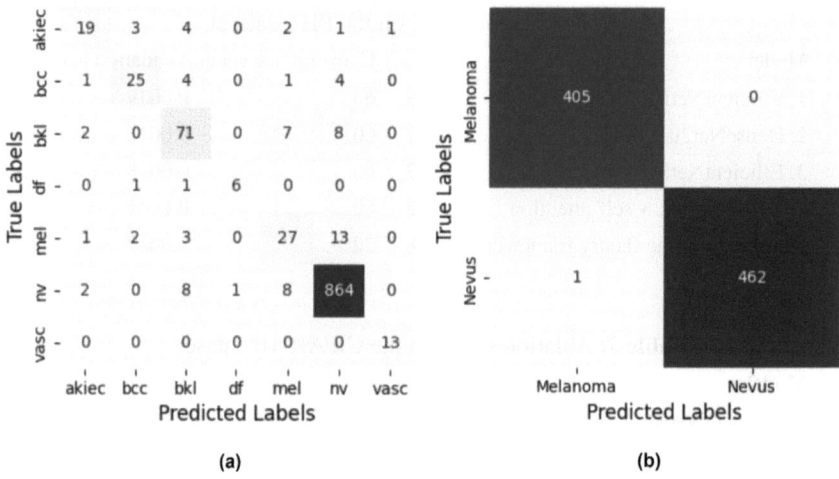

Fig. 7. Confusion matrices on (a) HAM100000 and (b) PH2 datasets.

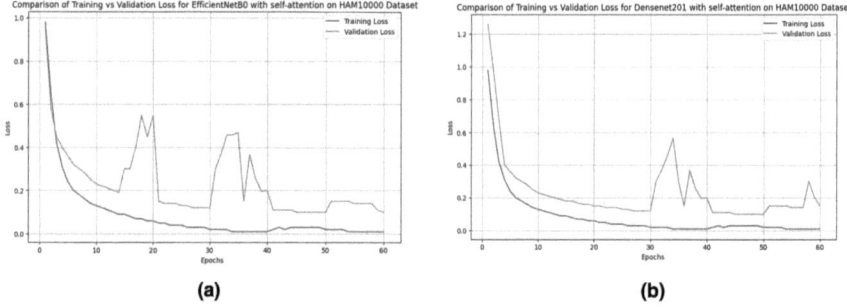

Fig. 8. Comparison of training vs validation losses on the HAM100000 dataset using (a) EfficientNetb0 with self-attention, and (b) DenseNet201 with self-attention.

Fig. 10 and Fig. 11. Through the use of dynamic learning rate modifications based on the *Improvement Factor*, the framework enables each model to alter its performance in reaction to the advancement of the others. This interaction minimizes the need for computational resources while promoting a quick and effective learning process. By combining the unique contributions of each model, the use of Shapley values for the weighted fusion of model predictions improves the classification accuracy and produces better diagnostic results.

The confusion matrices for both datasets are provided in Fig. 7. Training and validation loss curves are compared for the HAM10000 dataset in Fig. 8 and for the PH2 dataset in Fig. 9. Here, in Table 5, Table 6 and Table 7 we have shown accuracies and validation losses on two skin lesion datasets and one benchmark dataset (CIFAR-10)

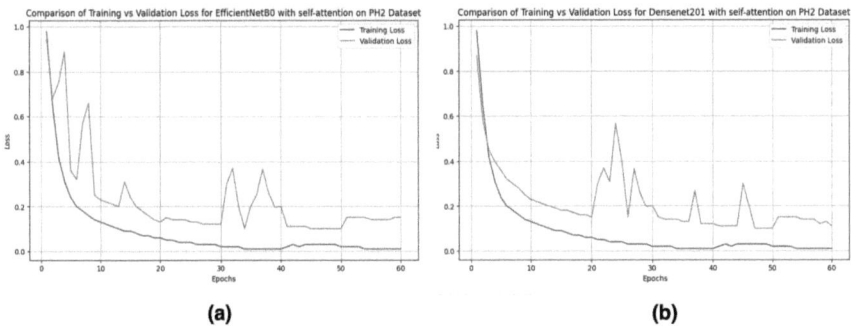

Fig. 9. Comparison of training vs validation losses on the PH2 dataset using (a) EfficientNetb0 with self-attention, and (b) DenseNet201 with self-attention.

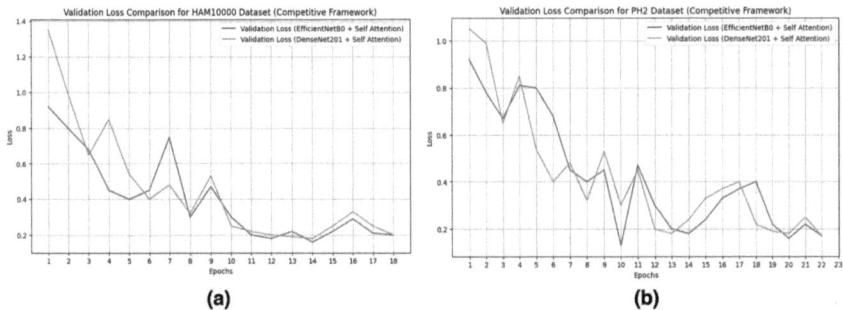

Fig. 10. Visualization of gradual convergence of validation losses on (a) HAM10000 and (b) PH2 datasets using the proposed approach.

using base models, aided with and without self-attention modules and our proposed methodology. It is prominent from the results that our proposed methodology helps the models to converge on their validation losses over a very less number of epochs, thereby making it robust and resource efficient. Experimental results on HAM10000 and PH2 datasets confirm the efficacy of this approach, with the proposed framework achieving superior performance metrics compared to several existing models, notably attaining an accuracy of 92.93% on the HAM10000 dataset and 99.65% on the PH2 dataset.

Table 3 highlights the performance of the model across various thresholds (0.0001, 0.0005, and 0.001) on four different datasets: HAM10000, PH2, MSID and CIFAR-10. Among the evaluated thresholds, the model demonstrates superior performance with a threshold value of 0.001, achieving the highest accuracy, precision, recall, and F1-score across all datasets. Specifically, on the HAM10000 dataset, the model attains an accuracy of 92.93% and a balanced F1-score of 92.89%. Similarly, for the PH2 dataset, the performance peaks with a remarkable accuracy of 99.88% and an F1-score of 99.63%. On the MSID dataset, the model achieves 96.59% accuracy and 97.11% for the F1-score, which are significantly better compared to the other thresholds.

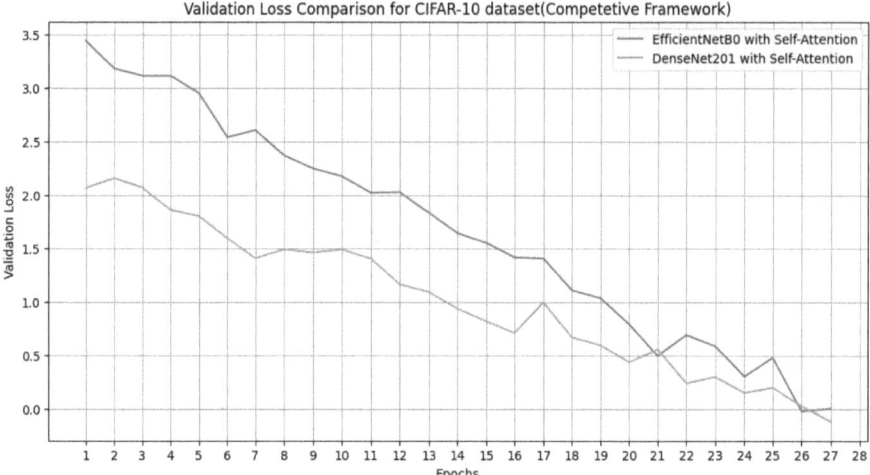

Fig. 11. Visualization of gradual convergence of validation losses on the CIFAR-10 dataset using the proposed approach.

This consistent improvement in all metrics at the 0.001 threshold demonstrates the robustness and reliability of the model when tuned to this value. The choice of this threshold ensures optimal trade-offs between precision and recall, indicating the models capability to maintain high classification accuracy while minimizing false positives and false negatives. Consequently, the threshold value of 0.001 has been selected as the default for the final model, given its ability to generalize effectively across diverse datasets.

6 Conclusion and Future Scopes

The non-cooperative game theory-based framework introduced in this research for co-training of deep learning models in order to provide an effective solution for the classification of skin lesions. By integrating principles of game theory into the training process, the framework achieves faster convergence, reduced computational costs, and enhanced classification accuracy. The competitive learning mechanism and Shapley value-based ensemble method underpin the robustness and efficiency of this approach, positioning it as a compelling advancement in medical image analysis.

However, we admit that the framework has some limitations that need to be addressed for broader applicability and improved performance. Firstly, the reliance on manually set thresholds, such as the standard deviation based threshold for the stopping criterion may limit the model's ability to generalize across different datasets. In the future, an automated threshold optimization method could be implemented, using meta-learning or reinforcement learning to dynamically adjust the thresholds based on the models performance across various datasets.

Secondly, while the adaptive learning rate mechanism helps to accelerate training, it is still relatively simplistic, relying on a linear adjustment based on the Improvement Factor. More sophisticated adaptive learning strategies, such as those utilizing meta-gradients or second-order optimization techniques, could be employed to further refine the learning rate adjustments, enhancing convergence speed and stability. By addressing these limitations, the framework can be made more robust and effective, thereby improving its utility for real-world medical imaging applications.

Acknowledgment. We are grateful to the CMATER Research Lab of the Department of Computer Science and Engineering, Jadavpur University, India for providing vital infrastructural assistance.

References

1. Arun, C.B., Anusha, M., Rao, T., Kodipalli, A.: Analysis of skin lesion classification using computational models. In: 2024 Asia Pacific Conference on Innovation in Technology (APCIT), pp. 1–7. IEEE (2024)
2. Bala, D., et al.: MonkeyNet: a robust deep convolutional neural network for monkeypox disease detection and classification. Neural Netw. **161**, 757–775 (2023)
3. Benyahia, S., Meftah, B., Lézoray, O.: Multi-features extraction based on deep learning for skin lesion classification. Tissue Cell **74**, 101701 (2022)
4. Bhowal, P., Sen, S., Yoon, J.H., Geem, Z.W., Sarkar, R.: Choquet integral and coalition game-based ensemble of deep learning models for COVID-19 screening from chest X-ray images. IEEE J. Biomed. Health Inform. **25**(12), 4328–4339 (2021)
5. Bhowmik, S., Sarkar, R., Das, B., Doermann, D.: GiB: a game theory inspired binarization technique for degraded document images. IEEE Trans. Image Process. **28**(3), 1443–1455 (2018)
6. Dinitra, N.T., Wijayanto, I., Akhyar, F., Safitri, A.S., Firdhaust, R.S.: Enchancing skin cancer classification using EfficientNet-based architectures. In: 2024 International Conference on Data Science and Its Applications (ICoDSA), pp. 415–420. IEEE (2024)
7. El-Khatib, H., Popescu, D., Ichim, L.: Deep learning-based methods for automatic diagnosis of skin lesions. Sensors **20**(6), 1753 (2020)
8. Foahom Gouabou, A.C., et al.: Computer aided diagnosis of melanoma using deep neural networks and game theory: application on dermoscopic images of skin lesions. Int. J. Mol. Sci. **23**(22), 13838 (2022)
9. Gajera, H.K., Nayak, D.R., Zaveri, M.A.: A comprehensive analysis of dermoscopy images for melanoma detection via deep CNN features. Biomed. Signal Process. Control **79**, 104186 (2023)
10. Gessert, N., Nielsen, M., Shaikh, M., Werner, R., Schlaefer, A.: Skin lesion classification using ensembles of multi-resolution EfficientNets with meta data. MethodsX **7**, 100864 (2020)
11. Ghosh, M., Ghosh, K.K., Bhowmik, S., Sarkar, R.: Coalition game based feature selection for text non-text separation in handwritten documents using LBP based features. Multimed. Tools Appl. **80**(2), 3229–3249 (2021)
12. Goodfellow, I., et al.: Generative adversarial nets. Adv. Neural Inf. Process. Syst. **27** (2014)
13. Gururaj, H.L., Manju, N., Nagarjun, A., Aradhya, V.N.M., Flammini, F.: Deepskin: a deep learning approach for skin cancer classification. IEEE Access (2023)

14. Hait, M., Das, R., Ali, A., Chaudhari, S.S., Djemal, K.: Skin cancer classification using levy stable based ensemble and its real-time implementation on OpenVINO toolkit. In: 2023 Twelfth International Conference on Image Processing Theory, Tools and Applications (IPTA), pp. 1–6. IEEE (2023)
15. Hosny, K.M., Kassem, M.A., Foaud, M.M.: Skin cancer classification using deep learning and transfer learning. In: 2018 9th Cairo international Biomedical Engineering Conference (CIBEC), pp. 90–93. IEEE (2018)
16. Iqbal, I., Younus, M., Walayat, K., Kakar, M.U., Ma, J.: Automated multi-class classification of skin lesions through deep convolutional neural network with dermoscopic images. Comput. Med. Imaging Graph. **88**, 101843 (2021)
17. Krizhevsky, A.: Learning multiple layers of features from tiny images. Technical report (2009)
18. Mendonça, T., Ferreira, P.M., Marques, J.S., Marcal, A.R., Rozeira, J.: PH 2-a dermoscopic image database for research and benchmarking. In: 2013 35th Annual International Conference of the IEEE Engineering in Medicine and Biology Society (EMBC), pp. 5437–5440. IEEE (2013)
19. Mridha, K., Uddin, M.M., Shin, J., Khadka, S., Mridha, M.F.: An interpretable skin cancer classification using optimized convolutional neural network for a smart healthcare system. IEEE Access (2023)
20. Mushtaq, S., Singh, O.: A deep learning based architecture for multi-class skin cancer classification. Multimed. Tools Appl. 1–23 (2024)
21. Nayak, T., et al.: Detection of monkeypox from skin lesion images using deep learning networks and explainable artificial intelligence. App. Math. Sci. Eng. **31**(1), 2225698 (2023)
22. Ozkan, I.A., Koklu, M.: Skin lesion classification using machine learning algorithms. Int. J. Intell. Syst. Appl. Eng. **5**(4), 285–289 (2017)
23. Paschos, G.: Perceptually uniform color spaces for color texture analysis: an empirical evaluation. IEEE Trans. Image Process. **10**(6), 932–937 (2001)
24. Rastegar, H., Giveki, D.: Designing a new deep convolutional neural network for skin lesion recognition. Multimed. Tools Appl. **82**(12), 18907–18923 (2023)
25. Ray, A., Sarkar, S., Schwenker, F., Sarkar, R.: Decoding skin cancer classification: perspectives, insights, and advances through researchers lens. Sci. Rep. **14**(1), 30542 (2024)
26. Roy, D., Pramanik, R., Sarkar, R.: Margin-aware adaptive-weighted-loss for deep learning based imbalanced data classification. IEEE Trans. Artif. Intell. (2023)
27. Sahoo, S.R., Dash, R., Mohapatra, R.K.: Fusion of deep and wavelet feature representation for improved melanoma classification. Multimed. Tools Appl. 1–27 (2024)
28. Sarkar, S., Majumder, S., Mandal, D., Gulvanskii, V., Kaplun, D., Sarkar, R.: Skin lesion classification using a decentralized peer-to-peer federated learning framework. In: 2024 20th International Symposium on Medical Information Processing and Analysis (SIPAIM), pp. 1–4. IEEE (2024)
29. Sarkar, S., Ray, A., Kaplun, D., Sarkar, R.: A combination of soft attention-aided CNN models using Dempster-Shafer theory for skin cancer classification. In: Alikhanov, A., Tchernykh, A., Babenko, M., Samoylenko, I. (eds.) CPAMCS 2023. LNNS, vol. 1044, pp. 410–421. Springer, Cham (2023). https://doi.org/10.1007/978-3-031-64010-0_38
30. Shen, D., Ling, H., Pham, K., Blasch, E., Chen, G.: Computer vision and pursuit-evasion game theoretical controls for ground robots. Adv. Mech. Eng. **11**(8), 1687814019872911 (2019)
31. Shen, S., et al.: A low-cost high-performance data augmentation for deep learning-based skin lesion classification. BME Front. (2022)
32. Tan, T.Y., Zhang, L., Lim, C.P.: Intelligent skin cancer diagnosis using improved particle swarm optimization and deep learning models. Appl. Soft Comput. **84**, 105725 (2019)

33. Tschandl, P., Rosendahl, C., Kittler, H.: The HAM10000 dataset, a large collection of multi-source dermatoscopic images of common pigmented skin lesions. Sci. Data **5**(1), 1–9 (2018)
34. Uysal, F.: Detection of monkeypox disease from human skin images with a hybrid deep learning model. Diagnostics **13**(10), 1772 (2023)
35. Vaswani, A.: Attention is all you need. Adv. Neural Inf. Process. Syst. (2017)
36. Waheed, Z., Waheed, A., Zafar, M., Riaz, F.: An efficient machine learning approach for the detection of melanoma using dermoscopic images. In: 2017 International Conference on Communication, Computing and Digital Systems (C-CODE), pp. 316–319 (2017). https://doi.org/10.1109/C-CODE.2017.7918949
37. Winter, E.: The Shapley value. Handb. Game Theory Econ. Appl. **3**, 2025–2054 (2002)
38. Yao, P., et al.: Single model deep learning on imbalanced small datasets for skin lesion classification. IEEE Trans. Med. Imaging **41**(5), 1242–1254 (2021)

LoRA-Based Summarization of Data Privacy Clauses in Terms and Conditions Documents Aligned with India's 2023 Digital Personal Data Protection Act

Preet Kanwal, Amish Gupta(✉), Sai Mohananshu Jujjavarapu, and Prasad B. Honnavalli

Department of Computer Science, PES University, Bengaluru, India
{preetkanwal,prasadhb}@pes.edu, amishgupta2909@gmail.com

Abstract. The increasing complexity and legal jargon found in the terms and conditions documents of various organizations pose significant challenges for users attempting to understand potential privacy risks. To address this issue, we developed a method to automatically summarize these documents, particularly focusing on clauses that could lead to privacy breaches. Furthermore, our work aligns with the principles outlined in India's Digital Personal Data Protection Bill, passed in 2023, ensuring that our approach is not only effective but also compliant with emerging privacy regulations. Our approach involves fine-tuning a large language model, Mistral 7B, using a custom dataset derived from the TOS; DR dataset. We employed the Low-Rank Adaptation technique to optimize the model's performance while ensuring computational efficiency. On inference, our model achieved an average BERTScore of 88.414%. The results of our experiments demonstrate that our method can produce concise and semantically accurate summaries that effectively highlight potential privacy concerns, offering users a clearer understanding of the terms they agree to.

Keywords: Data privacy · Automated summarization · Natural language processing · Low-Rank Adaptation(LoRA) · Large Language Model (LLM) · Digital personal data protection bill

1 Introduction

In this digital age where every aspect of our lives involves some form of a software application either in the form of a mobile app or a web app, our personal information slips away from our fingertips quite as easily as with the click of a button which is the agreement to the terms and conditions before using the application. These documents are often filled with vast amounts of legal jargon and vague information that prove overwhelming to the reader which hinders their comprehension of the document. A survey done on user interaction with policy documents shows ignorance towards reading such tedious and lengthy texts [15]. How can the latest advancements in natural language processing (NLP) and artificial neural networks help overcome such a hurdle and simplify such a cumbersome task for a user?

The advent of self-supervised (or pre-trained) language models, like Googles' BERT model (Bidirectional Encoder Representations from Transformers) [4], has lead to major improvement in natural language processing (NLP). Google introduced BERT as a bidirectional transformer model that analyzes words in text by accounting for both their left-to-right and right-to-left contexts. Unlike GPT models, which focus on text generation, BERT is designed to enhance computer systems' ability to comprehend text. Its strength lies in natural language understanding (NLU) tasks, such as sentiment analysis, where it excels in capturing the nuances of language for more accurate interpretation. Self-supervised language models are pre-trained on vast collections of general texts, including sources like Google Books and Wikipedia. This pre-training has led to substantial improvements across multiple fine-tuning tasks involving domain-specific datasets. The success of these models has sparked numerous applications and extensions, greatly advancing the field of natural language processing.

However, while general pre-training on large corpora has consistently improved performance across various NLP tasks, domain-specific pre-training, such as using a legal corpus, has not shown significant gains. Despite expectations that tailoring models to specialized fields would yield better results, the impact of domain-specific pre-training in the legal domain appears to be minimal [25]. Terms and conditions documents typically employ language that is analogous to that used in legal documents.

Proper evaluation of text summarization is a task that has always been challenging. Human evaluation is the default standard norm in everyday instances but it is time-consuming and requires the availability and effort of a qualified individual as per the task at hand. The procurement of services of such people might be costly. As a result, automatic evaluation or automatic text summarization (ATS) has been quite effective with the advent of Large Language Models (LLMs). This in turn provides solutions to problems that involve textual and comprehensive understanding of corpus has become ever the more viable [12]. In this context, automatic evaluation metrics (for example, ROUGE [13]) play an important role. These metrics, which are based on reference summaries and n-gram matching, are used in various types of summarization. However, surface-level word matching cannot accurately reflect the quality of the summary and in tasks like ToS (Terms of Service) summarization, the legal jargon presents a new problem and requires domain-specific fine-tuning. Additionally, it is challenging to evaluate the factual accuracy of the summary without utilizing the source document. Evaluation metrics based on pre-trained models such as BERTScore [24] have achieved better performances to that of the old standards which are human judgments. Large language models (LLMs) like GPT-3 [3,8,23] have the ability of in-context learning and instruction tuning that allow LLMs to align with human evaluation [16]. This enables LLMs to replicate or imitate the behavior of human evaluators, who generally evaluate summaries based on their experience, knowledge of the domain, and understanding examples and instructions.

In this paper, to build our model which summarizes data privacy clauses in terms and conditions documents according to the Digital Personal Data Protection Bill passed in 2023 in India [5], we have used Low-Rank Adaptation (LoRA) [10] which is a form of Parameter Efficient Finetuning (PEFT) for LLMs. We have implemented a modified version of the Mistral 7B model that has been trained on a custom dataset that has been manually prepared. The dataset has been prepared in four stages with the help of larger-scaled LLMs like Googles Gemini and ChatGPT v4.0. It provides a domain-specific dataset catered to terms and conditions retrieved via web scraping from the

ToSDR dataset [18]. This dataset enabled us to instruction-tune the Mistral model and let it undergo semi-supervised learning. We have benchmarked our model using the BERTScore [24] as our testing metric as it provides contextual embeddings contrary to word embeddings in its attention mechanism.

In the following sections of this paper, we present a comprehensive breakdown of our methodology, experimental design, results, and discussions. We emphasize both the strengths and limitations of our approach, as well as its potential applications across different domains.

2 Related Work

Automated summarization of data privacy clauses in terms and conditions has become increasingly important due to the growing concerns over data privacy and the complexity of legal documents. In this section, we categorize and discuss existing approaches to summarizing terms of service clauses, focusing on methodologies such as automatic text summarization, leveraging large language models (LLMs), and innovations in natural language processing (NLP) techniques. The following subsections highlight the effectiveness of these approaches in making legal content more accessible and comprehensible for users.

2.1 Deep Learning-Based Approaches

A recent study focused on identifying the relevant holding of a cited case through multiple-choice questions was made possible by the development of the custom dataset CaseHOLD [25]. This dataset represents a crucial task for legal professionals, offering both legal significance and considerable difficulty from an NLP standpoint. Notably, a BiLSTM baseline achieved an F1 score of only 0.4, underscoring the complexity of the task. Although a Transformer architecture like BERT pre-trained on general corpora such as Google Books and Wikipedia boosts performance, the most notable gains with the CaseHOLD dataset are achieved through domain-specific pre-training using a custom legal vocabulary. This approach yielded a 7.2% improvement in F1 score, reflecting a 12% performance increase over BERT. Additionally, consistent gains were observed across two other legal tasks. These findings suggest that domain-specific pre-training is beneficial when the task closely aligns with the pre-training corpus [25].

Another study, which focused on detecting unfair clauses in online terms of service, conducted an extensive comparison of various machine learning systems [14]. This included modern deep learning architectures for text categorization and a structured Support Vector Machine (SVM) for collective classification, which accounts for the sequence of sentences within a document. This approach aimed to improve the accuracy of clause detection by considering contextual relationships between sentences. They performed sentence-wide classification using various methods: SVMs, Long-Short Term Memory Networks (LSTMs), and collective classification using SVM-HMMs, which combine SVMs with Hidden Markov Models. The study also explored different sets of features for text categorization, employing a bag-of-words (BoW) model and word embeddings for sentence classification.

An improvement on this work was presented in a recent study that focused on unfair clause detection in terms of service across multiple languages [7]. This study extended the scope by not only using English terms of service but also including documents in different languages. The methods employed involved various NLP techniques such as Continuous Bag of Words (CBOW) and the use of advanced models like BERT and ELMo for text processing. The study concluded that applying machine translation at prediction time yields better results, marking it as an improvement over previous models like CLAUDETTE. This comprehensive comparison of different approaches highlights the effectiveness of using multilingual datasets and advanced NLP techniques in the detection of unfair clauses.

2.2 Domain-Specific Approaches

Recent studies have explored the benefits of prompt engineering for legal reasoning related tasks, particularly using datasets from the COLIEE 2021-22 competition [21]. These studies focus on reason-based prompting mechanisms and leverage zero through few-shot approaches using pre-trained large language models (LLMs) [1,3,19,20]. One popular framework utilized in this context is the IRAC (Issue, Rule, Application, Conclusion) method, which helps structure the legal reasoning process.

Additionally, some research has employed clustering techniques on COLIEE training data to perform entailment tasks based on bar exam questions [21]. However, this approach is more classification-oriented and does not align directly with the requirements of training an LLM for broader legal reasoning tasks.

Research on ensuring privacy policy compliance of wearables with IoT regulations has utilized privacy policies from major companies such as Apple, Samsung, Fitbit, Garmin, Withings, and Hexoskin. This research involves creating an ontology graph to identify key regulatory terms [6], which helps in pinpointing areas for privacy risk mitigation. While this approach is innovative and specifically tailored to IoT devices and wearables, its applicability may be limited for other contexts and requirements.

Recent research on the classification and annotation of relevant clauses in terms-and-conditions has employed the Contract Understanding Atticus Dataset [2]. This study utilizes few-shot prompting and achieved a high inter-coder agreement with a Cohen's Kappa score of 0.92 on 20% of the data. The annotation guidelines were iteratively refined by giving examples, edge cases, and definitions that are more precise. The study leverages Google's FLAN-T5 model to classify clauses into general categories. However, this approach focuses on the Italian language aspect of legal corpus clause categorization and uses a dataset to train and fine-tune the LLM.

2.3 Innovative Techniques

Recent studies have also introduced innovative techniques such as Polisis [9] which leverages the OPP-115 dataset and employs a two-stage approach for the analysis of privacy policies. In the unsupervised stage, domain-specific word embeddings are generated. The supervised stage involves training privacy text classifiers. Segments are split into tokens using PENN Treebank tokenization, and a multi-label cross-entropy loss function is used as the classifier's objective. Hyper-parameters for each classifier are optimized through a randomized grid-search. The evaluation metrics include top-1 precision, F1, recall, and precision scores. The system detects privacy clauses by identifying certain labels within the clause, and it compares model-generated labels with human annotations using Cohen's Kappa. Additionally, PriBot [9] is designed to answer free-form user questions from previously unseen privacy policies in real time, demonstrating the practical application of the system in enhancing user comprehension of privacy policies.

In contrast, another approach uses PrivacyCheck [22] which is an automatic summarize that uses data mining uses a corpus of 400 privacy policies to classify all clauses into three categories based on the risk: red, yellow and green for low, medium and high risk respectively. Implemented as a browser extension, PrivacyCheck exclusively focuses on privacy policies, providing users with immediate risk assessments of the clauses within these documents.

In the study Demystifying Legalese [17], the ToSDR dataset is used alongside models like RoBERTa, PrivBERT, Linear SVM, and Random Forest. Case classification is handled by Linear SVM, with the F1 score employed to address the dataset's imbalance. Cohens Kappa is used to measure the consistency of the annotations and the agreement level between annotators. This research utilizes a simplification strategy, labeling data based on annotations from the ToSDR website, rather than a summarization approach. The paper notes that summarization would lack necessary granularity for effective scoring. Additionally, it discusses the preference for BERT-family transformers over LLMs due to better operability in an end-to-end pipeline with a user interface.

3 Proposed Methodology

In this section, we present a detailed account of the methodology employed in our approach to summarize data privacy clauses in terms and conditions documents. We outline the key components of our method, including dataset preparation, choice of LLM and our fine-tuning process. We employ a multi-step process to prepare our dataset. Our approach integrates LoRA in our fine-tuning process to optimize the fine-tuning of our LLM. Detailed explanations and experimental procedures are provided in subsequent subsections.

3.1 Dataset Preperation

To fine-tune our LLM to perform the task of summarization of data privacy clauses in terms and conditions documents, we meticulously prepared our dataset, ensuring its suitability for the task at hand. We decided to make use of the TOS; DR dataset [18] for our problem statement. However, the TOS; DR dataset has some issues which made us perform some operations on it to make it suitable for the task at hand.

The TOS; DR dataset which we decided to use has a collection of over 9000 documents. The documents may be terms and conditions, cookie policies, privacy policies, refund policies etc. The limitation for our case in this dataset was that there were no summaries to compare our results with so that we could compute loss and finetune the LLM (Fig. 1).

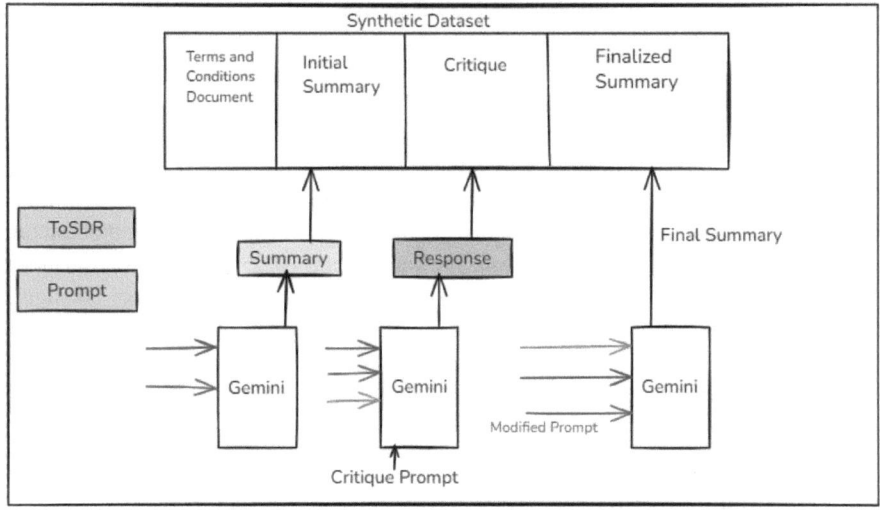

Fig. 1. Creation of the Synthetic Dataset.

To make the dataset usable, our first task was to generate a summarized document for each document in the dataset. Our goal is to make the model follow the Digital Personal Data Protection Bill [5] passed in India in 2023. To generate the summaries for all the documents, we decided to make use of Gemini.

By using Gemini and a well crafted prompt for our use case, we divide the process of generating the summary in a series of steps. The first step involves making a prompt that covers all the points conveyed by the Digital Personal Data Protection Bill. We designed our prompt by thoroughly reading and understanding the bill. Next we gave the document for which the summary has to be generated along with the prompt to Gemini and made it generate a summary. We did this for all the documents and saved the response for all the documents.

Following the above step, we need to ensure that the generated summaries follow the rules given to them in the prompt and whether they have been hallucinated or not. To check this we craft another prompt and give the original document, the new prompt and the summary to Gemini again to get a response from it regarding this. We once again save these results.

In the last step, we give the first summary generated, the changes suggested by Gemini in the previous step and ask Gemini once again to make the changes in the first summary generated. The summary generated by doing this is considered as our final summary that we use for fine-tuning our model. By doing these steps we ensure that all the summaries are following the rules and are not hallucinated.

3.2 Choice of LLM

To do our summarization we decide to leverage the powers of large language models based on the transformer architecture. Transformer architecture is the basis of modern language models. We decided to make use of the Mistral-7B-Instruct-v0.2-GPTQ . Mistral 7B [11] is an open-source large language model with seven billion paramenters. The specific version of the Mistral model that we have chosen is based on Mistral-7B-v0.1, a transformer model that possesses architecture choices such as Grouped-Query Attention, Sliding-Window Attention and Byte-fallback BPE tokenizer.

This model is a GPTQ model which means that it is quantized to make the model more efficient in terms of memory and computational resources. It has been fine-tuned on a diverse dataset specifically designed for instruction following tasks that encompass question answering, summarization, and various other interactive tasks. This model although provides improved efficiency and speed due to quantization it does not significantly sacrifice performance but instead maintains a balance between performance and resource utilization. The Mistral-7B-Instruct-v0.2-GPTQ possesses a context length of 8192 tokens which we found sufficient for handling terms and conditions and related corpus.

3.3 Fine-Tuning Process

For fine-tuning the language model we have made use of Low-Rank Adaptation (LoRA) [10] which is a form of parameter-efficient fine-tuning. LoRA (Low-Rank Adaptation) optimizes the fine-tuning process for large-scale models by focusing on a small, highly impactful subset of the model's parameters, rather than updating a broad range of weights as done in conventional fine-tuning techniques. Instead of directly modifying these weights, LoRA captures and tracks the changes, ensuring a more efficient and resource-saving fine-tuning approach. This method allows for greater flexibility and performance without the need for large-scale updates across the entire model,this can be referred to as "augmentation" of the weights of the large language model. LoRA achieves its efficiency by decomposing the large matrices representing weight changes into smaller, low-rank matrices that hold the "trainable parameters". LoRA introduces weight updates in a low-rank form by factorizing the change in weights, ΔW, as the product of two smaller matrices, B and A, i.e., $\Delta W = BA$. Instead of updating the full weight matrix W_0, the model applies these trainable low-rank matrices, significantly reducing the number of parameters while maintaining expressiveness. The final transformation thus becomes $h(x) = W_0 x + BAx$, where W_0 remains frozen, and only B and A are optimized during fine-tuning, as illustrated in Fig. 2. This decomposition reduces the computational burden, as only these smaller matrices are adjusted during training.

We have fine-tuned our model using a synthetic dataset created from a corpus of over 9000 documents from the TOS; DR dataset. The nature of the fine-tuning task at hand is that of a causal language modeling. In the initial steps, we set the model to training mode and enabled gradient checkpointing to reduce memory usage during backpropagation. We also prepared the model for quantized training to further reduce memory usage and improve training efficiency. To define the dimensionality of the adaptation we set the rank of the low-rank matrix to 8 and also defined a scaling factor of 32, controlling the extent of the adaptation. We also define a dropout rate for the LoRA layers, introducing regularization to prevent overfitting (Fig. 3).

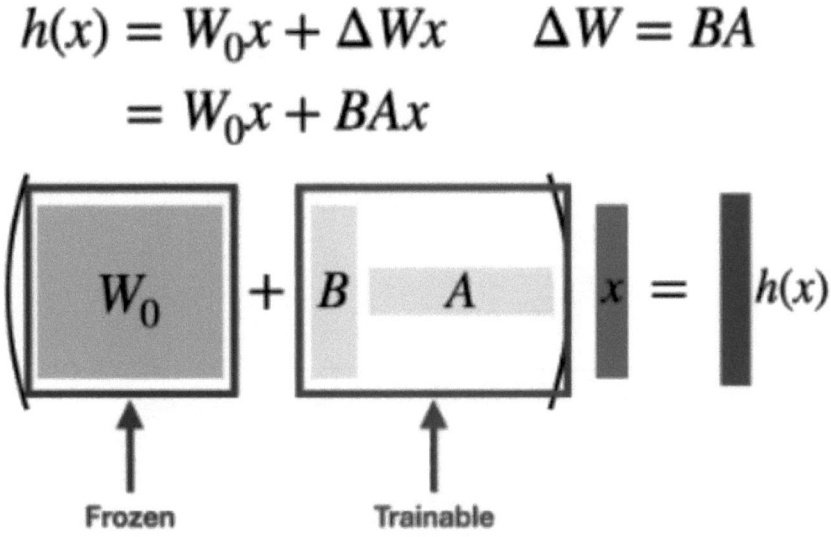

Fig. 2. Low-Rank Adaptation (LoRA).

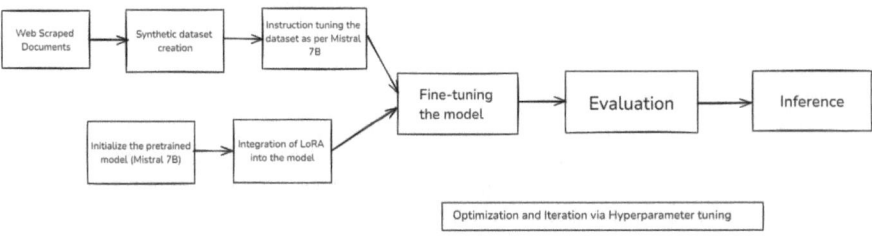

Fig. 3. System Architecture Implementing Low-Rank Adaptation (LoRA).

In the context of our model input regular truncation was not feasible due to the vast size of the terms and conditions and similar documents. Even utilizing chunking without a sliding window seemed to eat up system resources profoundly thus demanding an extensive number of Graphic Processing Units (GPUs) or more powerful GPUs with larger memory sizes to accommodate larger tensors produced by our model. To efficiently manage large text inputs and maintain contextual integrity for the language model without the eating up of system resources, we implemented a sliding window approach with a chunk size of 512 tokens. This technique ensures that the contextual understanding is preserved across chunks.

4 Experiments

Through extensive experiments and evaluations, we demonstrate the effectiveness and adaptability of our LoRA-based approach in summarizing data privacy clauses within

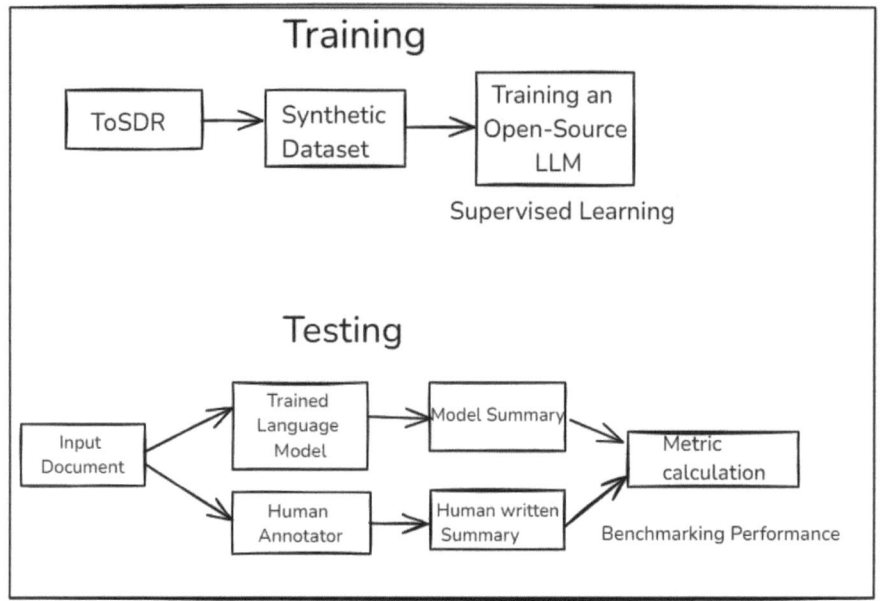

Fig. 4. Abstract Architecture Overview.

terms and conditions documents. We provide both qualitative and quantitative analysis to assess the performance of our method, highlighting its potential for various practical applications in legal document analysis, compliance checking, and consumer awareness. Furthermore, we discuss future research directions aimed at advancing the state-of-the-art in legal text summarization and promoting practical applications in data privacy and beyond (Fig. 4).

4.1 Qualitative Analysis

Going through a study on human-like summarization by Large Language Models (LLMs) [8], we have realised that utilising a metric that captures the essence of the text or in other words the semantic meaning of the text is paramount if we want to obtain something which is similar to something written by a human professsional in that particular domain.

To evaluate the qualitative performance of our LoRA-based summarization model, we employ BERTScore [24], a state-of-the-art evaluation metric that leverages contextual embeddings from pre-trained transformer models like BERT. Unlike traditional metrics that rely on exact word matching, BERTScore evaluates the degree of semantic similarity between generated summaries and the reference text by comparing the contextual embeddings. This approach allows for a more nuanced and context-aware evaluation, capturing the meaning and relevance of the summarized data privacy clauses more effectively. The use of BERTScore provides deeper insights into the model's ability to preserve critical information and convey it accurately in a summarized form, thereby validating the efficacy of our approach in real-world applications where semantic correctness is paramount.

To maintain the quality of our synthetic dataset we manually inspected the summaries generated by Gemini to ensure no hallucination to prevent the poisoning of our model and depletion of its performance and tuning. Thus we have managed to retain high quality summaries within our synthetic dataset which helped improve the overall performance of our fine-tuned model.

4.2 Quantitative Analysis

For quantitative evaluation, we employ the BERTScore metric [24], a widely recognized measure for assessing the semantic similarity between the generated summaries and the reference texts. It makes use of contextual embeddings rather than the normal word embeddings while handling each token in the input text. BERTScore ranges from 0 to 1, with higher values indicating closer semantic alignment between the generated summaries and the ground truth which in this case is a summary written by a human, thus reflecting the model's ability to accurately capture and convey the essential information in the data privacy clauses.

Using a quantized version of the Mistral-7B-Instruct-v0.2-GPTQ model we have managed to reduce the training time and compute resource utilization by a significant amount, to be precise we managed to reduce the training time per epoch by 50%. After training our model for 10 epochs, we compute the training and validation BERTScores to evaluate the performance of our LoRA-based summarization approach. The training BERTScore measures the semantic similarity on the training dataset, while the validation BERTScore assesses the model's generalization ability on a held-out validation dataset.

On inference our model produced an average accuracy of 88.414% when we tested it on 20 unseen documents. These results demonstrate the effectiveness of our approach in generating concise and semantically coherent summaries of data privacy clauses, preserving the essential meaning and context of the original text.

5 Conclusion and Future Work

In conclusion, our paper has introduced a novel approach to summarizing data privacy clauses within terms and conditions using Low-Rank Adaptation (LoRA) [10] techniques. By leveraging a LoRA-enhanced version of the Mistral-7B-Instruct-v0.2-GPTQ model, we have effectively generated concise and semantically coherent summaries that retain the essential information of the original clauses by proving the user with an account of the types of personal data collected and also warn the user about any clauses that may be in potential violation of the Indian Digital Privacy Bill. Our framework is designed in such a manner that with a few adjustments it can be adapted to be used for the analysis of various legal documents, providing users with summaries that are both accurate and contextually relevant and bypass the hassle of understanding legal jargon.

Our experimental results have demonstrated the effectiveness and robustness of our approach across a wide variety of terms and conditions and other related privacy policy documents, highlighting its potential for real-world applications in legal document analysis, compliance checking, and enhancing consumer awareness. Looking ahead, future research directions include the possible utilization of the architecture in the form of a tool that can be configured into a browser extension to facilitate a seamless and automatic summarization. We also hope to keep exploring more sophisticated summarization strategies, optimizing them for real-time applications, improving domain-specific adaptability, and incorporating user feedback mechanisms. By pursuing these avenues, our goal is to push the boundaries of legal text summarizing technology and promote practical applications in data privacy, ultimately contributing to better information transparency and accessibility for users worldwide.

Disclosure of Interests. The authors have no competing interests to declare that are relevant to the content of this article.

References

1. Alex, N., et al.: RAFT: A real-world few-shot text classification benchmark. arXiv preprint arXiv:2109.14076 (2021)
2. Bizzaro, P.G., et al.: Annotation and Classification of Relevant Clauses in Terms-and-Conditions Contracts. arXiv preprint arXiv:2402.14457 (2024)
3. Brown, T., et al.: Language models are few-shot learners. In: Advances in Neural Information Processing Systems 33, pp. 1877–1901 (2020)
4. Devlin, J., et al.: Bert: pre-training of deep bidirectional transformers for language understanding. arXiv preprint arXiv:1810.04805 (2018)
5. Digital Personal Data Protection Act, 2023. https://www.meity.gov.in/writereaddata/files/Digital%20Personal%20Data%20Protection%20Act%202023.pdf
6. Echenim, K.U., Elluri, L., Joshi, K.P.: Ensuring privacy policy compliance of wearables with IoT regulations. In: 2023 5th IEEE International Conference on Trust, Privacy and Security in Intelligent Systems and Applications (TPS-ISA). IEEE (2023)
7. Galassi, A., et al.: Unfair clause detection in terms of service across multiple languages. Artif. Intell. Law 1–49 (2024)
8. Gao, M., et al.: Human-like summarization evaluation with chatgpt. arXiv preprint arXiv:2304.02554 (2023)
9. Harkous, H., et al.: Polisis: automated analysis and presentation of privacy policies using deep learning. In: 27th USENIX Security Symposium (USENIX Security 2018) (2018)
10. Hu, E.J., et al.: Lora: low-rank adaptation of large language models. arXiv preprint arXiv:2106.09685 (2021)
11. Jiang, A.Q., et al.: Mistral 7B. arXiv preprint arXiv:2310.06825 (2023)
12. Jin, H., et al.: A comprehensive survey on process-oriented automatic text summarization with exploration of LLM-based methods. arXiv preprint arXiv:2403.02901 (2024)
13. Lin, C.-Y.: Rouge: a package for automatic evaluation of summaries. Text summarization branches out (2004)
14. Lippi, M., et al.: CLAUDETTE: an automated detector of potentially unfair clauses in online terms of service. Artif. Intell. Law **27**(2), 117–139 (2019). https://doi.org/10.1007/s10506-019-09243-2

15. Obar, J.A., Oeldorf-Hirsch, A.: The biggest lie on the internet: ignoring the privacy policies and terms of service policies of social networking services. Inf. Commun. Soc. **23**(1), 128–147 (2020)
16. Ouyang, L., et al.: Training language models to follow instructions with human feedback. In: Advances in Neural Information Processing Systems 35, pp. 27730–27744 (2022)
17. Soneji, S., et al.: Demystifying Legalese: An Automated Approach for Summarizing and Analyzing Overlaps in Privacy Policies and Terms of Service. arXiv preprint arXiv:2404.13087 (2024)
18. Terms of Service; Didnt Read. 2012. Project Website for Terms of Service; Didnt Read. https://tosdr.org/
19. Webson, A., Pavlick, E.: Do prompt-based models really understand the meaning of their prompts?. arXiv preprint arXiv:2109.01247 (2021)
20. Wei, J., et al.: Finetuned language models are zero-shot learners. arXiv preprint arXiv:2109.01652 (2021)
21. Yu, F., Quartey, L., Schilder, F.: Exploring the effectiveness of prompt engineering for legal reasoning tasks. Findings of the Association for Computational Linguistics: ACL (2023)
22. Zaeem, R.N., German, R.L., Barber, K.S.: Privacycheck: automatic summarization of privacy policies using data mining. ACM Trans. Internet Technol. (TOIT) **18**(4), 1–18 (2018)
23. Zeng, X., Song, F., Liu, A.: Similar Data Points Identification with LLM: A Human-in-the-loop Strategy Using Summarization and Hidden State Insights. arXiv preprint arXiv:2404.04281 (2024)
24. Zhang, T., et al.: Bertscore: evaluating text generation with bert. arXiv preprint arXiv:1904.09675 (2019)
25. Zheng, L., et al.: When does pretraining help? Assessing self-supervised learning for law and the casehold dataset of 53,000+ legal holdings. In: Proceedings of the Eighteenth International Conference on Artificial Intelligence and Law (2021)

Comparison of AI Speech-to-Text Systems and Their Application in Artillery Command and Fire Control Systems

Martin Blaha[1(✉)], Jaroslav Varecha[2], Jan Drábek[1], and Jiří Novák[1]

[1] University of Defence, Brno, Czech Republic
martin.blaha@unob.cz
[2] Akadémia Ozbrojených Síl Generála Milana Rastislava Štefánika, Liptovsky Mikulas, Slovakia

Abstract. This paper presents a comparative analysis of three leading AI speech-to-text (STT) systems: Descript.com, Google Vertex AI Studio (Chirp), and OpenAI Whisper. The objective of the study is to evaluate the accuracy, functionality, and potential applications of these technologies, with a particular focus on their integration into artillery command and fire control systems.

The analysis outlines the evolution of speech recognition technologies, from traditional methods based on Hidden Markov Models (HMMs) to modern deep neural networks, including Recurrent Neural Networks (RNNs), Convolutional Neural Networks (CNNs), and Transformer-based architectures. Practical testing was conducted on a dataset of English and Czech recordings with varying audio quality. The results indicate that Google Chirp achieves the highest accuracy in English transcriptions, while OpenAI Whisper demonstrates superior performance for the Czech language.

Additionally, the paper explores the optimization of STT systems for combat environments, including the use of Ant Colony Optimization (ACO) algorithms to minimize errors and enhance the relevance of transcriptions. The study also highlights security risks associated with deploying cloud-based STT services in military applications and emphasizes the advantages of on-premise solutions to ensure data protection.

Finally, the paper discusses strategies for modernizing defense capabilities through AI technologies. It advocates for increased investment in automated command and control systems, fire control, and situational awareness, emphasizing their crucial role in improving response times and the accuracy of artillery fire support.

Keyword: Speech-to-Text systems · Artificial intelligence · Data security · Fire control system

---

© The Author(s), under exclusive license to Springer Nature Switzerland AG 2025
A. Hadjali et al. (Eds.): DeLTA 2025, CCIS 2627, pp. 82–95, 2025.
https://doi.org/10.1007/978-3-032-04339-9_6

1 Introduction

In recent years, artificial intelligence (AI) has made remarkable progress, particularly in natural language processing applications such as speech-to-text (STT) conversion. These AI systems play a crucial role in various domains, including accessibility services, transcription applications, virtual assistants, and more. Among the wide range of available solutions, this study focuses on evaluating three leading STT systems: Descript.com, Google Vertex AI Studio (Chirp), and OpenAI Whisper [5, 8, 18].

This comparative analysis aims to provide an overview of these three state-of-the-art AI-powered STT systems. By assessing their functionality, accuracy, usability, and potential applications, we seek to offer deeper insights into their strengths and limitations. Understanding the differences and capabilities of these platforms is essential for businesses, developers, researchers, students, and end-users seeking optimal solutions for speech recognition tasks [7, 19].

Each of the selected AI systems represents cutting-edge technology developed by renowned organizations and research institutions. Descript.com leverages advanced machine learning algorithms and promises high accuracy and real-time transcription capabilities. Google Vertex AI Studio integrates powerful tools and scalable machine learning workflows, including speech recognition models trained on extensive datasets. OpenAI Whisper, built on OpenAI's proprietary deep learning architectures, claims exceptional adaptability to various linguistic contexts.

1.1 Technological Background

Speech-to-text technology has undergone significant development since its early use. This evolution has been primarily driven by advances in computational power and artificial intelligence. The journey of this technology began in the mid-20th century with simple speech recognition systems capable of identifying and transcribing only a limited set of spoken words. Early systems heavily relied on manually crafted linguistic rules and statistical approaches to interpret audio signals.

The integration of artificial intelligence techniques revolutionized speech recognition, enabling systems to adapt and learn from vast amounts of data. In the 1980s and 1990s, Hidden Markov Models (HMMs) became widely used in speech recognition, allowing for more accurate modeling of speech patterns and phoneme transitions.

The 21st century brought a paradigm shift in speech recognition with the emergence of deep learning, particularly Recurrent Neural Networks (RNNs) and Convolutional Neural Networks (CNNs). These models demonstrated superior performance in capturing complex dependencies in speech data, leading to a significant improvement in accuracy. More recently, Transformer-based architectures have further enhanced STT capabilities, allowing for even more efficient processing of contextual speech variations.

This study builds upon these technological advancements by comparing the effectiveness of state-of-the-art STT systems in diverse linguistic and environmental conditions, with a specific focus on their application in military command and fire control systems.

1.2 Modern AI-Based Speech-To-Text Systems: Methodology and Computational Considerations

Contemporary speech-to-text (STT) systems employ a combination of acoustic modeling, language modeling, and pattern recognition techniques. Acoustic models analyze audio signals and extract relevant features, such as phonemes and prosody, while language models incorporate linguistic context to enhance transcription accuracy. Pattern recognition algorithms then map these extracted features into text sequences.

Modern STT systems heavily rely on AI models trained on large-scale datasets, primarily using supervised learning techniques. Deep neural networks, including Recurrent Neural Networks (RNNs), Long Short-Term Memory (LSTMs), and Transformer architectures (e.g., BERT), have become the standard for modeling complex speech patterns.

Computational Infrastructure and Deployment. AI-driven STT models are typically deployed on high-performance computing infrastructures, including GPUs (Graphics Processing Units) and TPUs (Tensor Processing Units), to manage the computational demands of real-time transcription. Cloud-based solutions provide scalability and flexibility, allowing for dynamic resource allocation based on performance requirements.

The computational requirements for training and deploying these models vary depending on model complexity, dataset size, and the desired transcription quality. Training large models can be extremely resource-intensive and time-consuming, whereas inference (i.e., real-time speech transcription) requires optimized hardware accelerators for efficient processing.

Speech Processing Workflow. During speech-to-text conversion, the audio input undergoes preprocessing to extract key acoustic parameters, such as spectrograms and Mel-Frequency Cepstral Coefficients (MFCCs). These features are then fed into an AI model, which predicts the corresponding text output through sequence-to-sequence mapping.

Following transcription, post-processing techniques, including language modeling and error correction mechanisms, further refine the accuracy of the output. These enhancements ensure that AI-driven STT systems can deliver high-precision transcriptions even in complex acoustic environments, making them suitable for mission-critical applications such as artillery command and fire control systems [6, 12, 13, 15, 16, 24].

2 Optimization of Speech-To-Text for Artillery Command and Control Systems

Modern combat operations emphasize the need for speed and accuracy in processing large volumes of real-time data. Speech-to-text (STT) systems can significantly enhance the efficiency of artillery fire control by rapidly processing voice commands from commanders, forward observers, and sensor systems.

One of the key challenges in deploying STT systems in military environments is maintaining transcription accuracy under adverse conditions, including battlefield noise, signal interference, diverse accents, and speech variations. As demonstrated in this study,

Google Chirp excels in English transcriptions, whereas OpenAI Whisper performs better in Czech, making it particularly relevant for deployment within the Czech Armed Forces (AČR).

2.1 Algorithmic Optimization for Fire Support Management

The accuracy and reliability of STT systems can be further improved through optimization algorithms that filter and validate transcriptions, minimizing errors and enhancing relevance for military applications. One promising approach is Ant Colony Optimization (ACO), which has been successfully used for generating test scenarios under constrained conditions [4, 9].

A similar principle can be applied to automated military decision-making systems, particularly in:

- Filtering and optimizing voice commands – reducing the risk of misinterpreted orders.
- Adapting STT to dynamic battlefield conditions – adjusting to changing factors such as environmental noise and signal degradation.
- Target prioritization for fire support – ensuring faster and more accurate fire mission execution [10, 13, 23].

In our implementation, Ant Colony Optimization was used to simulate transcription optimization in high-noise scenarios. For example, the algorithm was applied to prioritize clearer audio segments and reduce command misinterpretation in simulated artillery fire missions, improving command processing reliability.

By integrating AI-driven STT solutions with algorithmic optimizations, modern artillery command and control systems can achieve greater efficiency, improved situational awareness, and enhanced response times, contributing to superior operational effectiveness in combat scenarios.

2.2 Adapting Speech-To-Text Systems to Variable Battlefield Conditions

The adaptation of speech recognition for military applications relies on the analysis of historical data and the optimization of transcription outputs using Ant Colony Optimization (ACO). By leveraging past operational data, STT models can be dynamically adjusted to improve recognition accuracy in combat environments.

Target Prediction and Prioritization. More efficient processing of voice commands can significantly enhance fire mission execution, leading to faster and more accurate target engagement. The application of ACO algorithms enables a reduction in transcription errors, similar to their 32.62% reduction in software testing costs, as demonstrated in previous studies. Implementing these principles could improve the accuracy and reliability of military STT solutions within the Czech Armed Forces' (AČR) command and fire control systems [23].

3 Optimization of Military Operations and Resource Allocation

Optimization in military operations extends beyond software algorithms to the strategic allocation and deployment of key military assets. The application of Microsoft Excel Solver in this study was demonstrated on a simplified simulation scenario where computational resources for STT were dynamically distributed across multiple artillery units. Solver provided rapid solutions for allocating limited processing power based on real-time tactical demands and environmental constraints [2, 8].

Similarly, optimization approaches can be applied to computational resource allocation in combat scenarios, as shown in Optimization of the Weighted Multi-Facility Location Problem Using MS Excel. Experiments using MS Excel Solver have demonstrated that even widely available software can provide high-quality solutions for complex optimization problems, including:

- Distribution of computational capacity across command centers.
- Efficient processing of multi-source sensor and STT data [4, 24].

This approach can be adapted for real-time STT computation distribution among artillery units, where multiple factors—target prioritization, signal strength, and computing availability—influence decision-making on optimal data routing [2, 3, 15].

The same optimization methods can be applied to STT processing, such as:

- Dynamic allocation of STT computing resources in combat environments.
- Adaptive distribution of STT models across command centers.
- Optimization of data transmission within military networks.

Integrating these optimization techniques with fire control systems can lead to enhanced command efficiency, reduced response times, and greater operational effectiveness of artillery units [15].

3.1 Security Considerations and On-Premise Deployment of STT Systems

Given the sensitivity of military communications, ensuring maximum data security and integrity is imperative. The use of cloud-based STT services poses potential security risks, making the deployment of on-premise models or specially optimized versions (e.g., Whisper deployed within a closed infrastructure) the preferred approach.

At the same time, automated fire control systems must account for potential failures, as demonstrated in the study A Manual Method of Artillery Fires Correction Calculation. The Czech artillery school employs an analytical correction calculation method, which can offer greater accuracy than graphical methods used within NATO standards.

Under degraded operational conditions, manual fire control calculations must be performed quickly and efficiently. STT systems can serve as a command support tool, facilitating voice-based input and verification of fire correction calculations, thereby reducing errors and shortening artillery unit response times [23].

Additionally, integrating ACO-based approaches could enhance real-time detection of unauthorized or anomalous inputs, improving situational awareness and cybersecurity in command and fire control systems.

3.2 Artillery Position Defense and STT Utilization for Combat Effectiveness

Another critical aspect is the ability of artillery units to ensure self-defense and protection of firing positions, particularly in the face of combined enemy attacks. The study Evaluation of the Effectiveness of a Firing Battery in Self-Defense and Protection in the Area of Firing Positions Using Constructive Simulation [9] emphasizes the need for allocating additional resources (personnel, equipment, and materials) to enhance the protection of artillery units in combat zones.

Experiments conducted in the MASA SWORD simulation environment indicate that the survivability of artillery units in reserved or maneuver areas (ARA/AMA) is crucial for successful mission execution [1, 10].

Integrating STT systems into command and fire control structures could significantly enhance artillery unit responsiveness in defensive operations. Automated processing of voice commands, threat identification based on acoustic signals (e.g., detecting approaching enemy vehicles or incoming fire), and direct integration with simulation and battle management systems would enable real-time tactical decision-making.

For example, linking STT with MASA SWORD constructive simulation would allow:

- Dynamic battlefield analysis.
- Automated generation of defensive countermeasure recommendations.
- Optimized resource allocation for fire position defense.

By combining speech recognition, AI-based threat detection, and constructive simulation tools, modern artillery units could improve their resilience and operational effectiveness in contested environments.

4 Overview of Selected Technologies

4.1 Descript.Com

Descript.com is a speech-to-text solution known for its real-time transcription capabilities and high accuracy. Leveraging advanced machine learning algorithms, Descript.com delivers robust performance across various audio sources and linguistic contexts.

Its intuitive interface and efficient processing make it a preferred choice for users seeking reliable transcription services. However, a notable drawback is its subscription-based pricing model, which may limit accessibility for some users.

Despite its user-friendly interface, Descript.com depends heavily on cloud infrastructure and a subscription model, which may limit its usability in field-deployed military systems where connectivity and licensing are constrained.

On the other hand, Descript.com offers the most user-friendly and well-organized interface among the three compared solutions, making it a strong contender for usability-focused applications [5].

4.2 Google Vertex AI

Google Vertex AI Studio is a comprehensive machine learning platform that supports speech-to-text functionalities. One of its key components is Google's Universal Speech Model, known as Chirp.

Chirp is distinguished by its impressive scale, featuring 2 billion parameters and achieving exceptional accuracy in speech recognition across multiple languages.

What makes Chirp unique is its innovative training methodology, which incorporates self-learning algorithms applied to an extensive dataset comprising:

- Millions of hours of audio recordings.
- Over 28 billion sentences in more than 100 languages.

While Chirp shows high accuracy, especially in English, its performance in less-represented languages such as Czech is more limited. Furthermore, full model customization options are restricted in cloud-based deployments, potentially limiting adaptability to military-specific language use cases [8].

Unlike traditional models, Chirp integrates unsupervised data from diverse linguistic sources, enabling it to effectively capture speech patterns and linguistic variations. This is achieved using a zero-shot learning framework, where Chirp generalizes to unseen language combinations without requiring labeled training data. This architecture allows it to perform reasonably well in under-resourced languages, although domain-specific fine-tuning is still limited.

Key Advantages of the Chirp Model:

- Advanced encoder architecture – utilizes modern techniques to process speech input across different languages.
- Fine-tuning with minimal supervised data – significantly improves accuracy in specific languages and dialects where conventional models often struggle.
- Exceptional English recognition accuracy (98%) and a 300% improvement in speech recognition across other languages.

Due to its adaptability and versatility, Chirp is a powerful tool for speech recognition tasks, ranging from short phrase transcription to processing extended audio files exceeding one minute.

4.3 OpenAI Whisper

OpenAI's speech-to-text solution is recognized for its adaptability and versatility in processing diverse linguistic variations and accents. At the core of this solution is Whisper, an automated speech recognition (ASR) system built on an extensive training dataset.

Whisper is trained on a large-scale multilingual and multi-purpose dataset, including 680,000 h of supervised training data sourced from the web.

This vast and diverse dataset provides Whisper with high resilience to accents, background noise, and specialized terminology.

Additionally, thanks to its comprehensive training approach, Whisper is capable of:

- Transcribing and translating speech across multiple languages.

- Converting spoken input into English, facilitating multilingual interactions [4].

To foster innovation and collaboration, Whisper's models and inference code are openly accessible, making it easier for researchers and developers to build practical applications and advance robust speech processing solutions.

Key Differences Compared to Other Approaches. Unlike some existing approaches that rely on smaller, more specialized training datasets or unsupervised pre-training of audio models, the strength of Whisper is built upon:

- Training on an extensive and diverse dataset.
- Lack of fine-tuning on specific data subsets, which enhances its generalization ability

As a result, Whisper may not outperform specialized models in benchmark tasks, such as LibriSpeech, but it delivers significantly better results in real-world scenarios without additional training.

For instance, in tests using diverse datasets, Whisper achieved up to a 50% reduction in transcription errors compared to other models.

Whisper's generalist training approach means it lacks task-specific fine-tuning for niche use cases. Although this enhances real-world robustness, it may result in lower performance in standardized evaluation datasets. This trade-off must be considered in military contexts where command-specific phrase accuracy is critical.

5 Performance and Accuracy

5.1 Practical Research

In terms of overall performance and accuracy, all three solutions demonstrate similar competitive results, with unique strengths in specific areas:

- Descript.com excels in real-time transcription with stable accuracy across various audio sources.
- Google Chirp offers best-in-class English transcription performance and seamless scalability within the Google Cloud ecosystem.
- OpenAI Whisper is superior in handling linguistic variations and accents, making it an optimal choice for multilingual applications [5, 7, 19].

Each system provides distinct advantages depending on the intended use case, whether for high-speed real-time transcription, large-scale cloud deployment, or multilingual processing. The following Table 1 gives a summary (next see Fig. 1). Each dataset consisted of five 1-min recordings at 44.1 kHz sampling rate, with controlled variations in background noise levels (low, moderate, high).

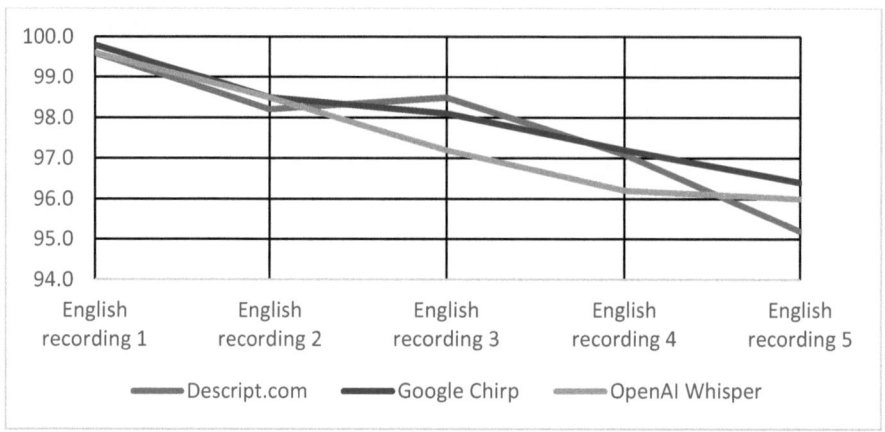

Fig. 1. Dependence of the percentage accuracy on the quality of the English recording.

Table 1. Comparison of AI Speech-to-Text System Performance at Different Audio Quality Levels – English recording.

	Descript.com	Google Chirp	OpenAI Whisper
English recording 1	99,6	99,8	99,6
English recording 2	98,2	98,5	98,5
English recording 3	98,5	98,1	97,2
English recording 4	97,1	97,2	96,2
English recording 5	95,2	96,4	96,0
Average	97,7	98,0	97,5

5.2 Results for Czech Audio Recordings

When focusing on Czech audio recordings, it was found that all three STT solutions officially support Czech. However, OpenAI Whisper achieved the best results among them.

- Descript.com exhibited the weakest performance, but still produced usable and reliable transcriptions.
- Overall transcription accuracy exceeded 97% for English and remained above 87% for Czech.
- More significant accuracy differences between the systems emerged primarily in lower-quality recordings.
- For Czech recordings, transcription accuracy declined more rapidly with decreasing audio quality compared to English recordings, highlighting the importance of high-quality input audio for achieving precise results.

Table 2. Comparison of AI Speech-to-Text System Performance at Different Audio Quality Levels – Czech recording.

	Descript.com	Google Chirp	OpenAI Whisper
Czech recording 1	95,2	95,6	96,1
Czech recording 2	91,3	92,1	94,2
Czech recording 3	87,6	88,1	94,2
Czech recording 4	85,0	87,2	92,2
Czech recording 5	80,2	83,1	85,6
Average	87,9	89,2	92,5

Each system provides distinct advantages depending on the intended use case, whether for high-speed real-time transcription, large-scale cloud deployment, or multilingual processing. The following Table 2 gives a summary (next see Fig. 2).

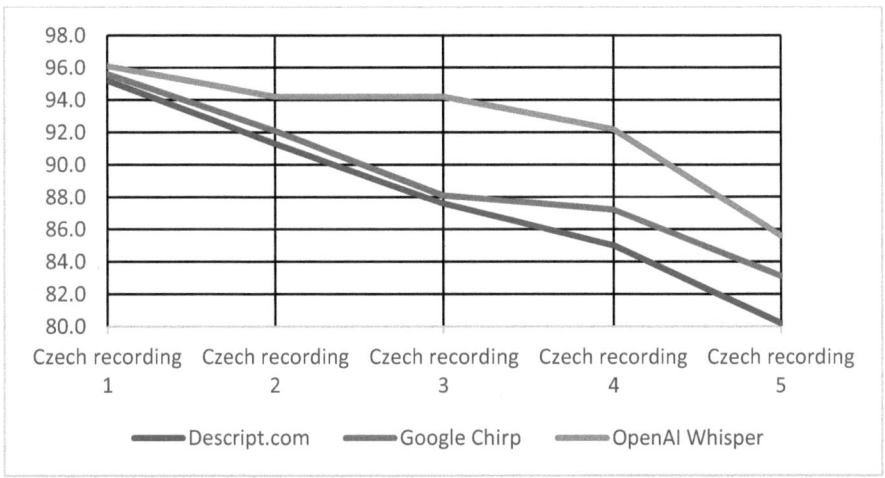

Fig. 2. Dependence of the percentage accuracy on the quality of the English recording.

Although Whisper achieved the highest accuracy for Czech recordings, no fine-tuning on domain-specific Czech military corpora was performed. This decision was influenced by the unavailability of extensive annotated Czech datasets and operational constraints that favor the use of pre-trained, black-box models.

5.3 Data Security Considerations

The use of speech-to-text (STT) tools in the military and security sectors presents significant risks, particularly regarding data protection and privacy.

Deploying these tools typically involves transmitting audio data to the service provider's cloud servers, raising critical concerns about the confidentiality and integrity of sensitive information.

Reliance on third-party cloud servers introduces the risk of data leaks or unauthorized access, potentially leading to national security threats and compromised operational secrecy.

Mitigating Security Risks. To counter these threats, a viable solution is the deployment of proprietary speech-to-text technology within an on-premise infrastructure.

- Military and security agencies can maintain full control over their data, minimizing the risk of exposure to external entities [11, 21].

5.4 Advantages of Proprietary (On-Premise) STT Solutions

Deploying self-managed STT solutions provides:

- Customization and integration with existing security protocols and infrastructure.
- Implementation of robust encryption mechanisms to protect sensitive data.
- Advanced access control and monitoring systems, ensuring a higher level of cybersecurity.

This approach enhances digital sovereignty, increases resilience to cyber threats, and protects against adversarial activities, making it a strategic necessity for military and intelligence applications.

6 Conclusion

This analysis aimed to provide an overview of three leading AI-powered speech-to-text (STT) solutions. Each platform offers unique features and capabilities, demonstrating strengths in different use cases.

Our practical testing highlighted the importance of evaluating these systems in real-world conditions, where factors such as audio quality and linguistic complexity significantly impact model performance:

- For English recordings, Google Chirp emerged as the most high-performing solution, achieving exceptional accuracy, particularly when processing low-quality audio.
- All tested AI solutions delivered competitive results, confirming their reliability across different application scenarios.
- For Czech recordings, OpenAI Whisper achieved the best performance, reinforcing its high efficiency in multilingual recognition.

Despite the strengths of these technologies, their deployment in sensitive domains, such as military and security applications, raises critical risks, particularly concerning data protection and privacy. Dependence on third-party cloud services for data processing raises concerns about unauthorized access and potential information leaks, which could compromise operational security and national defense integrity.

Strategic Investments in Automated Military Technologies. The findings of this analysis provide valuable insights for selecting optimal speech recognition solutions across various sectors. However, it is essential to acknowledge that AI technologies continue to evolve, and future developments may shift current performance trends.

The deployment of modern AI-driven technologies, including automated command and fire control systems, is critical to enhancing the defense capabilities of armed forces.

As outlined in the study An Alternative Model for Determining the Rational Amount of Funds Allocated to Defense of the Czech Republic in Conditions of Expected Risk, the current Czech Republic defense spending commitment of 2% of GDP may not adequately address real security threats or reflect the value of protected assets.

According to the authors, rational defense spending should range between 3.5% and 4.2% of GDP, based on threat probability and expected damage assessments [17, 22].

From this perspective, increased investments in automated command systems, fire control, and situational awareness technologies represent a cost-effective strategy to mitigate risks and enhance the effectiveness of defense measures.

The implementation of STT systems in artillery command and control offers:

- Faster command transmission and data processing – reducing response time in dynamic combat scenarios.
- Improved unit coordination – utilizing AI-driven transcription and communication analysis to enhance fire planning precision.
- Integration with modern simulation systems (e.g., MASA SWORD) – enhancing threat prediction and optimizing defensive operations based on advanced AI models [20].

This perspective demonstrates that modernizing defense capabilities through AI investments in combat decision-making and fire control systems is not merely a technological advancement—it is a fundamental component of strategic, effective, and rational defense planning.

Acknowledgments. The present work was supported by the Ministry of Defence of the Czech Republic, within the projects of the University of Defence, Faculty of Military Leadership "SV24-FVL-K107_DRA" and "DZRO-FVL22-LANDOPS".

Disclosure of Interests. The authors have no competing interests to declare that are relevant to the content of this article.

References

1. Blaha, M., Brabcova, K.: Communication environment in the perspective automated artillery fire support control system. In: Proceedings of the 10th WSEAS International Conference on Applied Informatics and Communications, and 3rd WSEAS International Conference on Biomedical Electronics and Biomedical Informatics, pp. 236–239 (2010)

2. Blaha, M., Brabcová, K.: Decision-making by effective C2I system. In: Proceedings of the 7th International Conference on Information Warfare & Security, pp. 44–50. Academic Publishing Limited, Seattle (2012). ISSN 2048-9870. ISBN 978-1-908272-29-4
3. Blaha, M., Potužák, L., Šustr, M., Ivan, J., Havlík, T.: Simplification options for more efficient using of Angular and Linear measuring Rules for Fire Control. Int. J. Educ. Inf. Technol. **15**(15), 28–34 (2021). ISSN 2074-1316. https://doi.org/10.46300/9109.2021.15.4
4. Bureš, M., Klíma, M., Blaha, M.: Ant colony optimization based algorithm for test path generation problem with negative constraints. In: IEEE International Conference on Software Quality Reliability and Security, pp. 701–712. IEEE Computer Society (2024). ISSN 2693-9185, ISBN 979-8-3503-6563-4. https://doi.org/10.1109/QRS62785.2024.00075
5. Descript: Speech to text tools. In: Descript Documentation. Descript. https://www.descript.com/tools/speech-to-text. Accessed 13 Jan 2025
6. Drozd, J., Procházka, J.: Konstruktivní simulace: účinný nástroj hodnocení operační efektivnosti v procesu plánování schopností. Vojenske Rozhledy-Czech Military Rev. **31**(2), 54–70 (2022). ISSN 1210-3292. IF 0,100. https://doi.org/10.3849/2336-2995.31.2022.02.054-070
7. Google Cloud: Getting Started with Chirp – Google's Universal Speech Model (USM) on Vertex AI. In: Google Cloud Blog. Google. https://medium.com/google-cloud/getting-started-with-chirp-the-googles-universal-speech-model-usm-on-vertex-ai-f54edaf4da93. Accessed 3 Jan 2025
8. Google Cloud: Speech-to-Text in Vertex AI. In: Google Cloud Documentation. Google. https://cloud.google.com/vertex-ai/docs/generative-ai/speech/speech-to-text. Accessed 20 Jan 2025
9. Havlík, T., Šustr, M., Ivan, J., Pekař, O., Mušinka, M.: Evaluation of the effectiveness of a firing battery in self-defense and protection in the area of firing positions using constructive simulation. J. Defense Model. Simul., 9 November 2024. ISSN 1548-5129. https://doi.org/10.1177/15485129241291579
10. Hujer, V., Slouf, V., Farlik, J.: Utility as a key criterion of a decision-making on structure of the ground based air defence. In: Mazal, J., et al. Modelling and Simulation for Autonomous Systems. MESAS 2021. Lecture Notes in Computer Science, vol. 13207, pp. 249–260. Springer, Cham (2022). ISSN 0302-9743. ISBN 978-3-030-98259-1. https://doi.org/10.1007/978-3-030-98260-7_15
11. Ivan, J., Šustr, M., Pekař, O., Potužák, L.: Prospects for the use of unmanned ground vehicles in artillery survey. In: Gini G., Nijmeijer H., Burgard W., Filev D. Proceedings of the 19th International Conference on Informatics in Control, Automation and Robotics (ICINCO), vol. 2022, pp. 467–475. Scitepress, Lisabon (2022). ISSN 2184-2809. ISBN 978-989-758-585-2. https://doi.org/10.5220/0011300100003271
12. Ivan, J., Šustr, M., Sládek, D., Varecha, J., Gregor, J.: Emergency meteorological data preparation for artillery operations. In: Gini, G., Nijmeijer, H., Filev, D.: Proceedings of the 20th International Conference on Informatics in Control, Automation and Robotics, vol. 1, pp. 250–257. ScitePress, Řím (2023). ISSN 2184-2809. ISBN 978-989-758-670-5. https://doi.org/10.5220/0012205500003543
13. Korecki, Z., Baláž, T., Krejčí, J., Racek, F.: Detection and localization system to determine the location of the laser source in the attack on the aircraft. In: Transport Means - Proceedings of the International Conference. Palanga, Lithuania: Kaunas University of Technology, vol. 1, pp. 140–146 (2023). ISSN 2351-7034. https://doi.org/10.5755/e01.2351-7034.2023.P1
14. Němec, P., Blaha, M., Pecina, M., Neubauer, J., Stodola, P.: Optimization of the weighted multi-facility location problem using MS excel. Algorithms **14**(7), 191 (2021). ISSN 1999-4893. https://doi.org/10.3390/a14070191
15. Němec, P., Šmerek, M., Pekař, O., Pecina, M.: Optimization of the complex multi-facility location problem using microsoft excel. AD ALTA: J. Interdisc. Res. **13**(1), 332–344 (2023). ISSN 1804-7890. IF 0,700

16. Neubauer, J., Vlkovský, M., Michálek, J.: Statistical modeling of cargo securing on selected military trucks and road surfaces. J. Defense Model. Simul. Appl. Methodol. Technol. **21**(3), 341–355 (2024). ISSN 1548-5129. IF 1,000. https://doi.org/10.1177/15485129241227012
17. Odehnal, J., Neubauer, J., Olejníček, A., Boulaouad, J., Brizgalová, L.: Empirical analysis of military expenditures in NATO nations. Economies **9**(3), 107 (2021). ISSN 2227-7099. https://doi.org/10.3390/economies9030107
18. OpenAI: Speech to text. In: OpenAI Documentation. OpenAI. https://platform.openai.com/docs/guides/speech-to-text. Accessed 3 Jan 2025
19. OpenAI: Whisper – robust speech recognition via large-scale supervised learning. In: OpenAI Research. OpenAI. https://openai.com/research/whisper. Accessed 13 Jan 2025
20. Palasiewicz, T., Rolenec, O., Kroupa, L., Maňas, P., Coufal, D.: Blast-induced deformations of the building entrance part caused by improvised shaped charges. In: Mazal, J., et al. Modelling and Simulation for Autonomous Systems. MESAS 2022. Lecture Notes in Computer Science, vol. 13866, pp. 109–130. Springer, Cham (2023). ISBN 978-3-031-31268-7. https://doi.org/10.1007/978-3-031-31268-7_7
21. Rolenec, O., Maňas, P., Palasiewicz, T.: Charakteristika násilných vstupů do objektů a experimentální posouzení možného vlivu rozletu střepin na bezpečnost výcviku při použití nálože na ploty. Vojenské rozhledy **33**(2), 147–166 (2024). ISSN 1210-3292. IF 0,300. https://doi.org/10.3849/2336-2995.33.2024.02.147-166
22. Šlouf, V., Blaha, M., Müllner, V., Brizgalová, L., Pekař, O. An alternative model for determining the rational amount of funds allocated to defence of the Czech Republic in conditions of expected risk. Obrana a strategie **2023**(1), 149–172 (2023). ISSN 1214-6463. IF 0,300. https://doi.org/10.3849/1802-7199.23.2023.01.149-172
23. Šustr, M., Ivan, J., Blaha, M., Potužák, L.: A Manual method of artillery fires correction calculation. Mil. Oper. Res. **27**(3), 77–94 (2022). ISSN 1082-5983. IF 0,700
24. Vališ, D., Hasilová, K., Vintr, Z., Rymarz, J.: City bus reliability measurement based on sparse field data supported by selected state space models. Transp. Res. Rec., August 2024. ISSN 0361-1981. IF 1,600. https://doi.org/10.1177/03611981241263563

Rhythm Fusion: Synchronizing Audio and Motion Features for Music-Driven Dance Generation

Nuha Aldausari[1(✉)], Gelareh Mohammadi[2], and David Cooper[3]

[1] Department of Computer Sciences, College of Computer and Information Sciences, Princess Nourah Bint Abdulrahman University, Riyadh, Saudi Arabia
nnaldausari@pnu.edu.sa
[2] School of Computer Science and Engineering, University of New South Wales, Sydney, Australia
g.mohammadi@unsw.edu.au
[3] Dolby Laboratories, Sydney, Australia
david.cooper@dolby.com

Abstract. Dance and music, as universal languages of emotion and expression, have been integral practices throughout human history, often associated with social and religious ceremonies. Manually animating a dancing person based on music is a challenging task that requires skill, time, and effort. However, with the help of an artificial intelligence (AI) model, dances can be generated automatically in response to music. Despite significant advancements in motion generation using AI techniques such as transformers, diffusion models, and GANs, challenges remain because existing frameworks primarily aim to produce movements that appear plausible in a general sense, rather than fully realistic. We developed our model with a specific goal: to understand the rhythmic link between music and motion and build specific components to learn the relationship between these features and then utilize that in the generative model.

To achieve this, we employ a state-of-the-art diffusion-style model to create dance sequences. We then introduce two sub-models: the Fusion Sync Classifier and Fusion Sync Enhancer. These sub-models, when integrated into the main model, Rhythm Fusion, ensure audio-video synchronization and facilitate the alignment and correlation between motion and music. Through the use of quantitative metrics, we show that our model outperforms other state-of-the-art models.

Keywords: Motion generation · Diffusion models · Audio-to-Keypoints mapping · Rhythm synchronization · Dance generation · Audio-video synchronization · Music-motion alignment

1 Introduction

As people immerse themselves in music, they often find themselves involuntarily tapping their feet or swaying to align their movements with the music's rhythmic structure. This spontaneous physical response to music reflects the human desire to interpret the music and convey their emotions. Dance that is synchronized to music can be used in various creative mediums, including film, video games, visual art, or even educational dance videos. For example, a video with a dance sequence could be used to teach individuals specific dance movements. Another example could be using tools to animate a movie actor to perform professional dance movements despite not being a professional dancer. Currently, to animate a character in response to music, an animator typically has to manually create keyframes according to the music using animation software such as Blender [12]. This animation process is time-consuming, demanding both skill and a deep understanding of dance and music. However, deep learning models have emerged as a solution to automate or assist this process and create choreography synchronized with music.

Audio-to-video generation models can produce video content by utilizing an audio signal as an input. Deep learning motion generation based on audio signals can be categorized by three main applications: speech synchronization, dance generation, and general motion generation. In speech synchronization, the objective is to map input speech signals to individuals delivering that speech, sometimes with an input image of the speaker. Dance generation applications involve mapping the input music to a person dancing to that music. In general motion generation, the goal is to map specific sounds to animations, such as translating the sound of ocean waves into visualizing ocean wave videos. In all these applications, the model is trained to generate a sequence of frames at either the pixel or keypoint level. This paper focuses on dance generation models based on music that primarily rely on keypoints. Here, keypoint generation surpasses pixel-level generation as the model learns to track fewer points along a sequence of frames. In addition, the datasets of the keypoints tend to have less diversity as there are a limited number of points in all the frames and across the dataset.

With the advancements in deep learning, dance generation models leverage diverse network architectures, including CNNs [1,15], RNNs, LSTMs [5,16,35], GANs [21,34], Transformers [22–24], and diffusion models [38]. With these architectures, some models employ separate encodings for audio and video, subsequently fusing these representations to generate the final dance [23,38]. Others opt for an autoregressive approach, predicting the dance code based on given music in the initial stage and refining it in the subsequent stage to generate the final dance [31]. Despite these varied architectures, all these models share a general objective, which is to translate the provided music into a cohesive sequence of continuous 2D or 3D human pose representations. However, they diverge in their specific goals. For instance, FACT, an early transformer-based model, aims to generate a lengthy sequence well-correlated with the input music, but a more recent model [40] has surpassed this model in these goals. EDGE focuses on the editing capabilities of the model, excelling in generating extended sequences, while Yang et al. [40] propose a model specifically designed for generating long-term dance, surpassing even the capabilities of EDGE and FACT. Bailando [31] seeks to satisfy dual objectives at spatial and temporal levels. Spatially, the model aims to capture only dance-like poses, while temporally, it targets a dancing sequence that follows the rhythm of the music. We share similar objectives with Bailando; however, our approach differs, as we avoid the complexities of difficult-to-train reinforcement learning methods [31] to achieve alignment between audio and motion. Moreover, in contrast to FACT, which attempts early fusion of music and motion to enhance the alignment between these modalities, we not only integrate these modalities early on but also employ pre-trained models that are specifically designed to classify the correlation between these modalities and actively work towards bringing them into closer alignment.

The Main Contributions of This Paper Can Be Summarized as Follows

- We developed a novel audio-to-keypoints model that leverages two pre-trained networks to align audio and motion features more effectively in a new way.
- We conducted quantitative evaluations to assess the performance of our network and compare it with three state-of-the-art models.
- We performed an ablation study to measure the individual effectiveness of each pre-trained network.

The rest of the paper is structured as follows. The following Sect. 2 provides an overview of the existing motion generation models with a specific focus on audio-to-motion generation. Section 3 outlines the architecture of our model with its two pre-trained models. In Sect. 4, we presented the outcomes of our proposed model, accompanied by a series of experiments to compare its performance with other models. In the last section, Sect. 8, a summary and concluding remarks are provided.

2 Related Work

Generative models are a category of deep learning models designed to generate new data points that conform to the distribution of the training data. Generative models can create diverse data types, such as images, videos, audio, and text, based on the training dataset. However, the result of video synthesis models remains in its early stages, particularly when juxtaposed with more established domains like images. This is because handling videos demands architectures that account for both spatial and temporal data. In addition, video generation might include other data, potentially incorporating an audio stream [2]. Commonly employed generative models in the current literature include Generative Adversarial Networks (GAN) [13], Variational Autoencoder (VAE) [19], flow-based models [30], and diffusion models [14,32,33]. GAN is one of the first generative models that demonstrated advanced results that closely resemble real training data, yet their adversarial training nature often leads to instability and less diverse outputs. In contrast, diffusion models have exhibited superior performance in the image domain compared to GANs [8]. First introduced by Sohl et al. in 2015 [32], diffusion models have undergone refinement by subsequent researchers such as Song et al. [33] and Ho et al. [14]. Diverging from the dual-network structure of GANs, which consists of a generator and a discriminator, diffusion models establish a Markov chain of diffusion steps, progressively introducing Gaussian noise to the original data at each forward diffusion process time step. This gradual introduction results in the loss of distinguishable features in the original data. In the reverse diffusion process, this procedure is inverted to reconstruct the data from the noise. In this paper, we utilize diffusion models as decoders to generate dance sequences, leveraging their superior performance.

Audio-to-video generative models find application in diverse domains such as speech, dance, and general motion generation. These models leverage various generative architectures like GANs, VAEs, transformers, diffusion models, or a combination thereof. In speech generation, the model processes audio inputs, occasionally accompanied by video inputs. When dealing solely with audio, the model generates video sequences based on the audio signal [3,4]. However, when both audio and video inputs are present, the application shifts to video retargeting or reenactment [29]. The primary goal of video retargeting or reenactment is to transfer motion trajectories from an input video to a generated one, often involving changing the person's identity. In the realm of video retargeting, some models use 2D or 3D keypoints as reference points [7,10,17,18,39,42,44,45], while others focus on the pixel level. For example, LipGAN [20] is a pixel-level face retargeting model that takes audio and a random face image as input, alongside another image featuring a desired pose with the lower half masked. LipGAN employs an audio encoder for audio signal processing and a visual encoder for the identity and masked pose images. Training involves both synchronized and unsynchronized audio-video pairs, integrating a reconstruction loss, contrastive loss, and adversarial loss. Similarly, Wav2lip [29], an extension of LipGAN, utilizes a pre-trained lip-sync expert during training. This lip-sync expert is a classifier that judges whether the pair of audio and video is from the same segment or not. This model is trained with sync loss and can be used as a pre-trained component with Wav2lip. In addition to sync loss, Wav2lip is incorporating reconstruction loss and a discriminator to assess video segment synchrony and quality. While Wav2lip shares both audio and visual encoders with LipGAN, it differentiates itself by working with audio-video segments rather than static images and short audio clips. In contrast to the pixel-level lip-sync model in Wav2lip, this paper proposes using a synchronization classifier at the keypoint level to determine whether a pair of audio and video are synchronized or not.

Motion generation refers to the process of creating realistic and dynamic movements, often within the context of computer vision, animation, or robotics. Models within this category have the capability to transfer motion between different skeletons without integrating audio signals as an input [6,41,43]. A preliminary attempt in incorporating audio signals for motion generation, as demonstrated in [35], utilizes traditional Long Short-Term Memory (LSTM) based autoencoders. In this approach, the model initiates by encoding the music signals, followed by the utilization of a predictor to generate the corresponding pose features. This attempt is succeeded by additional attempts, such as the work of Zhuang et al. [46], which incorporates WaveNet [27] from the speech generation domain to synthesise dance movements. Another model [11] employs semi-supervised methods, where the initial pretraining stage involves training the model on unlabeled data. During this stage, the model encodes Mel spectrogram representations and decodes them into Mel spectrogram, melody, and rhythm. In the subsequent stage, the pre-trained encoder and music embedding are utilized to generate the skeleton sequence. Ahn et al. [1] leverage a pre-trained music genre classifier to create music representations, followed by the use of dilated convolution layers for generating skeleton sequences. More recent works have used current generative models such as transformers and diffusion models. For example, FACT [23] utilizes the Full-Attention Cross-modal Transformer network to generate 3D dance motion conditioned on music. The generation process begins with audio feature extraction, followed by encoding using audio transformers. Motion features are encoded separately using motion transformers, and the resultant features are fused and decoded using a cross-modal transformer. The Bailando [31] model addresses the challenge of generating dance sequences synchronized with music differently. It introduces a choreographic memory to encapsulate meaningful dance units within a quantized codebook. These units are then combined into dance sequences coherent with the music using an actor-critic Generative Pre-trained Transformer (GPT). EDGE [38] stands out as a method tailored for editable dance generation. It employs a transformer-based diffusion model, incorporating Jukebox, a robust music feature extractor, to create realistic and physically plausible dances. These three models, FACT, Bailando, and EDGE, represent significant advancements in the field of 3D/2D dance generation. While existing models focus on editing dance or generating smooth motion trajectories, they face challenges in effectively synchronizing music with motion. Our model addresses these issues by placing a significant emphasis on achieving precise synchronization between the two. This emphasis is crucial, as unsynchronized motion with music is not only noticeable to the human eye but can also diminish the overall aesthetic quality of the generated dance sequence.

3 Methodology

In this work, we propose the Rhythm Fusion model, a diffusion-style generator that creates a sequence of dance skeletons from the input audio. This model employs a diffusion-style keypoint generator, that we augmented with two pre-trained networks known as Fusion Sync Classifier and Fusion Sync Enhancer. Fusion Sync Classifier plays a role in classifying whether the input audio and motion are synchronized, and Fusion Sync Enhancer aids the model in learning the relationship between audio and motion features. In the following subsections, we discuss each network separately.

3.1 Fusion Sync Classifier

The Fusion Sync Classifier is responsible for determining whether the input audio and motion are synchronized. To do this, the audio Mel spectrogram and the keypoints are first encoded separately using two encoders, as illustrated in Fig. 1. The architecture of each encoder comprises multiple convolutional layers with residual blocks. These two embeddings are then stacked to be inputted to the fully connected classifier. To train the classifier, we use two types of data samples: (1) synchronized pairs, where the audio Mel spectrogram is matched with the

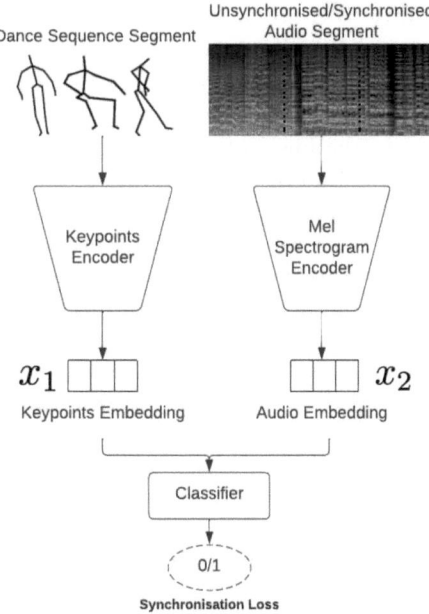

Fig. 1. The architecture of the Fusion Sync Classifier consists of two encoders for Mel spectrogram and keypoints sequence. The encoded signals are utilized to obtain the classification score.

correct corresponding dance keypoint sequence, and (2) unsynchronized pairs, where the audio is paired with an incorrect dance sequence. These two samples are associated with a Boolean variable y, which is used as the ground truth value to indicate whether a given pair is synchronized or not to be used in the loss function. To compute this loss, we employ binary cross-entropy after applying cosine similarity between the encoded embeddings. To calculate cosine similarity, a dot product is calculated between the ReLU-activated motion embedding x_1 and music embedding x_2 as shown in (1), resulting in a value ranging between 0 and 1 for each sample, indicating the probability that the input audio-video pair is synchronized. In (1), the value of ϵ is 1e-8 to prevent division by zero. We further use the binary-cross-entropy loss to compare the resulting probability p from (1) with the ground truth Boolean variable y, as shown in (2). The variables N, p_i, y_i represent batch size, predicted probability for sample i, and ground truth label that represents (synchronized/unsynchronized) for sample i.

$$p = \frac{x_1 \cdot x_2}{\max\left(\|x_1\|_2, \epsilon\right) \cdot \max\left(\|x_2\|_2, \epsilon\right)} \quad (1)$$

$$\mathcal{L}_{\text{Sync Classifier}} = -\frac{1}{N} \sum_{i=1}^{N} y_i \cdot \log\left(p_i\right) + (1 - y_i) \cdot \log\left(1 - p_i\right) \quad (2)$$

3.2 Fusion Sync Enhancer

The primary objective of this network is to learn the correlation between audio and motion. The overall architecture comprises two-stream encoder-decoder networks, with each stream encoding the respective signal separately, and it is shown in Fig. 2. The encoders in the Fusion Sync Enhancer consist of multiple convolutional layers followed by Causal Temporal Layers. On the motion side, the keypoint sequence is encoded in one network to produce the keypoints embedding, while on the audio side, the audio Mel spectrogram is first encoded and then the audio embedding is further encoded via a mapper network to learn the correlation between audio and motion signals to result in mapped audio embedding. A loss function is then employed to compare the encoded mapped audio embedding s_i and keypoints embedding k_i, minimizing the distance between these two signals, as detailed in Eq. (3). Following the encoding of these signals, two decoders are used for each signal. Two loss functions are applied, one is the reconstruction loss for the audio signals between generated audio $\hat{a}_i$ and input audio a_i in (4) and the other is the reconstruction loss for motion signals between generated keypoints sequence $\hat{m}_i$ and ground truth sequence m_i as in (5) to facilitate the convergence of the entire network.

$$\mathcal{L}_{\text{Embedding loss}} = \frac{1}{N} \sum_{i=1}^{N} (s_i - k_i)^2 \tag{3}$$

$$\text{MSE}_{\text{AudioReconstructionLoss}} = \frac{1}{N} \sum_{i=1}^{N} (a_i - \hat{a}_i)^2 \tag{4}$$

$$\text{MSE}_{\text{KeypointsReconstructionLoss}} = \frac{1}{N} \sum_{i=1}^{N} (m_i - \hat{m}_i)^2 \tag{5}$$

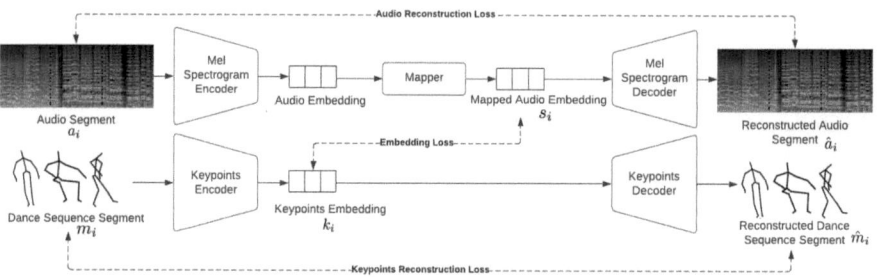

Fig. 2. Fusion Sync Enhancer: The network consists of two streams for audio and dance sequences. In each stream, there is an encoder and a decoder. The audio stream has an additional encoding step using a model called a mapper. The overall network has two reconstruction losses for each stream and an embedding loss to minimize the distance between these signals.

3.3 Rhythm Fusion

Rhythm Fusion is the overall architecture that incorporates components of the Rhythm Sync Enhancer and Rhythm Sync Classifier to improve synchronization within the model. The core of Rhythm Fusion adopts a Denoising Diffusion Probabilistic Model (DDPM) [14], and its architecture is illustrated in Fig. 3. The forward diffusion process is provided in (6), where $\mathbf{x}_1 \sim q(\mathbf{x})$ is a sequence of keypoints of a dance sampled from the training data points. This process creates a series of diffusion steps, $\mathbf{x}_1, \ldots, \mathbf{x}_T$, each progressively affected by additional noise, where T denotes the total number of diffusion steps. The extent of each diffusion step is regulated by a predefined variance schedule $\{\beta_t \in (0,1)\}_{t=1}^T$. The reverse diffusion process is according to (7), and can reconstruct the training sample from a Gaussian noise input, $\mathbf{x}_T \sim \mathcal{N}(\mathbf{0}, \mathbf{I})$. Thus, the reverse process can generate the dance sequence $\mathbf{x}_\theta(\mathbf{x}_t, m, t)$ condition on the audio representation input m, and timestep t, but for short notation, we refer to the entire input with $\mathbf{x}_\theta$. The audio representation used in our model includes the encoded Mel spectrogram, derived from waveform conversion, and additional representations such as Envelope, MFCC, and Chroma are extracted using Librosa Python package. In addition, the audio representations include peak and beat in one hot encoding format. At each time step, the input undergoes a linear layer, followed by a transformer decoder layer, and another linear layer to generate the sequence of skeletons.

$$q(\mathbf{x}_t \mid \mathbf{x}_{t-1}) = \mathcal{N}\left(\mathbf{x}_t; \sqrt{1-\beta_t}\mathbf{x}_{t-1}, \beta_t \mathbf{I}\right) \tag{6}$$

$$q(\mathbf{x}_{t-1} \mid \mathbf{x}_t, \mathbf{x}_1) = \mathcal{N}\left(\mathbf{x}_{t-1}; \tilde{\boldsymbol{\mu}}(\mathbf{x}_t, \mathbf{x}_1), \tilde{\beta}_t \mathbf{I}\right) \tag{7}$$

We adopted loss functions commonly used in diffusion-style generators [38, 40] which is a "simple" loss function $\mathcal{L}_{\text{simple}}$ introduced by Ho et al. [14]. In addition, we adopted Auxiliary loss functions $\mathcal{L}_{\text{Auxiliary}}$ [37, 38] to enhance the physical realism of kinematic motion. Specifically, these additional losses are designed to promote the alignment of three key aspects of physical realism: joint positions, velocities, and Contact Consistency Loss. The pre-trained models, Fusion Sync Classifier and Fusion Sync Enhancer, are integrated into our architecture. From the Fusion Sync Classifier, we employ the audio encoder to encode the input signal, and after generating the skeletons, we encode this signal. Our classifier determines whether these signals are synchronized. During the training of the Fusion Sync Classifier from scratch, we input both synchronized and unsynchronized pairs. However, after incorporating this network into Rhythm Fusion, we exclusively feed the network with synchronized data for faster convergence. From Fusion Sync Enhancer, we incorporate the audio encoder and the mapper for the audio input signal. We also add the keypoint encoder for the generated dance to minimize the distance between audio and motion representations. We added the equations (2) and (3) from Fusion Sync Classifier and Fusion Sync Enhancer, respectively. The overall loss function for the Rhythm Fusion model is described in (8).

$$\mathcal{L} = \mathcal{L}_{\text{simple}} + \lambda_{\text{Auxiliary}} \mathcal{L}_{\text{Auxiliary}} + \lambda_{\text{Sync Enhancer}} \mathcal{L}_{\text{Sync Enhancer}} \\ + \lambda_{\text{Sync Classifier}} \mathcal{L}_{\text{Sync Classifier}} \quad (8)$$

4 Experiment Details

4.1 Dataset

In our study, we leverage the AIST++ dataset introduced in [23], one of the most comprehensive publicly available datasets. AIST++ is a 3D human dance motion dataset, encompassing a pair of 3D motions with the musical audio file. One of the common representations of this dataset is 24 Skinned Multi-Person

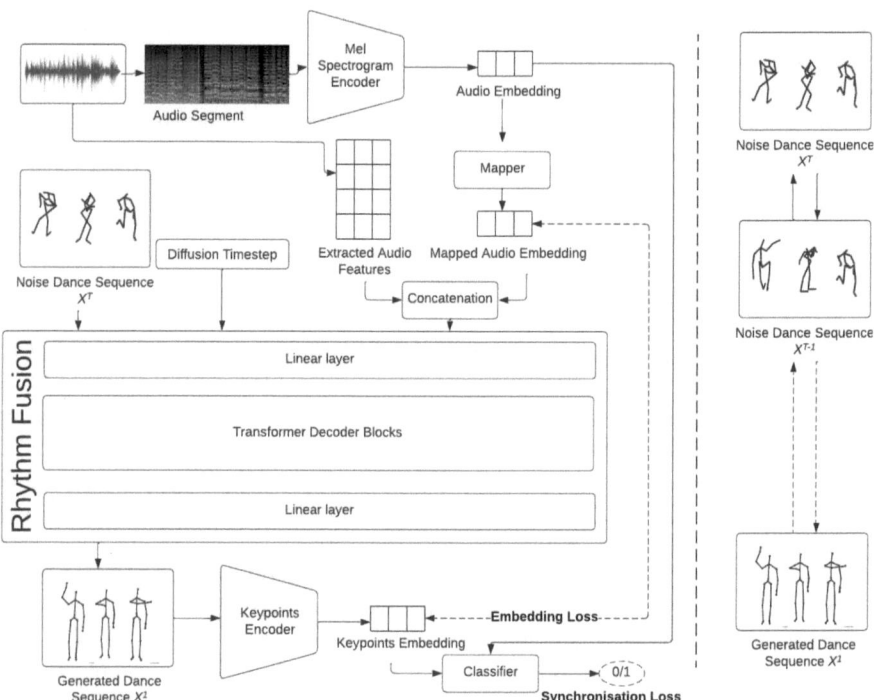

Fig. 3. The overall architecture of Rhythm Fusion includes components of Rhythm Sync Enhancer and Rhythm Sync Classifier. On the right side, there is an overview of our model that follows a diffusion process, denoising dance sequences starting from time $t = T$ to $t = 1$. On the left side, there is a detailed description of a diffusion process (Rhythm Fusion) at one time step. The input consists of noisy skeletons, diffusion timestep, and music features, while the output is a synthesized dance sequence. The generated dance is further encoded to be used in the optimization process as a loss function.

Linear Model (SMPL) pose parameters [25]. AIST++ has 1408 sequences, 30 subjects, and a diverse range of 10 dance genres that span both fundamental and sophisticated choreographies. We follow the approach of dance representation proposed in [38]. Each joint utilizes the 6-DOF rotation representation supplemented by a single root translation, denoted as $w \in R^{24*6+3=147}$. Binary contact labels ($b \in \{0,1\}^{2*2=4}$) for the heel and toe of each foot are incorporated. Consequently, the comprehensive pose representation is expressed as $x = \{b, w\} \in R^{4+147=151}$. Our diffusion-based framework is learning the synthesis of sequences comprising N frames, denoted as $x \in R^{N*151}$. In all our models, videos are trimmed to 5 s with a frame rate of 30 FPS, resulting in N being 150. The paired audio corresponds to the audio within the same 5-second video.

5 Quantitative Evaluation

We employ quantitative evaluation to assess the performance of our model and compare it against FACT, Bailando, and EDGE, which are currently the top state-of-the-art models. In our quantitative assessments, we utilize three key metrics across all models. Table 1 provides a summary of the quantitative metrics, which indicates that our model outperforms the state-of-the-art models on all metrics. The metrics are detailed below.

The first metric is the Physical Foot Contact score (PFC) [38], which assesses the physical plausibility of the generated content. PFC is an alternative to the feet sliding metric but in dance. Feet sliding checks if the feet are always in constant contact with the ground, but this is not the case for dances where sliding is part of the routine. Thus, PFC is based on two simple observations. The first observation is that when a dancer is moving forward/backward, they can only push off the ground using their feet. This means either at least one foot is staying still or the dancer's body isn't changing speed. The second observation is that jumping upwards requires pushing off the ground with the dancer's feet.

The next metric is the Fréchet Inception Distance (FID), which measures the performance by quantifying the distance between real and generated motion features. The extracted motion features could be based on kinetic features [28] identified as 'k' or geometric features [26] identified as 'g'. The kinetic feature is created by considering motion velocities and energies, thereby capturing the physical attributes of dance. In contrast, the geometric feature is formulated by leveraging various movement patterns, providing insights into the quality of choreography. This is done by tracking specific body points in the motion sequence. We follow FACT implementation [23] to calculate FID. We compared the ground truth dance motion sequences in the AIST++ test set with 40 generated motion sequences, that are randomly selected. As indicated in the second and third columns in Table 1, our generated motion sequences represent a distribution much more similar to the ground-truth motions in comparison to the other baseline models. We additionally assess the capacity of our model to produce varied dance motion patterns when presented with different input music. The diversity can be determined by the average Euclidean distance in the feature space, which can be the geometric feature space and the kinetic feature space, and it is denoted as 'g' and 'k', respectively. Based on the results from Table 1, our model shows marginally improved diversity compared to other models.

Lastly, the Beat score evaluates the correlation between motion and music, specifically the similarity between visual beats and music beats. The musical beats are obtained through Librosa, a Python package, while the kinematic beats are determined as the local minima of the kinetic velocity. The beat score measures the mean temporal gap between each musical beat and its nearest corresponding beat in the dance sequence [31]. The findings in the last row in Table 1 indicate that our proposed model surpasses previous state-of-the-art models across all three metrics.

6 Qualitative Evaluation

Although we initially considered juxtaposing still frames to illustrate qualitative performance, we found that such static comparisons offer limited insight into the dynamic aspects of our approach. Therefore, to provide a more comprehensive demonstration, we have included video examples in our GitHub repository. These videos more effectively showcase the systems behavior in real-world scenarios, allowing readers to better evaluate its qualitative performance.

We have included videos that compare our model with FACT, Bailando, and EDGE in the form of 2D representations. These videos offer clearer comparisons of synchronization and motion quality than static frame images. In these comparisons, the FACT model displays some irregularities, such as jittering. Both the FACT and Bailando models show less alignment with the musical beat, while EDGE and our model demonstrate better synchronization. Our model achieves superior alignment through the inclusion of two key components: the Fusion Sync Classifier and the Fusion Sync Enhancer, which improve the correspondence between audio and motion trajectories. Additionally, our model closely follows the beat and appears similar to the ground truth, further highlighting its improved synchronization performance.

Table 1. Quantitative evaluation with PFC, FID, and Beat scores. Our model demonstrates superior results in all metrics.

Model	PFC↓	FID↓	Beat Score↑
FACT	2.25	23.11	0.22
Bailando	1.75	24.82	0.23
EDGE	1.53	23.08	0.27
Our model	**0.98**	**22.89**	**0.29**

7 Ablation Study

We conducted an ablation study in which we removed each pre-trained network one at a time and assessed the model's performance using PFC, FID, and Beat scores. The results of this ablation study are presented in Table 2.

In the first ablation study, we removed the pre-trained Fusion Sync Classifier from the overall architecture. Specifically, we excluded the Mel Spectrogram encoder and the keypoints encoder of the Fusion Sync Classifier. Additionally, we omitted the synchronization loss detailed in (2) from the total loss function shown in (8). It appears that this modification had a relatively small impact on the overall performance, as shown in the first row in Table 2. One reason for this might be that making decisions on an entire sequence of dance with audio music may result in a score that does not necessarily represent each beat but rather an average score. Therefore, this score, (2), may not be a very informative loss function, especially when comparing a dance sequence with music from the same song but a different segment.

We also conducted another ablation study where we removed components of the pre-trained Fusion Sync Enhancer. In other words, we excluded the Mel Spectrogram encoder, keypoints encoder, and the embedding loss from the total loss in (8). We observed that this had a more pronounced effect on the overall proposed architecture. The reason behind this is that the pre-trained Fusion Sync Enhancer not only attempts to reconstruct the input signal and compare it with the ground truth but also generates an embedding (audio/ keypoints) that is correlated with the other embedding (audio/keypoints). This is a crucial aspect of the overall architecture because we aim to generate dance sequences inspired by the ground truth and synchronized with the music. It is worth mentioning that when we trained the Fusion Sync Enhancer without the mapper, the encoded features might not align perfectly in every beat, and the model did not converge. This is because when we extracted the audio and visual beats, as defined in [31], and analyzed the audio and dance beats to see if they aligned in each frame, we found a significant number of instances where the audio beat occurred first, followed by the visual beat in the subsequent frame. This resembles human-like behaviour, as it mirrors the response after hearing the audio beat. Aligning all visual beats with audio beats would result in a robotic-like dance, which is not the case for the AIST++ dataset.

The last row in Table 2 represents the model with both pre-trained models, which demonstrates the superiority of the model over having only one of the pre-trained models across all three metrics. Therefore, to achieve optimal performance, it is crucial to utilize both pre-trained networks.

Table 2. An ablation study was conducted on our proposed model, involving the removal of one pre-trained network in each row of the table.

Model	PFC↓	FID↓	Beat Score↑
- w/o Fusion Sync Classifier & sync. loss	2.25	24.35	0.21
- w/o Fusion Sync Enhancer & embed. loss	1.75	23.16	0.23
Our model	**0.98**	**22.89**	**0.29**

8 Conclusion and Future Research

In this paper, we introduced Rhythm Fusion, a novel diffusion-style framework that leverages two subnetworks, the Fusion Sync Classifier and the Fusion Sync Enhancer, to explicitly model and reinforce the correlation between audio and motion signals. These subnetworks were specifically designed to address a critical gap in previous work: learning alignment between music and dance movements.

Compared to state-of-the-art methods, FACT, Bailando, and EDGE, our model provides two key advantages. First, our Fusion Sync Enhancer learns a shared embedding space that captures the rhythmic and structural relationships between motion and music, leading to more coherent and precise synchronization. Second, the Fusion Sync Classifier continuously evaluates audio–video pairs to ensure that each generated dance sequence remains well aligned with the input music, enforcing global temporal consistency. As a result, we achieve superior performance across multiple metrics such as physical plausibility, diversity, and beat alignment, thereby closing an important gap in music-driven dance synthesis.

In addition to outperforming other approaches, our model brings forth three key contributions: (1) a dedicated mechanism for learning audio–motion correlations via a mapper network, (2) an integrated classification module that explicitly enforces synchronization, and (3) an end-to-end diffusion-style generator that captures both local rhythmic cues and overall motion dynamics. These architectural choices collectively enable us to generate dance sequences that are not only visually appealing and realistic but also demonstrate tighter adherence to musical beats.

While the goal of this project is to align audio with video features, a future approach could be refining the alignment by focusing on specific attributes such as beats or motion strength. This specific synchronization can be first extracted from the audio and video files and then used as input to improve the performance of the model. Specific feature synchronization is first achieved in the context of generating background music [9]. However, the feasibility of this approach relies on the availability of a suitable dataset that has MIDI files associated with the songs, as emphasized by [9]. Another promising future direction involves expanding the training scope to include non-human rhythmic events that there is a desire to match to music, such as keypoint representations of videos of fireworks. As far as our current knowledge extends, there is no music-keypoints dataset beyond dance-related datasets and speech-oriented datasets. However, if such a dataset were to be collected, it would open doors to various applications, one of which is music visualization. For instance, our model could be trained to map music to an unstructured finite set of points, subsequently generating dynamic motion patterns using these points. These keypoints can then be rendered into moving objects in an artistically engaging manner, further extending the utility and creative potential of the model.

References

1. Ahn, H., Kim, J., Kim, K., Oh, S.: Generative autoregressive networks for 3D dancing move synthesis from music. IEEE Robot. Autom. Lett. **5**, 3501–3508 (2020)
2. Aldausari, N., Sowmya, A., Marcus, N., Mohammadi, G.: Video generative adversarial networks: a review. ACM Comput. Surv. (CSUR) 1–25 (2022)
3. Aldausari, N., Sowmya, A., Marcus, N., Mohammadi, G.: Diverse audio-to-video gan using multiscale image fusion. In: Australasian Joint Conference on Artificial Intelligence, pp. 29–42. Springer (2022)
4. Aldausari, N., Sowmya, A., Marcus, N., Mohammadi, G.: Cascaded siamese self-supervised audio to video GAN. In: Proceedings of the IEEE/CVF Conference on Computer Vision and Pattern Recognition, pp. 4691–4700 (2022)
5. Alemi, O., Françoise, J., Pasquier, P.: GrooveNet: real-time music-driven dance movement generation using artificial neural networks. Networks **8**, 26 (2017)
6. Chan, C., Ginosar, S., Zhou, T., Efros, A.: Everybody dance now. In: Proceedings of the IEEE/CVF International Conference on Computer Vision, pp. 5933–5942 (2019)
7. Chen, L., Cao, C., Torre, F., Saragih, J., Xu, C., Sheikh, Y.: High-fidelity face tracking for AR/VR via deep lighting adaptation. In: Proceedings of the IEEE/CVF Conference on Computer Vision and Pattern Recognition, pp. 13059–13069 (2021)
8. Dhariwal, P., Nichol, A.: Diffusion models beat GANs on image synthesis. Adv. Neural. Inf. Process. Syst. **34**, 8780–8794 (2021)
9. Di, S., et al.: Video background music generation with controllable music transformer. In: Proceedings of the 29th ACM International Conference on Multimedia, pp. 2037–2045 (2021)
10. Doukas, M., Zafeiriou, S., Sharmanska, V.: Headgan: one-shot neural head synthesis and editing. In: Proceedings of the IEEE/CVF International Conference on Computer Vision, pp. 14398–14407 (2021)
11. Duan, Y., et al.: Semi-supervised learning for in-game expert-level music-to-dance translation. arXiv Preprint arXiv:2009.12763 (2020)
12. Foundation, B. Blender (software). https://www.blender.org/
13. Goodfellow, I., Pouget-Abadie, J., et al.: Generative adversarial nets. In: Advances in Neural Information Processing Systems, pp. 2672–2680 (2014)
14. Ho, J., Jain, A., Abbeel, P.: Denoising diffusion probabilistic models. Adv. Neural. Inf. Process. Syst. **33**, 6840–6851 (2020)
15. Holden, D., Saito, J., Komura, T.: A deep learning framework for character motion synthesis and editing. ACM Trans. Graph. (TOG) **35**, 1–11 (2016)
16. Huang, R., Hu, H., Wu, W., Sawada, K., Zhang, M., Jiang, D.: Dance revolution: long-term dance generation with music via curriculum learning. arXiv Preprint arXiv:2006.06119 (2020)

17. Huang, P., Yang, F., Wang, Y.: Learning identity-invariant motion representations for cross-id face reenactment. In: Proceedings of the IEEE/CVF Conference on Computer Vision and Pattern Recognition, pp. 7084–7092 (2020)
18. Kim, H., et al.: Neural style-preserving visual dubbing. ACM Trans. Graph. (TOG) **38**, 1–13 (2019)
19. Kingma, D., Welling, M.: Auto-encoding variational bayes. arXiv Preprint arXiv:1312.6114 (2013)
20. KR, P., Mukhopadhyay, R., Philip, J., Jha, A., Namboodiri, V., Jawahar, C.: Towards automatic face-to-face translation. In: Proceedings of the 27th ACM International Conference on Multimedia, pp. 1428–1436 (2019)
21. Lee, H., et al.: Dancing to music. In: Advances in Neural Information Processing Systems, vol. 32 (2019)
22. Li, B., Zhao, Y., Zhelun, S., Sheng, L.: Danceformer: music conditioned 3D dance generation with parametric motion transformer. In: Proceedings of the AAAI Conference on Artificial Intelligence, vol. 36, pp. 1272–1279 (2022)
23. Li, R., Yang, S., Ross, D., Kanazawa, A.: AI choreographer: music conditioned 3D dance generation with AIST++. In: Proceedings of the IEEE/CVF International Conference on Computer Vision, pp. 13401–13412 (2021)
24. Li, J., et al.: Learning to generate diverse dance motions with transformer. arXiv Preprint arXiv:2008.08171 (2020)
25. Loper, M., Mahmood, N., Romero, J., Pons-Moll, G., Black, M.J.: SMPL: a skinned multi-person linear model. In: Seminal Graphics Papers: Pushing the Boundaries, vol. 2, pp. 851–866 (2023)
26. Müller, M., Röder, T., Clausen, M.: Efficient content-based retrieval of motion capture data. In: ACM SIGGRAPH 2005 Papers, pp. 677–685 (2005)
27. Oord, A., Dieleman, S., Zen, H., et al.: Wavenet: a generative model for raw audio. arXiv Preprint arXiv:1609.03499 (2016)
28. Onuma, K., Faloutsos, C., Hodgins, J.: FMDistance: a fast and effective distance function for motion capture data. In: Eurographics (Short Papers), pp. 83–86 (2008)
29. Prajwal, K.R., Mukhopadhyay, R., Namboodiri, V.P., Jawahar, C.V.: A lip sync expert is all you need for speech to lip generation in the wild. In: Proceedings of the 28th ACM International Conference on Multimedia, pp. 484–492 (2020)
30. Rezende, D., Mohamed, S.: Variational inference with normalizing flows. In: International Conference on Machine Learning, pp. 1530–1538 (2015)
31. Siyao, L., et al.: Bailando: 3D dance generation by actor-critic GPT with choreographic memory. In: Proceedings of the IEEE/CVF Conference on Computer Vision and Pattern Recognition, pp. 11050–11059 (2022)
32. Sohl-Dickstein, J., Weiss, E., Maheswaranathan, N., Ganguli, S.: Deep unsupervised learning using nonequilibrium thermodynamics. In: International Conference on Machine Learning, pp. 2256–2265 (2015)
33. Song, Y., Ermon, S.: Generative modeling by estimating gradients of the data distribution. In: Advances in Neural Information Processing Systems, vol. 32 (2019)
34. Sun, G., Wong, Y., Cheng, Z., Kankanhalli, M., Geng, W., Li, X.: DeepDance: music-to-dance motion choreography with adversarial learning. IEEE Trans. Multimedia **23**, 497–509 (2020)
35. Tang, T., Jia, J., Mao, H.: Dance with melody: an LSTM-autoencoder approach to music-oriented dance synthesis. In: Proceedings of the 26th ACM International Conference on Multimedia, pp. 1598–1606 (2018)

36. Tang, T., Jia, J., Mao, H.: Dance with melody: an LSTM autoencoder approach to music oriented dance synthesis. In: Proceedings of the 26th ACM International Conference on Multimedia, pp. 1598–1606 (2018)
37. Tevet, G., Raab, S., Gordon, B., Shafir, Y., Cohen-Or, D., Bermano, A.: Human motion diffusion model. arXiv Preprint arXiv:2209.14916 (2022)
38. Tseng, J., Castellon, R., Liu, K.: Edge: editable dance generation from music. In: Proceedings of the IEEE/CVF Conference on Computer Vision and Pattern Recognition, pp. 448–458 (2023)
39. Wu, W., Zhang, Y., Li, C., Qian, C., Loy, C.: Reenactgan: learning to reenact faces via boundary transfer. In: Proceedings of the European Conference on Computer Vision (ECCV), pp. 603–619 (2018)
40. Yang, S., Yang, Z., Wang, Z.: LongDanceDiff: Long-term Dance Generation with Conditional Diffusion Model. arXiv Preprint arXiv:2308.11945 (2023)
41. Yang, Z., Zhu, W., et al.: TransMoMo: invariance-driven unsupervised video motion retargeting. In: Proceedings of the IEEE/CVF Conference on Computer Vision and Pattern Recognition, pp. 5306–5315 (2020)
42. Zakharov, E., Shysheya, A., Burkov, E., Lempitsky, V.: Few-shot adversarial learning of realistic neural talking head models. In: Proceedings of the IEEE/CVF International Conference on Computer Vision, pp. 9459–9468 (2019)
43. Zhou, Y., Wang, Z., Fang, C., Bui, T., Berg, T.: Dance dance generation: motion transfer for internet videos. In: Proceedings of the IEEE/CVF International Conference on Computer Vision Workshops (2019)
44. Zhou, H., Liu, J., Liu, Z., Liu, Y., Wang, X.: Rotate-and-render: unsupervised photorealistic face rotation from single-view images. In: Proceedings of the IEEE/CVF Conference on Computer Vision and Pattern Recognition, pp. 5911–5920 (2020)
45. Zhang, J., et al.: Freenet: multi-identity face reenactment. In: Proceedings of the IEEE/CVF Conference on Computer Vision and Pattern Recognition, pp. 5326–5335 (2020)
46. Zhuang, W., Wang, C., et al.: Music2dance: music-driven dance generation using wavenet. arXiv Preprint arXiv:2002.03761 (2020)

Leveraging Synthetic Data for Deep-Learning-Based Road Crack Segmentation from UAV Imagery

Andriani Panagi[ORCID] and Christos Kyrkou[✉][ORCID]

KIOS Research and Innovation Center of Excellence, University of Cyprus, Cyprus, Nicosia, Cyprus
{panagi.andriani,kyrkou.christos}@ucy.ac.cy
https://www.kios.ucy.ac.cy/

Abstract. One of the critical tasks in monitoring of road infrastructure is the identification of road cracks. Recent efforts have been made in utilising Unmanned Aerial Vehicles (UAVs) to automate this task without interfering with the road network traffic and infrastructure. However, high-quality annotated datasets that allow the development of reliable deep learning models for this purpose, are scarce. Synthetic data generation offers a promising alternative to mitigate this issue by reducing annotation costs and enhancing the dataset's variability. This paper represents a comparative study of two state-of-the-art deep learning models - UNet with EfficientNet and UNet with MobileNet- for road crack segmentation trained with three loss functions (Dice Loss, Focal Loss, and Weighted Binary Cross-Entropy Loss (WBCEL)) by using synthetic datasets generated with three different ways. Performance evaluation indicates that the best results were achieved using UNet with a MobileNet encoder, trained with WBCEL and synthetic images, yielding an mIoU of 63.52% tested on real crack imagery. A more granular analysis underlines how both synthetic data realism and the choice of the loss function impact segmentation accuracy. Our preliminary study concludes that well-designed synthetic data and appropriate loss functions have the potential to allow better generalization of the model to real-world scenarios.

Keywords: Synthetic image generation · Semantic segmentation · UAV imagery · Road crack segmentation · Convolutional neural networks

1 Introduction

Monitoring and maintaining road networks is critical for ensuring continuous transport, minimising disruptions, and fostering and supporting economic productivity. Conversely, decayed infrastructure can have disastrous consequences, including increased accident risks, higher vehicle maintenance costs, and transportation delays. Road defects, such as cracks, not only contribute to inefficiencies in commerce and transportation but also threaten human lives. The World Health Organization (WHO) reports that 1.19 million traffic accidents occur annually, injuring 20 to 50 million people [1], with road damage often playing a vital role. Uneven surfaces can lead to loss of vehicle control, reduced traction between the tires and the road, and increased driver stress and attention, ultimately leading to increased chances of accidents.

Traditional methods for detecting road damage typically involve human visual inspection or vehicle-mounted cameras for infrastructure monitoring [7,28]. Human inspectors are able to estimate not only the visible damage but also contextual circumstances, such as surroundings or structural issues that may occur. However, human in the loop solutions often lead to delays and limited throughput. Meanwhile, vehicle-mounted cameras can cover a wide range of road networks relatively fast and acquire high-resolution images that can be analysed automatically or manually inspected when needed. Nevertheless, these methods have several limitations: manual inspections are highly labour-intensive, time-consuming, and costly, particularly when applied to large sections of road networks. Additionally, they put workers' safety at risk, especially when they are on busy highways or in severe weather conditions. Vehicle-mounted systems, while automated, can generate a vast amount of data which makes the processing of it extremely resource-intensive. Traditional methods are therefore less effective when used in proactive maintenance and repair strategies as they may be inconsistent and occasionally fail to detect vital yet minor road defects, which ultimately minimise their effectiveness.

Identifying road damages using UAVs provides an alternative and time-effective method for infrastructure monitoring [25] without disrupting road network operations. One of the advantages of UAVs is their ability to cover large, hard-to-reach areas such as highways and rural roads in less time, without interfering with traffic flow compared to traditional methods. Equipped with high-resolution cameras and sensors, UAVs can capture detailed imagery of the road surface from multiple angles, therefore improving the likelihood of detecting subtle defects such as cracks. Moreover, their deployment is more cost-effective and adaptable to various monitoring scenarios.

However, using UAVs for automated road inspection presents several challenges. Cracks typically occupy only a small percentage of image pixels, which requires deep learning-based segmentation models to be highly precise in detecting even the smallest defects for enabling quick responses. Additionally, acquiring training data and manually annotating the images is often a complex and time-consuming task that requires significant human and infrastructure resources. Other factors, such as weather conditions, safety concerns and regulatory restrictions, further complicate the data acquisition process.

A more efficient alternative is the use of synthetic image data. Synthetic image data refers to artificially generated images through computer algorithms or simulations, rather than those captured by traditional imaging devices. These images closely replicate real-world visuals and are useful for training machine learning models in object detection, segmentation, and recognition tasks, providing a scalable solution for model development. Moreover, the generation of synthetic data automatically creates pixel-accurate annotations, reduces annotation costs, and accelerates the experimentation cycles.

While synthetic data generation has been shown to improve detection performance [29], its effectiveness compared to real data, the choice of model architecture, and the impact of various synthetic data generation techniques remain open questions especially in the context of road crack segmentation from UAV imagery. This paper aims to address these gaps by conducting a comparative study of segmentation models (UNet with EfficientNet and MobileNet encoders), synthetic data generation methods, and loss functions (Dice Loss, Focal Loss, and Weighted Binary Cross-Entropy) for road crack segmentation using UAV imagery. The generalisation ability of models trained on synthetic data is evaluated on a real dataset. The key contributions of this paper are summarised as follows:

- We propose a framework for generating synthetic road crack images by first segmenting road regions from UAV-captured images with a top down view and then superimposing crack patterns using three different techniques to ensure realistic integration with the road surface (Fig. 1).
- We evaluate and compare the performance of different deep learning segmentation model configurations trained on the synthetic datasets, evaluating them with various metrics to assess the impact of three different loss functions on segmentation accuracy.
- We derive the best combination of data generation, backbone network and loss function for road crack segmentation for UAV imagery.

Fig. 1. Overlaying crack images on road infrastructure background images captured from a UAV with a top-down view. This involves ensuring that the crack images are appropriately sized and positioned to blend seamlessly into the road surface. The end result is the road image with its binary mask.

2 Literature Review

Over the past years, extensive research has been conducted on general crack detection, segmentation, and classification using various approaches. These include image processing techniques [10,17,26], model-based methods [8,14,16] and synthetic data generation [12,19]. Initially, using conventional image processing, A. Mohan and S. Poobal present a thorough analysis of 50 studies focused on the automatic identification of concrete surface cracks and their depth using image processing methods [17]. In addition to describing the various image processing methods according to the kind of image utilized, the survey also does a thorough study, paying particular attention to the accuracy and error rates of each approach.

However, such methods have been replaced by machine learning approaches that offer improved accuracy and efficiency. Specifically deep learning techniques dominate in crack detection. This can be seen in both [8,14], where Convolutional Neural Networks (CNNs) were employed for pixel-level semantic segmentation of road crack damages. These studies concentrate on the development of deep neural networks for crack detection on road surface images, which is a widely adopted approach among researchers. Additionally, a comparative study presented in [24] evaluates state-of-the-art deep learning models for semantic crack detection in construction materials. The authors evaluate the performance of various existing algorithms in capturing fine details for crack segmentation and suggest that generating realistic synthetic image data could be a potential solution to mitigate data scarcity issues during model training.

2.1 Relevant Synthetic Data Generation Methods

Kanaeva and Ivanova in [12] generated an image dataset by imposing road crack images on top of backgrounds taken from publicly available datasets such as KITTI and Cityscapes. They used this dataset to train UNet [21] and Mask R-CNN [9] models. When tested on real images of cracked roads, these models provided an mIoU score of 47%. However, the mismatch of background and overlaid image perspective adopted in their methodology is a limitation to the realism of the synthesised dataset. In addition, Rill-García et al. [19] proposed a new tool, named "Syncrack", for creating images with accurate labels of cracks to improve the accuracy of pixel-level prediction. For generating images, the synthesised crack shapes were overlaid onto custom-made background images. To produce a fully synthetic dataset, noisy annotations were generated using the pixel-accurate crack masks. The segmentation models, such as UNet, that were trained on this dataset, performed similarly to those that were trained on real world datasets, providing greater precision, particularly in detecting the crack width.

Aditionally synthetic data augmentation for dam crack detection is proposed in [29]. The paper presented and introduced a concept similar to the one developed for this research but this time for crack detection on the surface of dams. The method used the generation of synthetic crack images and training a deep neural network, yielding good results, which further demonstrates how synthetic data increases the ability to train models when real data may not be accessible or sufficient.

Other methods, such as Generative Adversarial Networks (GANs), have been popular recently, since the success of these models has motivated many researchers. A semantically-driven generative adversarial network for generating crack images is called CrackGauGAN proposed in recent work [3]. This network can generate realistic-looking crack images that are high in fidelity with diverse data distributions. This not only outperforms the latest semantic layout-based GANs but also shows greater improvement in generating crack images.

Although the aforementioned methods propose state-of-the-art improvements in the synthesis of images, they do not adequately address key factors such as background noise and the unique challenges posed by top-down UAV imagery, which is crucial for road surface analysis. Moreover, while individual studies have demonstrated improvements in segmentation using synthetic data, there is limited comparative analysis on how different synthetic data generation techniques impact model performance in real-world scenarios. Our study proposes a comprehensive comparative research, evaluating multiple segmentation models and synthetic data configurations to determine the most effective approach for robust road crack segmentation with UAV imagery. By directly comparing model performance across different datasets and loss functions, we aim to provide insights into how synthetic-to-real transfer can be optimized for UAV-based infrastructure monitoring.

2.2 Existing Crack Segmentation Datasets

Crack segmentation datasets have significantly evolved and expanded over time. Most of them are employed as benchmarks for tasks involving segmentation, detection, and classification. Nevertheless, there are some limitations to the existing crack segmentation datasets. One drawback is that models are unable to learn much about how the crack might relate to the surrounding environment because most images of the already existing datasets lack a contextual background. Scale inconsistencies, limited variability, and differences in viewpoints—along with most images being captured by vehicle-mounted cameras instead of UAVs—are common issues in these datasets, reducing their generalization power in real-world applications. In the following, we provide an overview of some existing datasets which can be leveraged to create synthetic images from a UAV perspective (also summarized in Table 1).

- **CRACK500 Dataset:** The CRACK500 Dataset [30] consists of more than 500 high-resolution pavement images collected by using a smartphone as the data sensor. Each image was annotated by multiple annotators. The dataset contains pavement cracks of variable widths.
- **Crack Segmentation Dataset by Roboflow:** There are 4029 images of concrete damage in the dataset [20]. The images in the dataset have undergone augmentations such as rotation, saturation, and brightness adjustments.

Table 1. Summary of crack segmentation datasets.

Dataset	View	Captured With	Number of Images	Image Size
CRACK500 [30]	Top view	Smartphone camera	500	3264 × 2448
Crack Segmentation Dataset by Roboflow [20]	Top view	–	4029	416 × 416
CrackTree200 [31]	Top view	–	206	800 × 600
CFD [23]	Top view	Smartphone Camera	118	480 × 320
GAPs [6]	Top view	Vehicle Mounted Camera	1969	1920 × 1080
DeepCrack [15]	Top view	–	537	544 × 384
CrackSeg9k [13]	Multiple views	Multiple Methods	9255	400 × 400
OmniCrack30k [2]	Multiple views	Multiple Methods	30K	Multiple image sizes

Although the dataset consists solely of concrete crack damage, the cracks can still be applied to pavement scenarios, as they have been processed in a way that allows them to effectively mimic pavement cracks after overlaying.
- **CrackTree200 Dataset:** The CrackTree200 Dataset consists of 206 pavement images, each with dimensions of 800×600 pixels, featuring a variety of crack types annotated at pixel level. The dataset captures images in complex asphalt environments, characterized by challenges such as shadows, occlusions, low contrast, and noise [31].
- **CFD Dataset:** The dataset includes 118 annotated images of road cracks, each of 480×320 pixels [23]. These images were captured on urban roads in Beijing and exhibit a significant presence of noisy pixels, such as oil spots and water stains. Additionally, some images were taken under poor lighting conditions, further complicating the annotation process.
- **GAPs Dataset:** A total of 1969 grayscale images with a resolution of 1920×1080 pixels are included in [6]. The images have been annotated manually. The surface material depicted in the images includes pavement from three different German federal roads.
- **DeepCrack Dataset:** This dataset [15] entails 537 images with manual annotation maps. Each image is made available to a pixel-wise segmentation map, which presents to be a mask exactly covering the crack regions. All of the images are of a fixed size of 544×384 pixels.
- **CrackSeg9k Dataset:** With 9255 images that aggregate other smaller open-source datasets, the dataset [13] is one of the biggest, most varied, and reliable crack segmentation collection. It consists of 10 sub-datasets namely Crack500 [30], DeepCrack [15], SDNET2018 [5], CrackTree200 [31], GAPs [6], Volker, Rissbilder [18], CFD [23], Masonry [4], and Ceramic [22], with images being preprocessed and resized to 400×400.

- **OmniCrack30K:** It is the largest benchmark for crack segmentation containing nearly 30K images from over 20 datasets [2]. The dataset features cracks found in various materials, including asphalt, ceramic, concrete, masonry, and steel. Although UAV images are included in this dataset, they do not depict road cracks.

The existing datasets primarily focus on segmenting and detecting cracks in structures or roads using close-up images captured by ground-based or vehicle cameras, often neglecting contextual information. In contrast, our dataset provides images with annotations based on aerial imagery (captured with a UAV), incorporating contextual information from the surrounding areas near the roads. This broader perspective enhances its applicability for real-world crack detection tasks.

3 Methodology

3.1 Synthetic Image Data Generation

Creating high-quality road crack segmentation datasets requires pixel-level annotations, which is a highly time-consuming process as each pixel corresponding to a crack must be accurately labeled in the image. Alternatively, generating synthetic image data using auxiliary datasets can substantially reduce the time and effort required for manual annotation, making it a highly efficient and scalable solution.

Obtaining relevant background images that will be used as the basis for superimposing the crack damage is crucial for the generation of the synthetic images (Fig. 2). In our case, top-down realistic road infrastructure images were captured by a UAV (with an altitude of 120 m) to provide a comprehensive view of the surface. Along with these, images of cracks and their corresponding binary masks were sourced from existing annotated datasets, though these originated from different contexts. The binary masks were used to isolate the cracks, allowing their integration onto our own road images. To ensure diversity in crack damages we combined multiple crack images sourced from different publicly available datasets such as Crack500 [30], DeepCrack [15], CrackTree200 [31], GAPs [6], Volker, Rissbilder [18], CFD [23], and the Crack Segmentation Dataset by Roboflow [20].

The process of overlaying cracks onto background images starts with first isolating the road segment. This ensures that cracks are applied only to the relevant road area, resulting in a more realistic outcome. A segmented mask of the road from the background images is used to guide the synthetic crack deployment. Valid positions (e.g., all (x,y) pixel points of the image that belong to the road area) for crack placement are determined to avoid unrealistic positioning, such as cracks appearing on non-road areas. The crack image is adjusted to match the color and scale of the road, ensuring a natural integration into the background. In detail, an inverse mask is created from the original binary mask of the crack. This inverse mask is used to black out the area where the

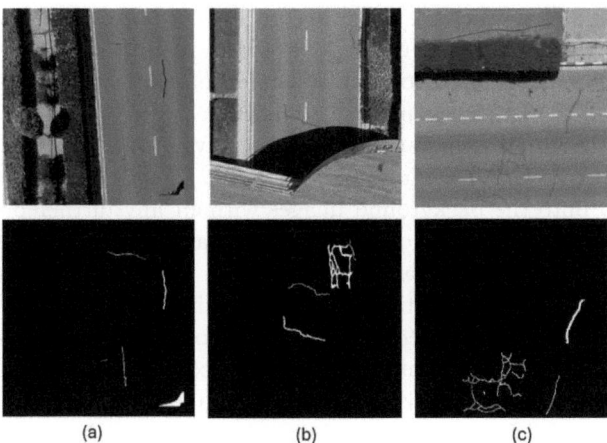

Fig. 2. Examples of crack image overlays: (a) The image displays the superimpose of cracks without any pre-processing performed to the crack overlay. (b) The crack images are scaled and then overlaid. (c) Both scaling and blending techniques are used during the deployment of cracks on the background images. Alongside each image, the corresponding binary mask is generated. The scaling and blending techniques used in (b) and (c) improve the visual coherence of the cracks with the background, making them appear more natural with the road.

crack will be applied, preserving the rest of the region in the background image. The inverse mask is applied to the region of interest (ROI)[1] in the background, ensuring that only the background area outside the crack region remains visible, while the area where the crack will be placed is blacked out. The crack image is further isolated by applying the binary mask to it, keeping only the crack region visible and blacking out the rest. The crack region is then resized to match the size of the background ROI, ensuring consistent dimensions between the crack and the background areas. Finally, the crack region is added to the blacked-out ROI from the background, effectively overlaying the crack on the background. This combined region is then placed back into the original background image at the specified position, resulting in a seamless integration of the crack into the background.

During this process, certain images containing vehicles on the road were identified. Since overlaying cracks on the cars would result in unrealistic damage of the road, these images were removed from the dataset so that the accuracy is maintained. Additionally, cracks were superimposed on road images where the

---

[1] The region of interest refers to the specific portion of the background image where the crack will be placed. It is defined by the coordinates (x, y), which represent the top-left corner of the area, and the dimensions h (height) and w (width) of the crack image. By slicing the background image from y to y+h (vertical range) and from x to x+w (horizontal range), the function extracts the exact area of the background where the crack will be overlaid, ensuring that the crack fits within this region.

roads had no natural cracks. The entire process of generating synthetic images with superimposed cracks is fully automated in a form of a pipeline. Human intervention is only required to manually remove images where cracks are mistakenly applied to cars, ensuring realistic results.

Synthetic images are generated in different ways of varying complexity. In the first case (referred to as the *Raw Cracks Dataset*), crack damage was superimposed directly onto the background images without any further pre-processing. In the second case, the crack overlays (referred to as *Cracks with Scaling Dataset*) were subsequently scaled to pixel values between the range 100–150 in order to better match the road color statistics. In the third case, referred to as *Cracks with Scaling and Blending Dataset*, in addition to scaling, a blending technique is used by doing a weighted sum of the initial pixel value with the crack image value, to ensure that the crack damages mix seamlessly with the road surface background.

3.2 UNet as the Baseline Network

The UNet architecture [21] comprises a down-sampling and an up-sampling path, where the encoder layers in the down-sampling direction reduce the spatial resolution of the input while capturing contextual information. In the up-sampling direction, the decoder layers reconstruct the segmentation map by decoding the encoded data and incorporating information from the down-sampling path through skip connections.

For crack segmentation, UNet processes a full-resolution input image (e.g., 512×512 pixels). During the down-sampling phase, the encoder progressively reduces the spatial dimensions of the image while increasing the number of channels by repeatedly applying 3×3 convolutional layers and max-pooling operations. This enables the encoder to capture hierarchical feature representations of the input. The up-sampling path takes the feature map from the bottleneck and reconstructs it to match the original input size using 2×2 up-convolution layers, followed by two 3×3 convolutional layers. These operations increase the spatial resolution while decreasing the number of channels. Skip connections from the down-sampling path are integrated into the up-sampling path to help the decoder accurately localize and refine features. Ultimately, the output is a binary segmentation map in which each pixel is classified as either a crack or part of the background.

3.3 Encoder Variants of the Network

For the encoder part, two variants were chosen: EfficientNet [27] having 7 million parameters and MobileNet V3 [11] having 2 million parameters. EfficientNet was chosen due to its scalable architecture, balancing accuracy w.r.t. efficiency, handling high-resolution image data with a large parameter space. The MobileNet V3 network was chosen for its lightweight design with less parameters that could be more suitable for on board processing of a UAV, as it provides faster computation and effective memory usage. The weights were initialized from ImageNet pre-trained models, to leverage transfer learning enabling better feature extraction capabilities and aid in faster convergence for our task. The encoder of the UNet model has a depth of 5^2 and the decoder has channels of 256, 128, 64, 32, and 16^3. To assess class imbalance and evaluate model performance, the models were trained using three distinct loss functions: the Dice Loss, the Focal Loss, and the Weighted Binary Cross Entropy Loss.

4 Experimental Set-up

4.1 Evaluation Metrics

We evaluate performance using various metrics, taking into account class imbalance, including Recall, Precision and F1-score, mean IoU (mIoU), and pixel accuracy during experimentation. The F1-Score is the harmonic mean between precision (the percentage of predicted positive pixels that are true positive) and recall (the percentage of actual positive pixels that were correctly identified). A model with either higher precision will have a low false positive rate or higher recall will retrieve most of the cracks. By examining region agreement, mIoU considers the spatial relationships between pixels, in contrast to these metrics that focus on classification accuracy. In particular, mIoU quantifies the degree to which the actual crack and predicted crack shapes match; a higher mIoU score indicates greater consistency. This is crucial for road crack segmentation tasks since high accurate pixel classification alone is insufficient to indicate good performance as cracks are often a small fraction of pixels in the image.

However, the aforementioned generic segmentation metrics such as mIoU and F1-score may not fully capture the quality of crack segmentation in real-world applications. For example, crack width consistency, total crack segmented area, and crack continuity are critical parameters in most real-world applications, e.g., road maintenance inspection. Although our evaluation is application-independent, future work could be to incorporate other measures that are directed towards the nature of cracks, such as width variation or continuity measurement to better reflect the structural integrity of the detected cracks.

4.2 Loss Functions

During the training of the models, we use 3 loss functions that address class imbalance issues. We assess their effectiveness in helping the different models learn reliable representations. The following loss functions were employed in the experimentation phase:

---

[2] A number of stages used in the encoder. Each stage generates features two times smaller in spatial dimensions than the previous one (e.g., for depth 0 we will have features with shapes [(N, C, H, W),], for depth 1 - [(N, C, H, W), (N, C, H // 2, W // 2)] etc. Here N represents the batch size, C the channels of the image, and H and W the height and the width of the image).

[3] List of integers which specifies the channels parameter for convolutions used in the decoder.

- **Weighted Binary Cross-Entropy Loss (WBCEL):** Same as the standard binary cross entropy loss but in order to combat class imbalance, it incorporates weights (w_0,w_1) for each class while calculating the loss between the true labels and the predicted probabilities.

$$\text{WBCEL}(y, \hat{y}) = -\frac{1}{N} \sum_{i=1}^{N} w_1 y_i \log(\hat{y}_i) + w_0(1 - y_i) \log(1 - \hat{y}_i) \quad (1)$$

- **Focal Loss:** It is a variant of the conventional cross-entropy loss that focuses more on examples which are hard to classify by down-weighting the loss contribution from cases which are well classified. Given below, γ is the focusing parameter that changes the loss contribution of the well-classified samples. α is the weight component which will help in handling the class imbalance.

$$\text{Focal Loss}(y, \hat{y}) = -\frac{1}{N} \sum_{i=1}^{N} \Big(\alpha y_i (1 - \hat{y}_i)^\gamma \log(\hat{y}_i)$$
$$+ (1 - \alpha)(1 - y_i)\hat{y}_i^\gamma \log(1 - \hat{y}_i) \Big) \quad (2)$$

- **Dice Loss:** Derived from the dice coefficient (which ranges from 0 to 1, where 1 represents perfect overlap) and transforms it into a loss function that ranges from 0 to 1, where 0 corresponds to perfect overlap.

$$\text{Dice Loss}(y, \hat{y}) = 1 - \frac{2 \sum_{i=1}^{N} y_i \hat{y}_i}{\sum_{i=1}^{N} y_i + \sum_{i=1}^{N} \hat{y}_i} \quad (3)$$

where y_i is the ground truth and $\hat{y}_i$ is the prediction in all the above equations.

The weights in the loss functions were calculated based on the amount of pixels of each class in the training set of the binary mask images. Specifically, the class weights are computed as a ratio of the total pixels count to each class count, giving higher weights to the less frequent classes. After that, the class weights are normalized so that the smallest one is 1, to make sure the class with the fewest pixels gets the base weight.

4.3 Datasets Splits

During the experiments, three distinct synthetic image datasets were produced to explore various scenarios as it was stated in the previous section. Each dataset was split into training and validation sets, with invalid images (e.g., cracks superimposed on cars) being removed. Hence, in the first dataset, where cracks were overlaid onto background images without any additional processing, there are 436 training and 148 validation images. In the second dataset, where scaling was applied to the crack images before overlaying them onto the background, we ended up with 449 training and 145 validation images. The third dataset, where both scaling and a blending technique were applied, resulted in 465 training and 150 validation images. These three datasets have varying image counts in their training and validation sets, reflecting the distribution seen in previously described datasets in Sect. 2.2.

4.4 Implementation Details

All experiments were conducted using the PyTorch framework on an Nvidia GeForce Tesla V100-PCIE-32GB GPU. The size of the input images used was 512×512 pixels. During the training phase, the Adam optimiser was employed with a maximum learning rate of 0.0001 and a weight decay factor of 0.0001. Image augmentation such as rotations and brightness were applied to the training images to enhance learning. The training was performed for 150 epochs with a batch size of 8.

5 Results and Analysis

5.1 Overall Performance of Models on Real Crack Test Data

The experimental results, summarized in Table 2 and illustrated in Fig. 3, highlight the performance of the models after being trained on synthetic data and then tested on real crack images. The evaluation was conducted across the three generated datasets. Among the different models and configurations of losses and datasets, the UNet-MobileNet model, trained on the scaled and blended cracks dataset using WBCEL, achieved the best performance, with an mIoU of 0.6352, an F1-Score of 0.7166, and a crack pixel accuracy of 0.3326. Additionally, the UNet-MobileNet model trained on the same dataset with Focal Loss delivered comparable results across all metrics. In applications where comprehensive detection of all crack pixels is critical, such as defect analysis or road crack monitoring, the UNet-MobileNet model with WBCEL is the preferred choice, prioritizing higher Recall over slightly lower Precision to ensure thorough crack identification. Other scenarios such as the UNet-MobileNet with WBCEL on the cracks with scaling dataset perform slightly worse but outperform other configurations such as the same scenario but with the Dice Loss function.

Figure 3 consists of four subplots visualizing the mIoU, F1-score, Recall, and Precision values obtained on test images by the three datasets utilized. It can be seen that UNet-Mobilenet consistently achieves the highest mIoU values, demonstrating its superior performance in spatially localizing cracks on each sample. Although the encoder has fewer parameters it is able to capture contextual information and achieve better results than EfficientNet.

In addition, Fig. 4 gives an overview of the Precision and Recall of the scenarios presented on Table 2. The different configurations fall into two categories: one with low precision and low recall, and another with high precision and low recall. Models trained on raw cracks generally show high precision but suffer from low recall, especially with Dice Loss. Incorporating scaling improves recall slightly but retains moderate to high precision. The best performance is achieved with "Cracks with Scaling and Blending" dataset, where models using Focal Loss strike the best balance between precision and recall, notably UNet-MobileNet with Focal Loss (Precision = 0.84, Recall = 0.67) as stated above. In contrast, Dice Loss models consistently exhibit very high precision but lower recall, indicating a tendency to miss cracks while making accurate detections when cracks

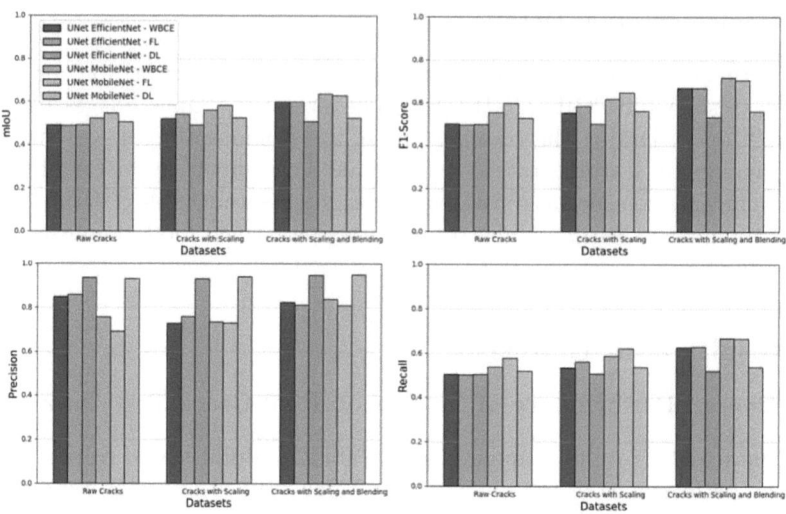

Fig. 3. Bar plots of evaluation metrics of crack segmentation models trained on synthetic data and tested on real crack images.

Table 2. Experimental results of models trained with synthetic data and tested on real crack images.

Dataset	Method	Parameters	Loss Function	mIoU	Pixel Accuracy (Crack)	Precision	Recall	F1-Score
Raw Cracks	UNet-EfficientNet	7M	WBCEL	0.4908	0.0077	0.8462	0.5034	0.5003
			Focal Loss	0.4877	0.0009	0.8565	0.5004	0.4943
			Dice Loss	0.4899	0.0043	0.9338	0.5026	0.4986
	UNet-MobileNet	2M	WBCEL	0.5209	0.0776	0.7563	0.5368	0.5529
			Focal Loss	0.5462	0.1628	0.6895	0.5769	0.5956
			Dice Loss	0.5048	0.0349	0.9299	0.5171	0.5259
Cracks with Scaling	UNet-EfficientNet	7M	WBCEL	0.5190	0.0654	0.7262	0.5340	0.5508
			Focal Loss	0.5405	0.1063	0.7590	0.5596	0.5833
			Dice Loss	0.4914	0.0075	0.9296	0.5040	0.5014
	UNet-MobileNet	2M	WBCEL	0.5600	0.1729	0.7335	0.5879	0.6166
			Focal Loss	0.5816	0.2389	0.7302	0.6210	0.6478
			Dice Loss	0.5244	0.0674	0.9377	0.5365	0.5598
Cracks with Scaling and Blending	UNet-EfficientNet	7M	WBCEL	0.5995	0.2453	0.8218	0.6262	0.6697
			Focal Loss	0.5994	0.2446	0.8119	0.6278	0.6690
			Dice Loss	0.5082	0.0369	**0.9468**	0.5205	0.5312
	UNet-MobileNet	2M	WBCEL	**0.6352**	**0.3326**	0.8385	**0.6665**	**0.7166**
			Focal Loss	0.6271	0.3297	0.8090	0.6637	0.7053
			Dice Loss	0.5240	0.0710	0.9488	0.5361	0.5578

are identified. Overall, using the cracks with scaling and blending dataset provides the most reliable results for crack segmentation.

Table 3. Comparison of model (UNet-MobileNet/WBCEL) trained with a real crack dataset and different initial weight configurations.

Initial Weights	Pretraining	Metrics on Real Cracks Dataset				
		mIoU	Pixel Accuracy (crack)	Precision	Recall	F1-Score
ImageNet	Synthetic	0.6550	**0.8202**	0.6891	**0.8918**	0.7469
Random	None	0.5821	0.7255	0.6218	0.8249	0.6632
ImageNet	None	**0.6787**	0.7888	**0.7180**	0.8811	**0.7701**
Random	Synthetic	0.6072	0.3899	0.7049	0.6966	0.6849

5.2 Performance of Model Trained on Real Crack Dataset with Different Initial Weight Configurations

Table 3 presents a comparison of different weight initialization strategies when training a UNet-MobileNet model with WBCEL on a real crack dataset. The experiments evaluate the impact of using pretrained ImageNet weights and synthetic pretraining. The first configuration involves initializing the model with ImageNet weights, followed by training on synthetic crack data before using these weights on the training of the model with real crack data. The second configuration starts with random weights and without any synthetic pretraining. The third configuration uses ImageNet-initialized weights but is trained directly on real crack data, skipping synthetic pretraining. Lastly, the fourth configuration initializes first the model with random weights, undergoes synthetic pretraining and then uses these pretrained weights to train the model on real crack images. The results shown in the table are from the evaluation of the different configurations on real crack data. The experiments confirm that pretraining, in any form, is beneficial. Models initialized with ImageNet weights consistently perform better, making them the preferred choice when available. However, for architectures lacking ImageNet pretrained weights, synthetic pretraining proves useful as it further enhances crack detection accuracy. These findings highlight the value of domain-specific synthetic pretraining, suggesting that its impact could be even greater with a larger pretraining dataset.

5.3 Effect of the Loss Functions

Table 2 demonstrates that the choice of loss function significantly impacts the performance metrics of the UNet models. Across all datasets, Focal Loss provides a balanced trade-off between precision and recall, resulting in higher mIoU, pixel accuracy, and F1 scores compared to WBCE and Dice Loss. WBCE achieves moderately high recall and pixel accuracy but relatively lower precision, leading to lower mIoU and F1 scores overall. In contrast, Dice Loss yields the highest precision across most experiments but suffers from notably lower recall, reducing its ability to detect all cracks consistently. This behavior is particularly evident in "Raw Cracks" and "Cracks with Scaling" datasets, where Dice Loss results

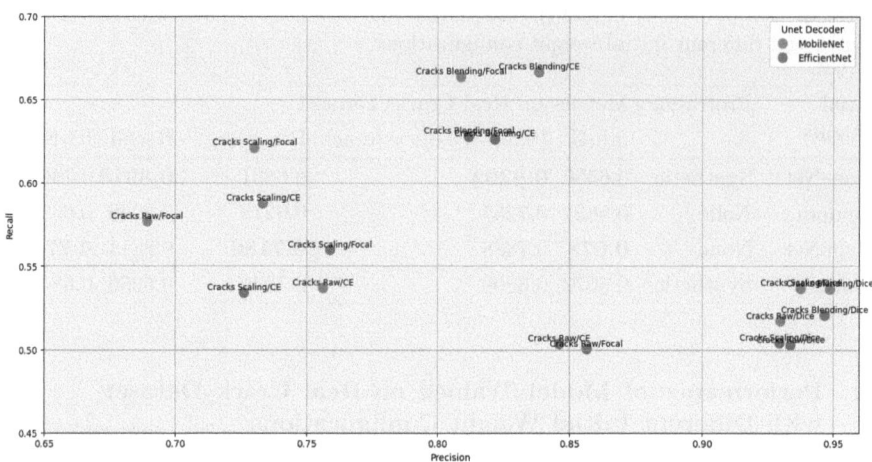

Fig. 4. Precision-Recall plot of models trained with synthetic data.

in precision values exceeding 0.92 but recall values below 0.55. Overall, while WBCE offers more balanced recall, Focal Loss stands out by consistently achieving superior mIoU, pixel accuracy, and F1 scores, making it the most effective loss function for crack segmentation tasks.

5.4 Visualization

Figure 5 visually demonstrates the crack segmentation results on real test data when using the models that were trained on "Cracks with Scaling and Blending" synthetic dataset. All models exhibit some falsely detected pixels, where some of them couldn't even predict the mask. UNet-MobileNet with either WBCEL or Focal Loss deliver the best predictions among the four images.

5.5 Further Fine-Tuning of Models on Real Crack Images

The best seven models in terms of accuracy and mIoU presented in Table 2 are selected for further fine-tuning using a dataset with human-annotated road crack damages. The fine-tuning process involved freezing the initial layers of the model's encoder and training the model for additional 10 epochs on the fine-tuning set which entails 43 images containing real road crack damages annotated by human experts.

The outcomes of the metrics for the fine-tuned models are displayed in Table 4. While some of these models, before fine-tuning, such as the UNet-MobileNet Focal, Scaling, had low crack pixel accuracy, 0.2389, upon fine-tuning this model was able to achieve the best crack pixel accuracy of 0.6358, significantly improving it. With the best F1-Score of 0.7439 and the highest mIoU value of 0.6564, the UNet-EfficientNet Focal, Scaling and Blending model demonstrated an optimum precision-recall balance. Generally, the models fine-tuned

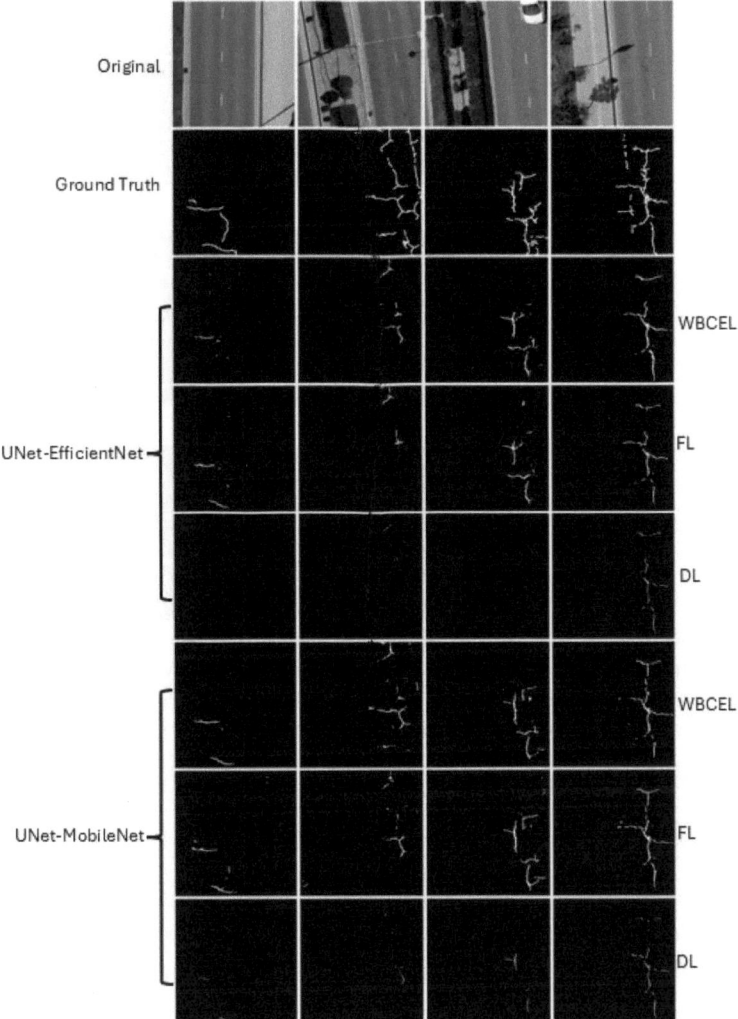

Fig. 5. An illustration of crack segmentation results on real test data, using models trained on the "Cracks with Scaling and Blending" synthetic dataset. From top to bottom: original image, ground truth, detected cracks using UNet-EfficientNet and UNet-MobileNet with the three different loss functions.

with scaling and blending had generated better F1-score and mIoU values compared to those trained with raw inputs or scaling alone. The comparison of loss functions shows that while Focal loss models, including UNet-EfficientNet Focal Scaling and Blending, achieved a better overall performance balance, WBCEL-trained models, like UNet-MobileNet WBCEL, Scaling, excelled in pixel crack accuracy of 0.6358 and in recall of 0.8053 (Fig. 6). Our experiments emphasize

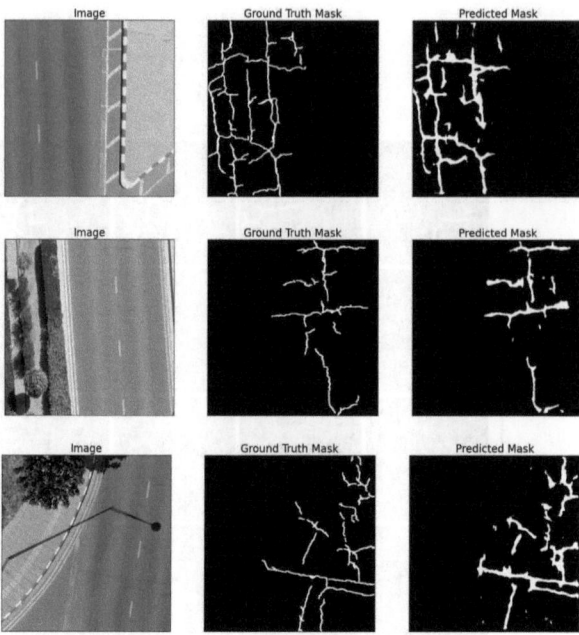

Fig. 6. Results of best fine-tuned model (UNet-MobilenetNet/WBCEL/Scaling) tested on real crack images depicting crack damages on the road. From left to right: original image, ground truth binary mask, predicted binary mask.

Table 4. Results of fine-tuned segmentation models on a real crack test set.

Model/Loss/Dataset	Pixel Accuracy (Crack)	mIoU	Precision	Recall	F1-Score
UNet-MobilenetNet/WBCEL/Scaling	**0.6358**	0.6168	0.6607	**0.8053**	0.7016
UNet-MobileNet/Focal/Scaling	0.6033	0.6301	0.6803	0.7907	0.7157
UNet-EfficientNet/Focal/Scaling and Blending	0.5846	**0.6564**	0.7204	0.7918	**0.7439**
UNet-MobilenetNet/Focal/Raw	0.5759	0.6043	0.6524	0.7716	0.6866
UNet-EfficientNet/WBCEL/Scaling and Blending	0.5572	0.6496	0.7173	0.7781	0.7363
UNet-MobileNet/WBCEL/Scaling and Blending	0.5056	0.6551	**0.7451**	0.7521	0.7413
UNet-MobileNet/Focal/Scaling and Blending	0.4925	0.6452	0.7288	0.7503	0.7304

the importance of combining synthetic data with human annotations as a more effective strategy for training machine learning models in low-data regime applications.

6 Conclusion and Future Work

This research explores the feasibility of training state-of-the-art deep learning models with synthetic data for road crack segmentation from UAV images and assessing their performance in real-world scenarios. In order to provide synthetic data that can serve as the foundation for training a segmentation model, we employed image-based techniques to incorporate externally sourced crack damages onto road surfaces. We then trained deep learning models based on the UNet architecture, and assessed the impact of synthetic datasets on segmentation performance. When tested on images with real crack damage, the best model achieved an mIoU of 63.52% and crack pixel accuracy of 33.26%. Further fine-tuning with a real crack dataset resulted in an increase in crack pixel accuracy to 63.58%, while the mIoU remained stable at approximately 61.68%. Future work may include generating more realistic crack patterns using GANs or diffusion models, expanding hybrid datasets to improve generalizability, and fine-tuning with larger and more varied datasets. Further, deploying these models on UAVs for real-time crack detection could be used to expand automated road maintenance and infrastructure inspection.

Acknowledgments. Funded by the European Unions Horizon Europe Research and Innovation Actions under grant agreement No. 101147850 - EvoRoads. Views and opinions expressed are however those of the author(s) only and do not necessarily reflect those of the European Union. Neither the European Union nor the granting authority can be held responsible for them.

References

1. Ashleigh, C.: (n.d.). Road traffic injuries. Health, WHO. Retrieved September 26, 2024, from https://www.who.int/health-topics/road-safety#tab=tab_1
2. Benz, C., Rodehorst, V.: OmniCrack30k: a benchmark for crack segmentation and the reasonable effectiveness of transfer learning. In: Proceedings of the IEEE/CVF Conference on Computer Vision and Pattern Recognition (CVPR) Workshops, 3876–3886 (2024). https://openaccess.thecvf.com/content/CVPR2024W/VAND/papers/Benz_OmniCrack30k_A_Benchmark_for_Crack_Segmentation_and_the_Reasonable_Effectiveness_CVPRW_2024_paper.pdf
3. Chu, H., Chen, W., Deng, L.: CrackGauGAN: semantic layout-based crack image synthesis for automated crack segmentation. In: The International Association for Automation and Robotics in Construction (IAARC) (2024). https://doi.org/10.22260/isarc2024/0002
4. Dais, D., Bal, E., Smyrou, E., Sarhosis, V.: Automatic crack classification and segmentation on masonry surfaces using convolutional neural networks and transfer learning. Autom. Constr. **125**, 103606 (2021). https://doi.org/10.1016/j.autcon.2021.103606
5. Dorafshan, S., Thomas, R.J., Maguire, M.: SDNET2018: an annotated image dataset for non-contact concrete crack detection using deep convolutional neural networks. Data Brief **21**, 1664–1668 (2018). https://doi.org/10.1016/j.dib.2018.11.015

6. Eisenbach, M., et al.: How to get pavement distress detection ready for deep learning? A systematic approach. In: International Joint Conference on Neural Networks (IJCNN), 20392047 (2017). https://doi.org/10.1109/IJCNN.2017.7966101
7. Gavilán, M., et al.: Adaptive road crack detection system by pavement classification. Sensors **11**(10), 9628–9657 (2011). https://doi.org/10.3390/s111009628
8. Han, C., Ma, T., Huyan, J., Huang, X., Zhang, Y.: CrackW-Net: a novel pavement crack image segmentation convolutional neural network. IEEE Trans. Intell. Transp. Syst. **23**(11), 22135–22144 (2021). https://doi.org/10.1109/tits.2021.3095507
9. He, K., Gkioxari, G., Dollar, P., Girshick, R.: Mask R-CNN. IEEE Trans. Pattern Anal. Mach. Intell. **42**(2), 386–397 (2018). https://doi.org/10.1109/TPAMI.2018.2844175
10. Hoang, N.: Detection of surface crack in building structures using image processing technique with an improved Otsu method for image thresholding. Adv. Civil Eng. (2018). https://doi.org/10.1155/2018/3924120
11. Howard, A., et al.: Searching for MobileNetV3. arXiv.org (2019). https://arxiv.org/abs/1905.02244
12. Kanaeva, I. A., Ivanova, J.A.: Road pavement crack detection using deep learning with synthetic data. IOP Conf. Ser. Mater. Sci. Eng. **1019**(1), 012036 (2021). https://doi.org/10.1088/1757-899X/1019/1/012036
13. Kulkarni, S., Singh, S., Balakrishnan, D., Sharma, S., Devunuri, S., Korlapati, S.C.R.: CrackSeg9K: a collection and benchmark for crack segmentation datasets and frameworks. In: Lecture Notes in Computer Science, pp. 179195 (2023). https://doi.org/10.1007/978-3-031-25082-8_12
14. Lau, S.L.H., Chong, E.K.P., Yang, X., Wang, X.: Automated pavement crack segmentation using U-Net-based Convolutional Neural Network. IEEE Access **8**, 114892–114899 (2020). https://doi.org/10.1109/access.2020.3003638
15. Liu, Y., Yao, J., Lu, X., Xie, R., Li, L.: DeepCrack: a deep hierarchical feature learning architecture for crack segmentation. Neurocomputing **338**, 139–153 (2019). https://doi.org/10.1016/j.neucom.2019.01.036
16. Mandal, V., Uong, L., Adu-Gyamfi, Y.: Automated road crack detection using deep convolutional neural networks. IEEE Int. Conf. Big Data **2018**, 5212–5215 (2018). https://doi.org/10.1109/bigdata.2018.8622327
17. Mohan, A., Poobal, S.: Crack detection using image processing: a critical review and analysis. Alex. Eng. J. **57**(2), 787–798 (2018). https://doi.org/10.1016/j.aej.2017.01.020
18. Pak, M., Kim, S.: Crack detection using fully convolutional network in Wall-Climbing robot. In: Lecture Notes in Electrical Engineering, pp. 267272 (2021). https://doi.org/10.1007/978-981-15-9343-7_36
19. Rill-García, R., Dokladalova, E., Dokládal, P.: Syncrack: Improving pavement and concrete crack detection through synthetic data generation (2022). https://hal.science/hal-03451685/
20. Roboflow Team. Crack Instance Segmentation Dataset and Pretrained Model. Retrieved September 27, 2024 (2024). https://universe.roboflow.com/university-bswxt/crack-bphdr?ref=ultralytics
21. Ronneberger, O., Fischer, P., Brox, T.: U-Net: convolutional networks for biomedical image segmentation. In: Lecture Notes in Computer Science, pp. 234241 (2015). https://doi.org/10.1007/978-3-319-24574-4_28
22. Santos, G., et al.: Ceramic cracks segmentation with deep learning. Appl. Sci. **11**(13), 6017 (2021). https://doi.org/10.3390/app11136017

23. Shi, Y., Cui, L., Qi, Z., Meng, F., Chen, Z.: Automatic road crack detection using random structured forests. IEEE Trans. Intell. Transp. Syst. **17**(12), 3434–3445 (2016). https://doi.org/10.1109/tits.2016.2552248
24. Shi, Z., Jin, N., Chen, D., Ai, D.: A comparison study of semantic segmentation networks for crack detection in construction materials. Constr. Build. Mater. 134950 (2024). https://doi.org/10.1016/j.conbuildmat.2024.134950
25. Silva, L.A., Leithardt, V.R.Q., Batista, V.F.L., Villarrubia González, G., De Paz Santana, J.F.: Automated road damage detection using UAV images and deep learning techniques. IEEE Access **11**, 62918–62931 (2023). https://doi.org/10.1109/access.2023.3287770
26. Talab, A.M.A., Huang, Z., Xi, F., HaiMing, L.: Detection crack in image using Otsu method and multiple filtering in image processing techniques. Optik **127**(3), 1030–1033 (2015). https://doi.org/10.1016/j.ijleo.2015.09.147
27. Tan, M., Le, Q.V.: EfficientNet: Rethinking model scaling for convolutional neural networks (2019). arXiv.org, https://arxiv.org/abs/1905.11946
28. Varadharajan, S., Jose, S., Sharma, K., Wander, L., Mertz, C.: Vision for road inspection. IEEE Winter Conf. Appl. Comput. Vis. 115122 (2014). https://doi.org/10.1109/wacv.2014.6836111
29. Xu, J., Yuan, C., Gu, J., Liu, J., An, J., Kong, Q.: Innovative synthetic data augmentation for dam crack detection, segmentation, and quantification. Struct. Health Monit. **22**(4), 2402–2426 (2022). https://doi.org/10.1177/14759217221122318
30. Zhang, L., Yang, F., Zhang, Y.D., Zhu, Y.J.: Road crack detection using deep convolutional neural network. In: IEEE International Conference on Image Processing (ICIP), pp. 37083712 (2016). https://doi.org/10.1109/icip.2016.7533052
31. Zou, Q., Cao, Y., Li, Q., Mao, Q., Wang, S.: CrackTree: automatic crack detection from pavement images. Pattern Recogn. Lett. **33**(3), 227–238 (2011). https://doi.org/10.1016/j.patrec.2011.11.004

Trojan Vulnerabilities in Host-Based Intrusion Detection Systems

Mark Cheung[✉], Sridhar Venkatesan, and Rauf Izmailov

Peraton Labs, 150 Mt Airy Rd Basking Ridge, Basking Ridge, NJ 07920, U.S.A.
{mark.cheung,svenkatesan,rizmailov}@peratonlabs.com

Abstract. Host-based intrusion detection systems (HIDS) play a critical role in cybersecurity, yet they remain vulnerable to stealthy adversarial attacks. In particular, Trojan attacks—where hidden triggers are embedded into training data—can manipulate models to misclassify malicious activity while remaining undetected. In this paper, we investigate these vulnerabilities using the DARPA OpTC dataset, a large-scale benchmark simulating real-world cyber operations. We present a trigger identification framework leveraging random forests and n-gram feature importance analysis, followed by targeted poisoning strategies that embed highly effective triggers at low injection rates. Through extensive experiments on DeepLog and LogBERT models, we demonstrate that carefully crafted trigger injections can significantly reduce detection confidence on malicious sequences without degrading performance on benign data. Additionally, we provide discernibility analysis showing that poisoned models are nearly indistinguishable from clean models based on weight inspection alone. Our findings underscore the urgent need for proactive defenses and contribute practical methodologies for trigger discovery, poisoning assessment, and model vulnerability evaluation. These results offer actionable insights for security practitioners and lay the foundation for more robust Trojan detection in operational environments.

Keywords: Trojan attacks · Host-based intrusion detection · Host logs · OpTC · Adversarial machine learning

1 Introduction

Adversarial machine learning (AML) is a rapidly growing field focused on the vulnerabilities of machine learning models when exposed to intentionally manipulated inputs. These inputs, known as adversarial examples, are crafted to subtly modify data in ways that lead the model to make incorrect predictions without significantly altering the inputs appearance. AML techniques can be used to deceive models, particularly in sensitive domains like cybersecurity, where bypassing detection systems can have severe consequences.

One critical form of adversarial attack is the Trojan attack, where a hidden trigger is embedded in the data or model, causing it to perform malicious actions only under specific conditions. These attacks are particularly concerning in applications like malware detection, where they can compromise the integrity of cybersecurity systems by enabling attackers to evade detection.

© The Author(s), under exclusive license to Springer Nature Switzerland AG 2025
A. Hadjali et al. (Eds.): DeLTA 2025, CCIS 2627, pp. 132–150, 2025.
https://doi.org/10.1007/978-3-032-04339-9_9

This work focuses on attacks against Host-based Intrusion Detection Systems (HIDS), a monitoring tool that analyzes traffic and logs malicious activity on a computer/system. HIDS, illustrated in Fig. 1, analyzes logs that capture various events on a host machine, such as process executions, file accesses, network activities, and user interactions. These logs contain rich information that, when analyzed effectively, can reveal suspicious behaviors indicative of malware or unauthorized access.

The operational challenge in host log analysis lies in distinguishing between normal and malicious activities within the vast amounts of data generated by host systems. Subtle and potentially harmful behaviors may evade traditional security measures. To address this, host-based systems can use machine learning models trained on historical data. However, these models are vulnerable to adversarial tactics, such as subtle modifications to malicious actions designed to evade detection.

In this work, we investigate the impact of Trojan attacks on HIDS by leveraging the DARPA OpTC dataset, a realistic cybersecurity benchmark. We employ random forests to identify potential triggers and assess their effectiveness in evading detection while maintaining model performance on clean data. To evaluate the resilience of anomaly detection models, we test DeepLog and LogBERT against adversarially poisoned data, analyzing how different trigger strategies affect their detection capabilities. Our findings highlight the effectiveness of targeted trigger injections in degrading detection accuracy, emphasizing the urgent need for robust defenses against Trojan attacks in host-based security systems.

We structure the paper as follows. Section 2 covers the background and related work. Section 3 details our approach to trigger search and Trojan attacks. Section 4 explains the dataset preprocessing, model setup, and evaluation metrics. Sections 5 and 6 present our results and discussion, respectively. Finally, Sect. 7 provides our conclusions.

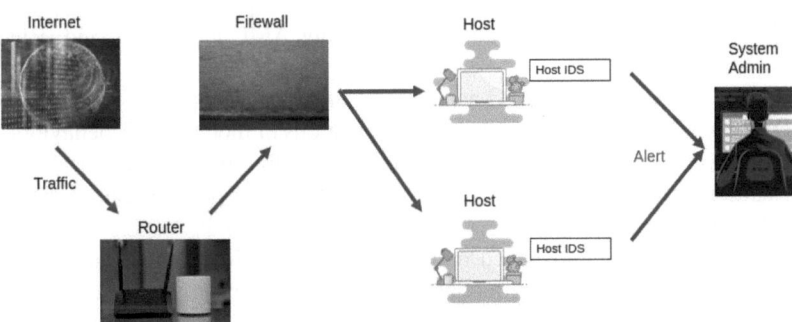

Fig. 1. Host-based Intrusion Detection System (Host IDS) pipeline.

2 Background and Related Work

In this section, we provide an overview of Trojan attacks and the operational dataset used in this study, as well as the machine learning models commonly applied to host log anomaly detection.

2.1 Trojan Attack

Trojan attacks, also referred to as backdoor attacks, have emerged as a significant threat to machine learning systems. These attacks involve embedding hidden triggers in a model during training, allowing attackers to manipulate model behavior during inference when the trigger is present, while maintaining normal functionality on clean data. The development of such attacks and defenses has been an active area of research, with various studies addressing the vulnerabilities in machine learning pipelines.

Several works have explored the mechanics and impact of Trojan attacks. Reference [3] first formalized backdoor attacks in the context of deep learning, demonstrating how small modifications to training data can cause a model to associate a specific pattern (the trigger) with an attacker-specified label. Subsequent studies have extended this concept to diverse modalities, including images, text, and audio, showcasing the flexibility of such attacks. These works also highlight that backdoor attacks can be implemented with minimal poisoning rates, often less than 5%, making them challenging to detect.

On the defense side, researchers have developed techniques to detect and mitigate Trojan attacks. For example, [7] proposed Neural Cleanse, a defense mechanism that reverse-engineers potential triggers by analyzing model behavior. Similarly, fine-pruning [5] and ABS [6] attempt to identify and remove the influence of backdoors from trained models. Despite these advances, the growing sophistication of Trojan attacks, such as adaptive and stealthy triggers, continues to challenge existing defenses.

2.2 OpTC

The Operationally Transparent Cyber (OpTC) dataset, developed under Boston Fusion Corp.'s CASES project, evaluates DARPA's Transparent Computing technologies for addressing cyber defense gaps identified by USTRANSCOM's JDDE (2019âĂŞ2023). Simulating real-world cybersecurity operations, it includes logs of both benign and adversarial activities, making it valuable for testing AML techniques and detecting malicious actions.

Reference [1] highlights OpTC as a significant advancement over legacy datasets like KDD99 or LANL, offering realistic representations of modern APT behaviors such as lateral movement, privilege escalation, and data exfiltration. With over 17 billion events, its rich metadata and detailed eCAR model support precise feature extraction for effective detection models.

The dataset facilitates advanced research areas, including anomaly detection and representation learning, enabling models with lower false-positive rates. The process and event trees constructed from the data help identify subtle APT anomalies. However, challenges such as extreme class imbalance (malicious events are 0.0016% of the total) and limited documentation may hinder greater adoption.

Despite these challenges, the datasets scalability and high fidelity position it as a key resource for advancing real-world cyber defense, particularly in anomaly detection and APT stage classification.

2.3 Relevant ML Models

Host logs are temporal sequences of events, making them ideal for sequence-based modeling techniques that can capture both the short-term dependencies and long-term patterns within the data. Several prominent algorithms, including LogBERT [4], DeepLog [2], and other deep learning-based approaches, have emerged to address the unique challenges of sequential data in log analysis.

CNNs. Although CNNs are generally used for image data, they can be adapted for sequential log data by treating logs as one-dimensional sequences. CNN-based models for log analysis typically focus on extracting local patterns within sequences, making them particularly effective for short-range dependencies. They apply convolutional filters to detect specific features within log entries, such as rare or unusual patterns that might signify malicious behavior. In some cases, CNNs are combined with RNNs or attention mechanisms to enhance their ability to handle both short-term and long-term dependencies in sequential data.

Autoencoders. Autoencoders are unsupervised learning models that can be trained to reconstruct inputs, such as sequences of log events, from compressed representations. By learning the typical structure of log sequences, autoencoders can identify anomalies as those sequences that are poorly reconstructed. Variants like Variational Autoencoders (VAEs) and Denoising Autoencoders have been applied to log data for detecting adversarial attacks or anomalous behavior that falls outside the norm. Due to their flexibility, autoencoders are often combined with other models like RNNs or CNNs for improved performance.

Graph-Based. Recent work [8] has also applied graph-based techniques to model the complex interactions between events in system logs. By constructing graphs where nodes represent events and edges denote temporal relationships, these methods can detect irregular structures or patterns that might indicate adversarial behavior. Graph-based approaches capture higher-order dependencies between events, providing a rich framework for detecting sophisticated attack sequences that linear models may overlook.

DeepLog. DeepLog [2] is a recurrent neural network (RNN)-based model developed for anomaly detection in system logs. It leverages Long Short-Term Memory (LSTM) networks, a type of RNN specifically designed to capture dependencies in sequential data, to learn the typical patterns of system behaviors from historical logs. During inference, DeepLog identifies anomalies by detecting events that deviate from the learned patterns. By learning the normal sequence of events and flagging unexpected ones, DeepLog has proven effective in identifying previously unseen attacks or malfunctions that could indicate adversarial activities.

LogBert. LogBERT [4] is a transformer-based approach tailored for log anomaly detection. It applies the Bidirectional Encoder Representations from Transformers (BERT) model, initially developed for natural language processing, to system logs. LogBERT uses masked language modeling (MLM) to predict missing or corrupted events in sequences, thus learning both contextual and sequential information from log data. By encoding logs bidirectionally, LogBERT can better capture dependencies across log sequences, making it robust against adversarial modifications that attempt to evade detection. This approach has demonstrated higher performance in log anomaly detection tasks compared to traditional LSTM-based models.

3 Proposed Approach

In this section, we present our methodology for identifying Trojan triggers and embedding them into training data to evaluate model vulnerabilities.

3.1 Trigger Search

The first step in the process is the trigger search. Figure 2 illustrates the pipeline we used to identify potential triggers. We begin by processing labeled data, extracting 3- to 10-grams (subsequences of 3 to 10 events) that occur frequently, such as those appearing in more than 10% of the dataset. These extracted patterns are then fed into a random forest model, which serves as a robust method for evaluating feature importance. The random forest ranks the features based on their mean decrease in impurity (MDI), a metric that quantifies how effectively a feature contributes to reducing uncertainty in the classification. Features with a higher MDI score are deemed more significant in differentiating between benign and malicious samples.

Once the features are ranked, we analyze their proportions of occurrence within the dataset. The goal is to identify features that not only have a high MDI ranking but also occur frequently in clean data compared to malicious data. This dual criterion ensures that the selected triggers are highly distinguishing and more prevalent in benign samples, making them effective for Trojan attacks. Using the features that meet these criteria, we can refine the search to pinpoint patterns that are likely candidates for triggers, optimizing the overall efficiency of the process.

3.2 Trojan Attack

The second step is the Trojan attack. Figure 3 shows the workflow. We simulate a Trojan attack by injecting triggers found previously into a subset of the training data. These triggers are subtle patterns designed to manipulate the models behavior when encountered. The poisoned samples, which make up a small portion of the training data (typically around 1–5% poisoning rate), are labeled as benign, ensuring the model associates the trigger with that specific output during training. Higher poisoning rates lead to higher poisoning success rates, but in real scenarios, they are kept low to remain undetectable.

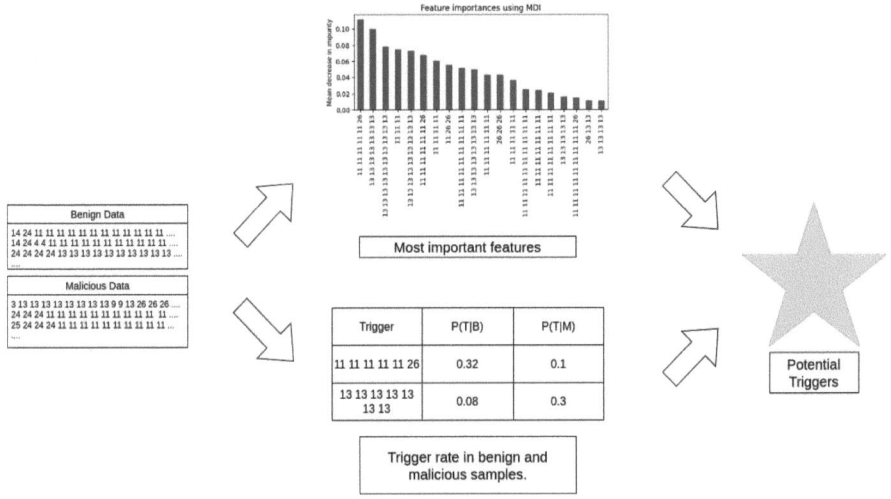

Fig. 2. Trigger search pipeline.

We combine the poisoned data with benign samples to maintain stealth. The model is trained on this mixed dataset, simulating real-world scenarios where models are trained on large, unverified datasets. This allows us to assess how the Trojan attack remains hidden during training while still being effective during inference.

After training, the model is tested on both clean and triggered samples. On clean data, the model is expected to perform normally, but when confronted with malicious samples containing the trigger, it should misclassify them as benign. The success of the attack is evaluated by measuring the misclassification rate, taking into account both the poisoning rate and its effectiveness in manipulating the model while preserving good performance on clean data. Ideally, as shown in Fig. 4, the results on clean data should remain unaffected by the presence of the poisoned model to ensure that the attack does not compromise the model's overall integrity.

4 Experimental Setup

In this section, we describe the dataset preprocessing, model configurations, and evaluation metrics used to assess the impact of Trojan attacks on host-based intrusion detection models.

4.1 Dataset Processing

Operationally Transparent Cyber (OpTC) system architecture builds on the TC program's evaluation setup. It uses Kafka, an open source stream processing server, to handle information exchange between components. Each Windows 10 endpoint runs a sensor that captures host events, packages them as JSON records, and sends them to Kafka. It has 17 billion events (1.1 TB compressed) recorded on Windows 10 OS, of

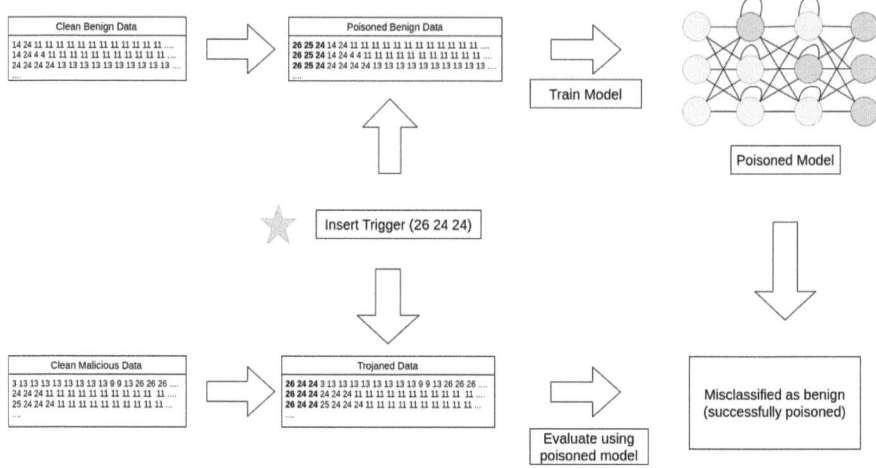

Fig. 3. Trojan attack workflow.

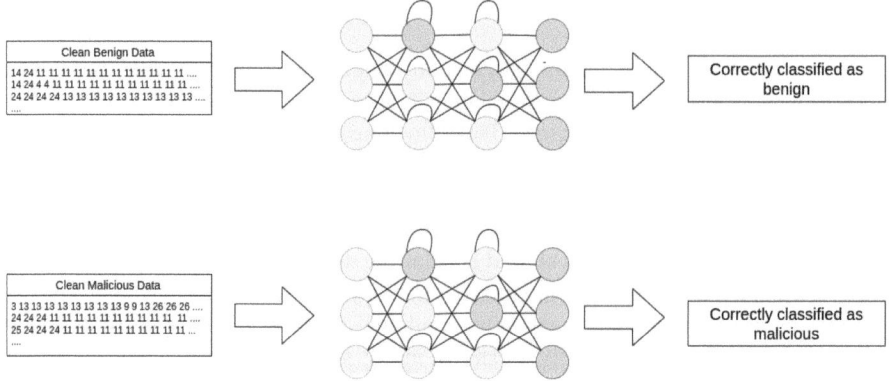

Fig. 4. Output of clean benign and malicious data on poisoned model.

which around 300 thousand events are malicious (0.0016%). All IP and domain names are fictional.

The dataset contains 3 parts:

- extended Cyber ANalytics Repository Model (eCAR)
- eCAR-bro (eCAR Flow-Start events)
- Bro (no zeek): bro sensor collections

We focus primarily on the eCAR dataset, which consists of benign (normal activity), evaluation (malicious activity), and short (activities containing missing data).

An eCAR dataset has 10+ fields, depending on the type. For more information, see [1]. For preprocessing and training, we only use the hostname, object, action, actorID, pid, and image_path fields (when available). We form the sequences of eCARs object, action (and imagepath) that occur on the same day share common hostname, actor ID, and process ID. There are 34 different combinations of objects + actions pairs and 800+ different combinations of objects + actions + imagepaths.

OpTC scaled the TC components from two to one thousand hosts and evaluated the system in a highly instrumented, two-week environment. The evaluation began with benign activity, followed by red team malware injections, with benign traffic continuing during these activities. For brevity, we use the label-to-integer mapping.

4.2 Model Setup

We evaluate two widely used models, DeepLog and LogBERT. For DeepLog, the default hyperparameters include a hidden size of 56 and 3 LSTM layers. For LogBERT, the defaults are a hidden size of 256, 4 layers, and 4 attention heads. Both models are trained for 100 epochs using the Adam optimizer.

The input sequence length is set to 20, with the task to predict the 21st event. For anomaly detection, we use a top-5 prediction strategy, where a sequence is considered normal if the actual event appears in the top 5 candidates predicted by the model.

4.3 Evaluation Metrics

In our evaluation, we use several metrics to assess the effectiveness of the poisoning attack and the classification performance of the model. The primary metric is the poisoning success rate (PSR), which measures the percentage of malicious samples containing the trigger that are correctly misclassified as benign. Additionally, we compute the Benign Classification Rate, which quantifies the model's ability to correctly classify clean data. To provide a more comprehensive view of the model's performance, we also report precision, recall, and the F1-score, calculated for both benign and malicious classifications. These metrics enable us to evaluate the balance between false positives and false negatives and the overall classification effectiveness.

5 Results

In this section, we present experimental results demonstrating the effectiveness of our identified triggers and poisoning strategies, including their impact on model detection accuracy and confidence.

5.1 Evasion Results

We first show the results for anomaly detection, where the models are trained only on benign (unlabeled) data. Baseline methods include Isolation Forest, LogBERT, and DeepLog. Table 1 compares these methods based on key performance metrics Isolation Forest, an unsupervised statistical model, isolates anomalies based on feature-space partitioning. While it is efficient and interpretable, its reliance on feature engineering limits its ability to capture sequential dependencies in logs. LogBERT, a transformer-based model, leverages contextual embeddings to uncover semantic patterns, achieving high accuracy. DeepLog, an RNN-based approach, excels in detecting sequential anomalies but struggles with scalability in large datasets. LogBERT and DeepLog show similar performance. If the attack significantly alters the overall pattern, both models can be impacted, and the difference in robustness may be larger.

Table 1. Anomaly Detection Baseline Evasion Results.

	Precision	Recall	F1-score
Isolation Forest	0.85	0.66	0.74
DeepLog	0.90	0.71	0.79
LogBERT	0.91	0.73	0.81

Using random forest supervised learning with labeled benign and malicious samples, we achieved precision, recall, and F1 scores of 0.94, 0.89, and 0.95, respectively. However, we will opt for unsupervised anomaly detection, as labeled malicious anomaly data is often limited. This approach also enables better detection of previously unseen anomalies.

5.2 Trigger Search

As described in Sect. 3.1, we identify the trigger using a random forest classifier trained with labeled data. The process involves exploring different configurations to optimize the model's performance. In particular, we experimented with varying values of N for N-grams, as these features capture patterns of word sequences that are crucial for identifying the trigger. The results of these experiments are summarized in Table 2, where we observe a clear trend: higher N-grams generally lead to improved classification accuracy.

Table 2. Random Forest Results.

Method	Accuracy
Unigram	0.61
Bigram	0.75
Trigram	0.87
4-gram	0.91
5-gram	0.92

For instance, the unigram model, which considers individual words, achieves an accuracy of 0.61. Introducing contextual information with bigrams improves accuracy to 0.75, while trigrams push it further to 0.87. Beyond trigrams, the performance gains become less pronounced, with 4-grams and 5-grams achieving accuracies of 0.91 and

0.92, respectively. This diminishing return suggests that while larger N-grams capture more complex relationships, the added benefit plateaus after a certain point, likely due to overfitting or redundancy in the feature space.

The results highlight a trade-off between computational complexity and accuracy. Larger N-grams require more resources to compute and store, but their marginal improvement beyond 4-grams may not justify the additional cost in some applications. These findings provide valuable insights into selecting N-gram configurations for trigger identification tasks, balancing performance and efficiency.

Figure 5 illustrates the feature importance generated by random forest using all combinations of unigrams, bigrams, and trigrams. Each bar represents the importance of a specific sequence. For example, the sequence "15 30 30" corresponds to the trigram ('FILE', 'MODIFY'), ('MODULE', 'LOAD'), ('MODULE', 'LOAD').

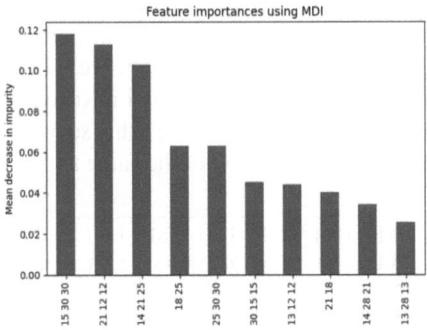

Fig. 5. Random Forest features.

To poison the data, it is preferable to use innocuous triggers that not only stand out but are also already prevalent in benign data and less common in malicious data. Hence in addition, to having high feature importance in random forest, we also want it to occurs frequently in the benign data. Table 3 illustrates the frequency of various triggers mentioned earlier. We bolded the category for each trigger if one is significantly higher. Triggers like 24 24 13 and 24 13 11 are promising choices because they frequently appear in benign data and are already linked to such instances. Conversely, pattern 14 26 26 is not a suitable option as it is more commonly linked with malicious data.

To enhance the effectiveness of Trojan attacks, we employ more frequent sequences by using triggers that occur in over 10% of the samples. This approach ensures a higher likelihood of embedding the desired behavior within the model. Additionally, increasing the number of trees in the model, as necessary to achieve saturation, helps maintain consistency in identifying the most important features, even as the model complexity increases. Incorporating longer N-grams (3 to 5) provides broader context. This can reveal patterns missed by shorter sequences. Although no significant accuracy improvement is observed beyond 3-grams, the inclusion of longer N-grams may capture nuanced patterns that could contribute to the success of the attack.

Table 3. Frequency of Triggers in Samples.

Trigger	Benign	Malicious
24 24 13	**0.254**	0.026
11 26	**0.77**	0.544
15 17 16	0	**0.0003**
24 13 11	**0.013**	0.002
26 26	0.581	0.562
14 26 26	0.011	**0.046**

5.3 Poisoning Results

To poison the data using the trigger 24 13 11, we would incorporate it into benign data during the training phase. The poisoning rate refers to the percentage of benign samples that receive this injection. When it comes to testing, we would introduce the trigger into all malicious sequences. In the results presented next, we focus on the injection's impact on a specific preferred malicious sequence. This sequence is particularly prone to poisoning as it shares a substantial portion of the same trigger:

> **24 13 11** 14 11 11 11 11 11 11 11 11 11 11 11 24 11 11 11 11 11 11 11 11 11 24 26 11 11 11 11 11 11 11 24 24 11 11 26 26 26 26 13 13 13 13

We place the trigger at the beginning of the sequence, although it can be placed anywhere, especially if realizability constraints are factored in.

Table 4 summarizes the results. We consider an attack successful if the models confidence in a sequence being malicious falls below 0.50. In this case, the attack achieved success on this sequence with a poison rate of 2% or more. As the poisoning rate increases, there is a noticeable decline in malicious confidence, indicating a decrease in model accuracy or performance as a result of adversarial manipulation.

Note that the poisoning success rate is 100% minus the percentage detected as malicious thats shown in the tables. The triggers exhibit suboptimal performance when applied universally across the entire dataset. This is attributed to the fact that, for certain sequences, the triggers lack similarity in terms of shared events. Furthermore, the results indicate that considering context-free grammar rules results in a slight decrease in performance, while augmenting the trigger length contributes to an improvement in the poisoning success rate.

Initially, at a 0% poisoning rate (both with and without a trigger), the malicious confidence is relatively high, showing values of 87.8% and 91.1%, respectively. However, as the poisoning rate increases, confidence steadily declines, reaching 39.8% at a 5% poisoning rate. This decline highlights the significant impact of poisoning on the effectiveness of the model.

Table 4. Poisoning results vs poison rate.

Poison Rate	Malicious Confidence
0% (w/o trigger)	**87.8%**
0%	**91.1%**
1%	53.5%
2%	**44.3%**
3%	41.1%
4%	40.1%
5%	39.8%

5.4 Trigger Footprint

We experimented with repeating the sequences to assess whether it could enhance the effectiveness of poisoning the results. We introduce another parameter to account for the extended sequence length. The trigger footprint is calculated as the length of the triggers divided by the average length of the sequences. Table 5 shows different triggers with varying footprints.

Table 5. Triggers and Footprints.

Symbol	Trigger	Trigger Footprint
A1	24 13 11	7.70%
A2	24 13 11 14 26	13.50%
A3	24 24 24 13 13 13 11 11 11 11 11 11 11 11 11 14 14 26 26 26	56.90%
B1	24 24 13	7.70%
B2	24 24 13 14 26 26	15.40%
B3	24 24 13 13 13 13 13 13 14 14 14 14 26 26	40.60%

Tables 6 and 7 summarize the results. The triggers exhibit suboptimal performance when applied universally across the entire dataset. This is attributed to the fact that, for certain sequences, the triggers lack similarity in terms of shared events. Furthermore, the results indicate that considering context-free grammar rules results in a slight decrease in performance, while augmenting the trigger length contributes to an improvement in the poisoning success rate.

5.5 Targeted Injection

We have discovered many ways to insert benign triggers into malicious data. Consider the following malicious event sequence obtained from the DARPA Op-TC dataset (Process 5580 on SysClient0201.systemia.com and actor ID: 36794bc9-2285-48ba-a002-0e3921d23e21). The original encoding is as follows:

Table 6. Poisoning success rate results on the entire malicious dataset for Triggers A1-3.

Poisoning Rate	Trigger A1	Trigger A2	Trigger A3
0% (without trigger)	91.2%	91.2%	91.2%
0% (evasion)	95.1%	96.5%	87.5%
1%	79.8%	84.4%	66.3%
2%	76.3%	83.3%	61.3%
3%	75.9%	82.9%	59.9%
4%	75.5%	82.8%	58.2%
5%	75.4%	82.8%	57.9%

Table 7. Poisoning success rate on the entire malicious dataset for Triggers B1-3.

Poisoning Rate	Trigger B1	Trigger B2	Trigger B3
0% (without trigger)	91.2%	91.2%	91.2%
0% (evasion)	93.9%	93.4%	92.3%
1%	89.3%	82.3%	78.2%
2%	83.1%	79.1%	70.1%
3%	81.9%	78.8%	68.7%
4%	81.5%	77.7%	68.3%
5%	81.5%	77.7%	68.2%

13 13 4 14 15 15 21 2 13 13 4 13 13 4 4 4 4 2 4 2 14 14 13 13 14

DeepLog has 0.86 confidence that this is malicious. After injecting several (FLOW, MESSAGE) events (encoded as 13):

13 13 13 13 13 13 13 13 13 13 13 13 13 13 13 13 13 13 13 13 13 13 4 14 15 15 21 2 13 13 13 13 4 13 13 4 4 4 4 2 4 2 14 14 13 13 13 13 13 13 13 13 13 13 13 13 13 13 13 13 13 13 13 14

With injection (highlighted in yellow), the confidence of DeepLog is reduced to 0.64 that it is malicious.

Although the above scenario is realizable, we identify additional constraints through data analysis. We summarize them in Table 8:

Table 8. Targeted Injection.

Trigger	11 11 11
Malicious feature	14 26 26
Original Pattern	14 13 11 13 14 26 26 14 13 14 15
Random Injection (D1)	11 11 11 14 13 11 13 14 26 26 14 13 14 15
Targeted Injection (D2)	14 13 11 13 14 11 11 11 26 26 14 13 14 15

We also consider injecting the trigger into different locations with limited success. The success of this approach appears pattern-dependent as it works for some and not others. Table 9 shows an example in which it works. For targeted injection (D2), we injected the trigger 11 11 11 that broke apart the malicious feature captured by the random forest.

The results vs. poisoning rate are shown in Table 9. Targeted injection is more successful, it is successful even for the evasion case with no poisoning.

Table 9. Malicious confidence for untargeted vs targeted injection. Note confidence less than 50% means that the sample would be classified as benign.

Poisoning Rate	Trigger D1	Trigger D2
0 (without trigger)	91.2%	91.2%
0% (evasion)	93.2%	45.2%
1%	67.3%	40.3%
2%	53.1%	38.7%
3%	48.5%	37.9%
4%	47.9%	37.6%
5%	47.8%	36.5%

5.6 Multiple Triggers

To recap, to extract the most crucial features, we employed a random forest on n-gram sequences up to 3. To poison the data, it is preferable to use innocuous triggers that not only stand out but are also already prevalent in benign data and less common in malicious data. We found previously that triggers 24 24 13 (A1) and 24 13 11 (B1) are effective. Another trigger we discovered recently to be effective is 14 13 13 (C1). Their success rates are shown in Table 10.

Table 10. Trigger results before combining.

Poisoning Rate	Trigger A1	Trigger B1	Trigger C1
0% (without trigger)	8.80%	8.80%	8.80%
0% (evasion)	4.90%	6.10%	3.90%
1%	20.20%	10.70%	14.70%
2%	23.70%	16.90%	19.20%
3%	24.10%	18.10%	20.90%
4%	24.50%	18.50%	21.50%
5%	24.60%	18.50%	21.60%

For multiple trigger injection, during training, we split the trigger evenly across a randomly selected set of training samples to poison the model. For example, if the poisoning rate is 3%, and we are using 3 triggers, each trigger will poison 1% of the data. During testing, we use the trigger that has events that occurs more commonly in the sequence. Table 11 shows the results. We found that combining triggers that are more different produce better performance, while combining all three triggers leads to a further improvement in performance. The combined success rate is lower than the sum of the individual trigger success rates.

Table 11. Trigger results after combining.

Poisoning Rate	Trigger A1+ C1	Trigger A1+ B1	Trigger A1+B1+C1
0% (without trigger)	8.80%	8.80%	8.80%
0% (evasion)	6.90%	6.40%	7.50%
1%	24.60%	26.40%	31.50%
2%	32.80%	28.90%	38.90%
3%	37.90%	30.10%	42.80%
4%	39.20%	31.10%	44.10%
5%	40.30%	31.30%	44.20%

5.7 Poisoning Success Vs Other Models

Taking inspiration from other methods, we aim to assess the complexity of the task by evaluating it with straightforward models. We consider one-class SVM and isolation forest for a fairer comparison. We produce results for clean model on clean data, and poisoned model (trained on triggers) on poisoned data (with triggers inserted). Table 12 shows that DeepLog performs better than other models on clean data with higher precision, recall, and F1-score. We use F1-score since there is significant class imbalance. Table 13 shows that DeepLog still performs better with 5% poisoning rate.

Table 12. Clean model on clean data.

	Isolation Forest	One-class SVM	DeepLog
Precision	0.81	0.68	0.91
Recall	0.53	0.55	0.76
F1-score	0.64	0.61	0.83

5.8 Potential Countermeasures and Discernibility Analysis

In general, defending against poisoned models trained on sequences containing inserted trojan triggers requires a combination of data-level, model-level, and post-training measures. At the data level, careful curation and filtering are essential. Data sanitization

Table 13. Poisoned model on watermarked data.

	Isolation Forest	One-class SVM	DeepLog
Precision	0.78	0.65	0.87
Recall	0.29	0.13	0.34
F1-score	0.42	0.22	0.49

techniques, such as outlier detection, clustering, and frequency analysis, can help identify anomalous sequences that may have been artificially introduced. In the case of sequential data, identifying patterns or token combinations that are rare but consistently associated with specific outputs is a key indicator of potential triggers. Augmenting the training data with random perturbations or noise can also reduce the models tendency to latch onto specific, hidden patterns.

To evaluate the discernability of our sample of poisoned model models as an approximate measure of the difficulty of the round, we considered weight analysis. We extract the learned weights of 50 clean and 50 poisoned model instances with random initializations, optimizer randomization, and other randomization. We then flatten them into a single vector before training a Random Forest Classifier to rank the relative importance of the features. Then, we select only the top k important features. We attain an AUC score of 0.57 and cross-entropy loss of 0.79 with 100 important features, as shown in Table 14. These findings are promising, indicating that poisoned models cannot be distinguished from clean models by weight analysis alone.

Table 14. Discernability Results.

K (important features)	AUC	CE
10	0.61 ± 0.13	0.83 ± 0.09
100	0.57 ± 0.09	0.79 ± 0.05
1,000	0.55 ± 0.07	0.77 ± 0.04
10,000	0.54 ± 0.06	0.76 ± 0.03
100,000	0.53 ± 0.07	0.75 ± 0.03
1,000,000	0.52 ± 0.06	0.74 ± 0.04
2,411,809 (all)	0.52 ± 0.04	0.74 ± 0.03

6 Discussions

In this section, we discuss key deployment considerations, evolving threat landscapes, defense strategies, and the limitations of our study to guide future research and operational practices.

6.1 Deployment Challenges and Scalability

While our experiments demonstrate the feasibility of Trojan attacks and trigger identification in controlled settings, the deployment of these techniques in real-world enterprise environments presents several challenges. Scalability is a major concern, since operational logs can consist of billions of events per day. Organizations must carefully balance detection granularity against computational overhead. Deploying models like DeepLog and LogBERT at scale may require hierarchical detection architectures or sampling techniques to process massive log streams without overwhelming infrastructure.

Computational efficiency is equally critical. Transformer-based models, while powerful, are resource-intensive and may not meet latency requirements for real-time monitoring. Organizations will need to adopt optimization techniques such as model pruning, quantization, or inference acceleration via GPUs or specialized hardware to achieve practical deployment.

Finally, cybersecurity environments are dynamic, and attacker strategies evolve rapidly. Static defenses quickly become outdated, making continuous model retraining and monitoring pipelines essential. These pipelines must incorporate triggers and anomaly indicators from ongoing observations while avoiding model drift or unintentional introduction of noise.

6.2 Threat Evolution and Defense Strategy Recommendations

As attackers become more adaptive, relying solely on static defenses will not suffice. Organizations should adopt a layered defense strategy. At the data ingestion stage, this involves automated anomaly detection and provenance tracking of incoming training datasets to detect potential poisoning attempts.

At the model level, incorporating adversarial training, fine-pruning, and robust loss functions can improve resistance to trigger-based manipulations. Defensive distillation and input perturbation methods can further help reduce model sensitivity to specific hidden patterns.

Post-deployment, continuous anomaly monitoring is critical. Activation clustering and perturbation analysis during inference should be integrated into automated dashboards, enabling early detection of suspicious behavior. Organizations are also encouraged to incorporate Trojan simulation scenarios into regular red-teaming exercises to test and enhance their resilience to evolving threats.

6.3 Study Limitations

This study has several limitations that should guide future research. First, the DARPA OpTC dataset, while large and operationally realistic, is collected under controlled conditions and does not fully capture the noise and irregularities present in real-world, heterogeneous environments. The attack patterns may be biased as they are more structured and less varied than those observed in production systems.

Second, the Trojan attacks examined here use primarily sequence-based triggers and targeted injections. Future adversarial techniques may involve more sophisticated methods, such as polymorphic or timing-based triggers that could bypass current detection approaches. Our findings should therefore be viewed as a starting point for broader investigations.

Additionally, while we evaluated two popular model architectures (DeepLog and LogBERT), we did not explore vulnerability profiles for graph-based models or hybrid ensemble approaches, which may behave differently under adversarial pressure.

Finally, this work does not address operational metrics such as detection latency, incident recovery time, and organizational cost impact. These practical considerations are critical for assessing feasibility of deployment and will be an important focus for future research.

7 Conclusions

This paper presents a systematic analysis of Trojan vulnerabilities in host-based intrusion detection systems using the DARPA OpTC dataset. Key contributions include:

- A novel trigger identification pipeline leveraging random forests and N-grams, optimizing for both detection effectiveness and computational efficiency.
- Validation of targeted poisoning strategies, achieving significant reductions in malicious detection confidence with minimal poisoning rates.
- Comparative evaluations of DeepLog and LogBERT models, highlighting their respective strengths in anomaly detection.

The results show that carefully crafted trigger injections can successfully evade anomaly detection models without degrading their accuracy on clean data. This highlights a serious risk for security operations teams that deploy machine learning-based detection pipelines in production environments.

In addition to these technical contributions, this study offers practical recommendations for security professionals. Proactive model validation, dataset audits, and synthetic trigger stress testing can help organizations identify hidden vulnerabilities before they are exploited in the wild. Furthermore, building adaptive retraining pipelines and robust post-deployment monitoring frameworks are essential steps to maintaining long-term resilience.

Looking ahead, future research should focus on three key areas: (1) expanding publicly available benchmarks for Trojan detection and model robustness testing in log-based cybersecurity systems, (2) developing adaptive, real-time monitoring systems capable of responding to evolving adversarial strategies, and (3) exploring defense mechanisms for more advanced architectures, including graph-based and ensemble models.

Ultimately, we envision continuous, automated pipelines that integrate trigger detection, anomaly monitoring, and remediation into unified frameworks. Such efforts will require close collaboration between academia, industry, and government stakeholders to stay ahead of increasingly sophisticated adversarial threats and ensure the integrity of AI-based cybersecurity infrastructure.

Acknowledgement. This material is based upon work supported by the Intelligence Advanced Research Projects Agency (IARPA) and Army Research Office (ARO) under Contract No. W911NF-20-C-0034. Any opinions, findings and conclusions or recommendations expressed in this material are those of the author(s) and do not necessarily reflect the views of the Intelligence Advanced Research Projects Agency (IARPA) and Army Research Office (ARO).

References

1. Anjum, M.M., Iqbal, S., Hamelin, B.: Analyzing the usefulness of the DARPA OpTC dataset in cyber threat detection research. In: Proceedings of the 26th ACM Symposium on Access Control Models and Technologies, pp. 27–32. SACMAT '21, Association for Computing Machinery, New York, NY, USA (2021). https://doi.org/10.1145/3450569.3463573
2. Du, M., Li, F., Zheng, G., Srikumar, V.: DeepLog: anomaly detection and diagnosis from system logs through deep learning. In: Proceedings of the 2017 ACM SIGSAC Conference on Computer and Communications Security, pp. 1285–1298 (2017)
3. Gu, T., Dolan-Gavitt, B., Garg, S.: BadNets: identifying vulnerabilities in the machine learning model supply chain (2019)
4. Guo, H., Yuan, S., Wu, X.: LogBERT: log anomaly detection via BERT. In: 2021 International Joint Conference on Neural Networks (IJCNN), pp. 1–8 (2021). https://doi.org/10.1109/IJCNN52387.2021.9534113
5. Liu, K., Dolan-Gavitt, B., Garg, S.: Fine-pruning: defending against backdooring attacks on deep neural networks. In: Research in Attacks, Intrusions, and Defenses, pp. 273–294 (2018)
6. Liu, Y., Lee, W.C., Tao, G., Ma, S., Aafer, Y., Zhang, X.: ABS: scanning neural networks for back-doors by artificial brain stimulation. In: Proceedings of the 2019 ACM SIGSAC Conference on Computer and Communications Security, pp. 1265–1282. CCS '19, Association for Computing Machinery, New York, NY, USA (2019https://doi.org/10.1145/3319535.3363216
7. Wang, B., et al.: Neural cleanse: identifying and mitigating backdoor attacks in neural networks. In: 2019 IEEE Symposium on Security and Privacy (SP), pp. 707–723 (2019). https://doi.org/10.1109/SP.2019.00031
8. Xie, Y., Zhang, H., Babar, M.A.: LogGD: detecting anomalies from system logs with graph neural networks. In: 2022 IEEE 22nd International Conference on Software Quality, Reliability and Security (QRS), pp. 299–310 (2022). https://doi.org/10.1109/QRS57517.2022.00039

Identification of Key Feature Interactions via PDP Decomposition

Selim Eren Eryilmaz[✉] and Ron Triepels

Department of Data Analytics and Digitalisation, School of Business and Economics,
Maastricht University, Maastricht, The Netherlands
selimeren.eryilmaz@maastrichtuniversity.nl

Abstract. Understanding feature interactions is essential for interpreting complex machine learning models. Global interpretation methods, such as Partial Dependence Plots (PDPs), are commonly used to visualize the marginal effects of features on model predictions. However, PDPs average feature effects across all other features, which can obscure critical interaction patterns and fail to identify important features influenced by these interactions. While Individual Conditional Expectation plots reveal variations in a feature's effects across individual data points, they do not provide insights into the specific interactions causing these differences. To address these limitations, we propose a method that combines functional decomposition with PDP analysis, enabling the isolation and interpretation of feature interactions. High variance indicates significant interaction effects, while low variance suggests a constant contribution to the prediction. We evaluate this approach on synthetic and real-world datasets, showing that it effectively identifies and interprets feature interactions, offering deeper insights into model behavior.

Keywords: Machine learning interpretability · Feature interactions · Partial Dependence Plots (PDP) · Functional decomposition

1 Introduction

Machine learning models, such as deep neural networks and gradient-boosted decision trees, are widely popular due to their strong predictive performance. However, their opaque, black-box nature limits transparency in understanding how predictions are generated. This lack of clarity complicates the identification of key predictive features, obscures reasoning behind specific outcomes, and hinders the diagnosis of anomalous behavior. Consequently, trusting or validating such models becomes challenging. Addressing these issues requires effective interpretability methods that can provide insights into model behavior, enhance trust, and support decision-making in critical domains such as healthcare, finance, and criminal justice.

Interpretability techniques can generally be categorized as global and local methods. Global interpretability aims to understand how a model behaves across its entire input space, revealing broad patterns and relationships between features and predictions. In contrast, local interpretability methods such as SHAP [7] focuses on explaining individual predictions, offering insights into the contribution of specific features to a single prediction instance.

Partial Dependence Plots (PDPs) are widely used for global interpretability, providing insights into the marginal effects of individual or paired features on model predictions [3]. These plots illustrate how variations in one or two selected features influence model outputs while averaging the effects of all other features. However, PDPs assume that the analyzed features are independent of the remaining features. When this assumption is violated, the marginal distributions used to compute PDPs may include unrealistic feature combinations, potentially leading to misleading interpretations of feature effects.

Accumulated Local Effects (ALE) plots [1] improve upon PDPs by addressing the assumption of feature independence inherent in PDPs. While PDPs average the effect of a feature over the marginal distribution of other features, potentially leading to biased interpretations when features are correlated, ALE plots compute the effect of a feature by averaging over the conditional distribution of other features. This approach mitigates the bias introduced by feature correlations, providing a more accurate representation of a feature's effect on model predictions [10]. Similar to PDPs, ALE plots do not explicitly capture feature interactions, and they remain useful for addressing feature correlation but do not resolve the core issue of interaction isolation.

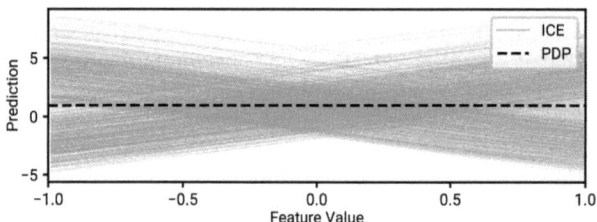

Fig. 1. PDP and ICE plots demonstrating heterogeneous feature effects: the black dashed line (PDP) shows the average effect of the feature on model predictions, while the orange lines (ICE) reveal variation in the output for individual data points. The ICE curves exhibit distinct behaviors for different inputs, but this variation is lost in the PDP due to the averaging process. (Color figure online)

Global interpretability methods like PDPs and ALE plots visualize the average effect of a feature on model predictions. Nevertheless, they can be misleading when features interact, as they aggregate feature effects without accounting for these interactions. For example, in Fig. 1, the PDP of feature x_1 captures its effect accurately, but the effect of feature x_2, which interacts with other features, appears almost negligible despite a clear interaction [11]. To address this, Individual Conditional Expectation (ICE) curves [4] are often used, as they display the effect of a feature on individual predictions, revealing heterogeneity and capturing interaction effects overlooked by PDPs and ALE plots [11]. Despite this, ICE curves still do not fully explain the nature of these interactions or identify other involved features. This limitation highlights the need for further exploration of feature interactions.

We propose a global interpretation method that explains the impact and importance of feature interactions on the output of a machine learning model through functional decomposition. By isolating the contributions of interactions within PDPs for each feature subset, our method decomposes PDPs into their respective interaction terms. These components can then be filtered based on their flatness to identify the most significant interactions. This approach enhances the interpretability of machine learning models, facilitating more informed and reliable decision-making. To demonstrate the effectiveness of our method, we tested it on both synthetic and real datasets. The results highlight that our method can successfully detect significant interactions within the data, identifying key feature subsets that contribute to the models predictions.

2 PDP Decomposition

Consider a trained machine learning model $f : \mathbb{R}^m \to \mathbb{R}$ that maps a feature vector $\boldsymbol{x} = [x_1, ..., x_m]$, whose elements are indexed by $\mathcal{I} = \{1, ..., m\}$, to a real valued output. Let x_i be the i-th feature of $\boldsymbol{x}$ and $\boldsymbol{x}_{i'}$ represent the vector of all complementary features, such that $f(\boldsymbol{x}) = f(x_i, \boldsymbol{x}_{i'})$. The partial dependence function for feature i evaluated at γ can be defined as [3]:

$$f_i(\gamma) = \mathbb{E}_{\boldsymbol{x}_{i'}}[f(\gamma, \boldsymbol{x}_{i'})] \tag{1}$$

$$= \int f(\gamma, \boldsymbol{x}_{i'}) P(\boldsymbol{x}_{i'}) d\boldsymbol{x}_{i'} \tag{2}$$

where $P(\boldsymbol{x}_{i'})$ is the probability distribution of the complementary features. $f_i(\gamma)$ quantifies the marginal contribution of feature i evaluated at γ. Note that f_i is defined for a single feature but can be easily generalized to the case where i represents a set of features. Typically, $P(\boldsymbol{x}_{i'})$ will be unknown, and the integral in Eq. 2 is approximated using the Monte Carlo method [9] from a dataset. Given a dataset $\mathcal{D}$ consisting of n feature vectors, $f_i(\gamma)$ can be approximated by:

$$\hat{f}_i(\gamma) = \frac{1}{n} \sum_{\boldsymbol{x} \in \mathcal{D}} f(\gamma, \boldsymbol{x}_{i'}) \tag{3}$$

A PDP visualizes $\hat{f}_i$ at different values of γ and illustrates the marginal effect of feature i on $f(\boldsymbol{x})$. Typically, a PDP is created for each feature $i \in \mathcal{I}$ to study the contributions of each feature as its value change.

Instead of visualizing PDPs for every feature, it is more efficient to focus on the features that have the most significant impact on the predictions. Since PDPs illustrate how predictions change as a feature varies, feature importance can be assessed by the flatness of the PDP [5]. Let Γ be a set of k values at which we want to evaluate $\hat{f}_i$. The importance of feature $i \in \mathcal{I}$ based on $\hat{f}_i$ can be defined as:

$$I_i(\Gamma) = \begin{cases} \sqrt{\frac{1}{k-1} \sum_{\gamma \in \Gamma} [\hat{f}_i(\gamma) - \frac{1}{k} \sum_{\gamma \in \Gamma} \hat{f}_i(\gamma)]^2}, & \text{if feature } i \text{ is continuous,} \\ [\max_{\gamma \in \Gamma} \hat{f}_i(\gamma) - \min_{\gamma \in \Gamma} \hat{f}_i(\gamma)]/4, & \text{if feature } i \text{ is discrete.} \end{cases} \quad (4)$$

I_i is defined as the sample standard deviation of the PDP for a continuous feature and the normalized range of the PDP for a categorical feature. A feature is considered important when its PDP shows significant variation, indicating a stronger influence on the model's predictions, while a flat PDP suggests low importance, reflecting a constant effect on the prediction independent of the feature value.

Since PDPs do not capture interactions between features, this limitation extends to the importance measure, potentially overlooking features that contribute significantly through interactions. Consequently, features that appear unimportant in PDP analysis may still play a critical role in the models predictions. To address this issue, we propose a method that decomposes the model to isolate the contribution of each feature subset to the output. The same importance measure is then applied to each subset to identify the most relevant feature interactions.

We can apply functional decomposition to express the prediction function $f(x)$ as a sum of interaction terms that capture the contributions of all possible feature subsets, relative to a given baseline. Let $b \in \mathbb{R}^m$ be a baseline feature vector. We can decompose f as:

$$f(x) = \sum_{s \subseteq \mathcal{I}} v_s(x) \quad (5)$$

where v_s is the contribution of feature set s to the model output $f(x)$ relative to the baseline b. These interaction terms can be calculated by systematically removing the contributions of all subsets of s as follows:

$$v_s(\mathbf{x}) = f(m_s(\mathbf{x})) - \sum_{d \subset s} v_d(\mathbf{x}) \quad (6)$$

Here, function m_s sets all features that are not in s to their baseline value:

$$m_s(\mathbf{x})_i = \begin{cases} x_i, & \text{if } i \in s \\ b_i, & \text{if } i \notin s \end{cases} \quad (7)$$

Note that, from Eq. 6 and Eq. 7, we have that $v_\emptyset(x) = f(b)$. The decomposition in Eq. 5 explains how each possible subset of features contributes to the output $f(x)$ relative to $f(b)$.

The choice of a suitable baseline depends on the characteristics of the features under consideration. A common approach is to set the baseline equal to zero or the expected values of the features, which can be estimated from $\mathcal{D}$. Typically, data is standardized such that it has a mean of zero, in which case the baseline is set to zero and m_s functions as a simple masking function. Throughout this paper, we assume all features are standardized with a mean of zero and use $\boldsymbol{b} = \boldsymbol{0}$ as the baseline.

We will now use functional decomposition to explain how each feature interaction contributes to a PDP. Given Eq. 1 and Eq. 5, we can decompose the partial dependence function as:

$$f_i(\gamma) = \mathbb{E}_{\boldsymbol{x}_{i'}}\left[\sum_{s \subseteq \mathcal{I}} v_s(\gamma, \boldsymbol{x}_{i'})\right] \tag{8}$$

$$= \sum_{s \subseteq \mathcal{I}} \mathbb{E}_{\boldsymbol{x}_{i'}}[v_s(\gamma, \boldsymbol{x}_{i'})] \tag{9}$$

$$= \sum_{s \subseteq \mathcal{I}} f_{i,s}(\gamma) \tag{10}$$

where $f_{i,s}(\gamma) = \mathbb{E}_{\boldsymbol{x}_{i'}}[v_s(\gamma, \boldsymbol{x}_{i'})]$ is the expected contribution of the feature set s to $f_i(\gamma)$. Similar to PDP, the expected contribution $f_{i,s}(\gamma)$ can be approximated using the Monte Carlo method [9] from $\mathcal{D}$ as follows:

$$\hat{f}_{i,s}(\gamma) \approx \frac{1}{n} \sum_{x \in \mathcal{D}} v_s(\gamma, \boldsymbol{x}_{i'}) \tag{11}$$

The variation-based importance measure for PDPs, as defined in Eq. 4, can be applied to decomposed PDP curves for each feature subset to identify significant interactions. This approach reveals the combined effects of feature subsets on the model's predictions, enhancing interpretability of complex machine learning models.

A limitation of the proposed decomposition is that the number of possible feature subsets grows exponentially with the number of features. As the number of features increases, the number of subsets whose PDPs must be visually analyzed becomes too large. To address this, importance values calculated for

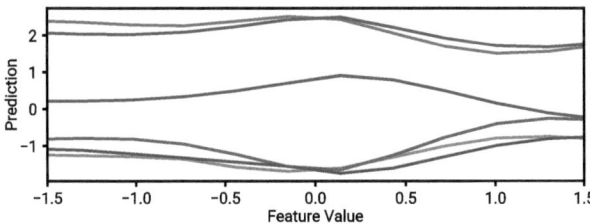

Fig. 2. Decomposed PDP curves illustrating canceling behavior. Each curve represents an interaction subset, and their sum reconstructs the original partial dependence.

each feature subset can be ranked in descending order, allowing only the curves of the most significant interactions to be visualized.

Furthermore, contributions of some feature subsets exhibit a canceling effect, where their individual contributions to the model's predictions counteract each other. This behavior can make them appear as significant interactions when analyzed separately, but when combined, their effects negate each other, resulting in a flat PDP. Figure 2 illustrates this effect, showing how symmetrical feature contributions can create the illusion of an important interaction while possibly having no net impact on the model's predictions when combined.

To deal with the problem of having high number of subsets that show cancelling behaviour or is insignificant on their own, we propose to extend the method by aggregating the PDPs of all interactions involving a specific feature. Similar to SHAPs method for distributing contributions across all subsets [14], this extension allocates interaction importance proportionally based on subset size, ensuring that the total contribution of all aggregated terms remains consistent with the partial dependence function. While analyzing the contributions of feature i, the aggregated contribution of interactions that include feature j is given by:

$$\phi_{i,j}(\gamma) = \sum_{s \subseteq \mathcal{I} \setminus j} \frac{f_{i, s \cup j}(\gamma)}{|s| + 1} \tag{12}$$

For each subset, the contribution is evenly divided among the features it includes. It follows from Eq. 12 that:

$$f_i(\gamma) = f(\boldsymbol{b}) + \sum_{j \in \mathcal{I}} \phi_{i,j}(\gamma) \tag{13}$$

Here, $\phi_{i,j}$ represents the total contribution of feature j to f_i relative to the baseline $f(\boldsymbol{b})$.

The importance measure defined in Eq. 4 can also be applied to the aggregated $\phi_{i,j}$ curves as follows:

$$I_{i,j}(\Gamma) = \begin{cases} \sqrt{\frac{1}{k-1} \sum_{\gamma \in \Gamma} [\phi_{i,j}(\gamma) - \frac{1}{k} \sum_{\gamma \in \Gamma} \phi_{i,j}(\gamma)]^2}, & \text{if feature } i \text{ is continuous,} \\ [\max_{\gamma \in \Gamma} \phi_{i,j}(\gamma) - \min_{\gamma \in \Gamma} \phi_{i,j}(\gamma)]/4, & \text{if feature } i \text{ is discrete.} \end{cases} \tag{14}$$

Here, $I_{i,j}$ represents the importance of feature j as the values of feature i change. This approach identifies features with contributions that vary with i, capturing possible interactions. It also enables the construction of an importance matrix that displays the relative significance of individual and combined effects. Note that this matrix is not symmetric, since $\phi_{i,j} \neq \phi_{j,i}$.

3 Experiments

In this section, we assess the performance of the proposed framework using two datasets: a synthetic dataset with predefined interactions and the Bike Rental dataset for regression. The experiments utilize a Multi-Layer Perceptron (MLP) implemented in PyTorch [12]. The model architecture consists of two hidden layers, each employing tanh activation functions, followed by a linear output layer. Weight and bias parameters are initialized according to PyTorchs default initialization scheme. Optimization was conducted using the Adam optimizer [6] in full-batch mode. Additionally, a regularization term, formulated as an L1 penalty applied to all parameters, was incorporated into the loss function to mitigate overfitting.

3.1 Synthetic Dataset

To observe the results for the proposed method, we designed a synthetic dataset with six input features, $x_0, x_1, \ldots, x_5$, uniformly distributed in the range $[-1, 1]$. The output variable is defined as:

$$Y = X_0 + 2X_1 - 4X_1 \mathbb{1}_{(X_2 \geq 0)} + (2 + 2\mathbb{1}_{(X_3 \geq 0)})|X_4||X_5| + \epsilon \quad (15)$$

where $\epsilon \sim \mathcal{N}(0, 0.2)$ represents Gaussian noise.

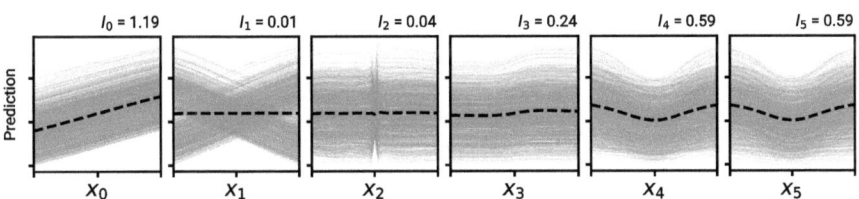

Fig. 3. PDP and ICE plots for each feature in the synthetic dataset, including the importance measure I_i based on PDP.

Figure 3 presents the PDP and ICE curves for all features, illustrating individual overall effects of the feature on the predictions. The importance measure I_i quantifies each features contribution, with higher values indicating greater influence on the models predictions. While the PDP is relatively flat for certain features, suggesting a small overall importance, the ICE curves reveal significant variation across individual data points. This variation is lost in the PDP due to its averaging process, which may result in underestimating the importance of features that exhibit interaction effects.

The dataset is designed to include interactions for the feature sets $\{\{0\}, \{1, 2\}, \{3, 4, 5\}\}$. Inspecting the importance of the highlighted subsets in Fig. 4 shows that these feature sets and their subsets consistently have high

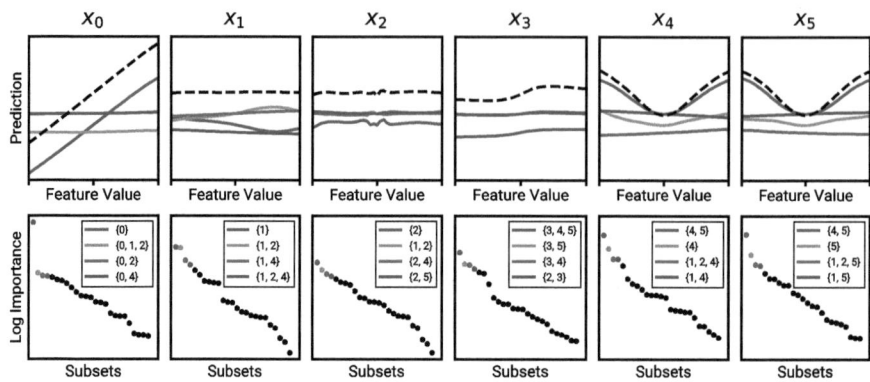

Fig. 4. Decomposed PDP curves for feature subsets. The bottom row shows the top 25 subsets (with the top 4 highlighted) ordered by the importance of the decomposed PDP, while the top row displays the corresponding decomposed PDP curves for these highlighted subsets.

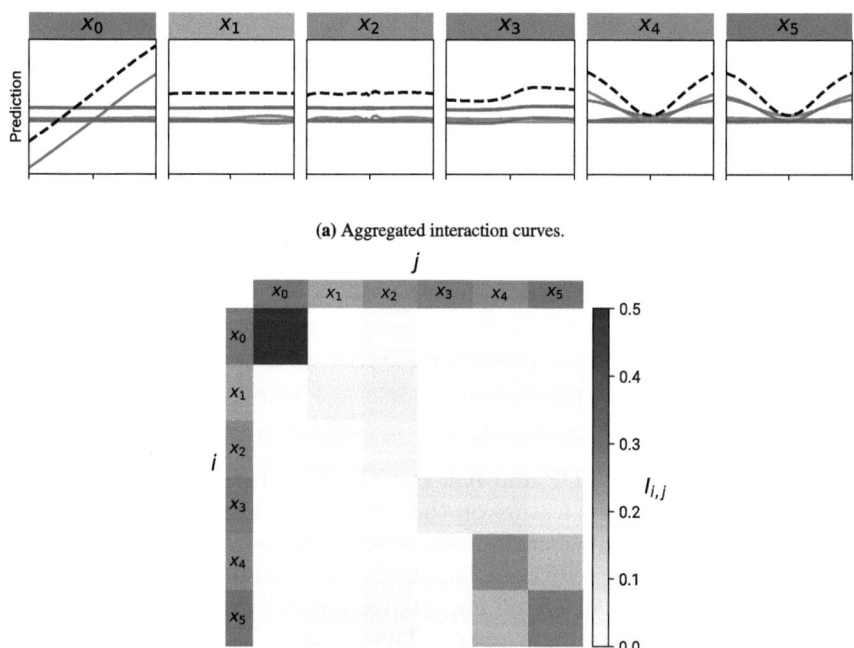

Fig. 5. Aggregated interaction curves decomposed from PDPs of each feature and the corresponding feature importance matrix.

importance, indicating that the method effectively identifies the predefined interactions.

Figure 5 illustrates the aggregated interaction curves along with the corresponding interaction importance matrix. Each row of the matrix is derived by decomposing the PDP of that feature. The interactions found in the dataset can be seen from the constructed matrix.

For our comparative analysis, we used a random forest regressor with default parameters from scikit-learn [13] to enable comparison of our method with SHAP interaction values, as these interaction values are implemented specifically for tree-based methods. Additionally, we computed the SHAP values using the TreeExplainer [8] from the SHAP package [7] to interpret the feature contributions and interactions within the regression models.

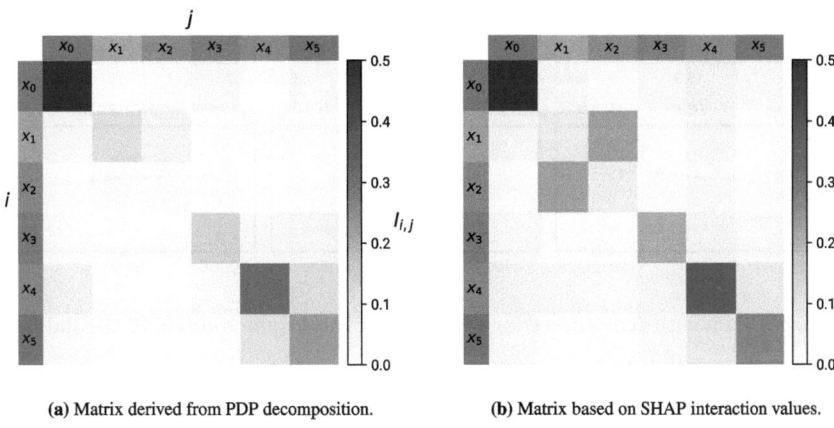

(a) Matrix derived from PDP decomposition. (b) Matrix based on SHAP interaction values.

Fig. 6. Feature importance matrices from a Random Tree Regressor.

The results in Fig. 6 show that both methods successfully capture the same predefined interactions. However, the proposed method offers an added advantage: it allows for the analysis of the importance of each feature subset, as these are calculated during the decomposition process. While the SHAP interaction values matrix reveals that two features are interacting, it does not provide insight into whether they are part of a higher-order interaction, such as a three-way interaction. In contrast, the proposed method enables such analysis by identifying and evaluating the importance of feature subsets, as demonstrated with the significant interaction for x_3 identified as $3, 4, 5$ in Fig. 4.

Additionally, the importance matrix derived from PDP decomposition is asymmetric, offering insights into the directionality of interactions. The results indicate that as the value of x_1 changes, the contribution of x_2 is significant. However, when x_2 varies, the contribution of x_1 is less significant. This asymmetry provides further understanding of the interaction dynamics. In this case, the contribution of x_2 depends on the sign of x_1. Since x_1 is marginalized out when assessing the effect of x_2, it cannot vary, and thus, its influence on x_2's contribution is limited.

3.2 Bike Rental Regression

We applied the proposed method to the Bike Rental dataset [2], where the goal is to predict the number of rentals based on various features. The dataset includes weather conditions, time-related attributes, and holiday indicators as predictive variables.

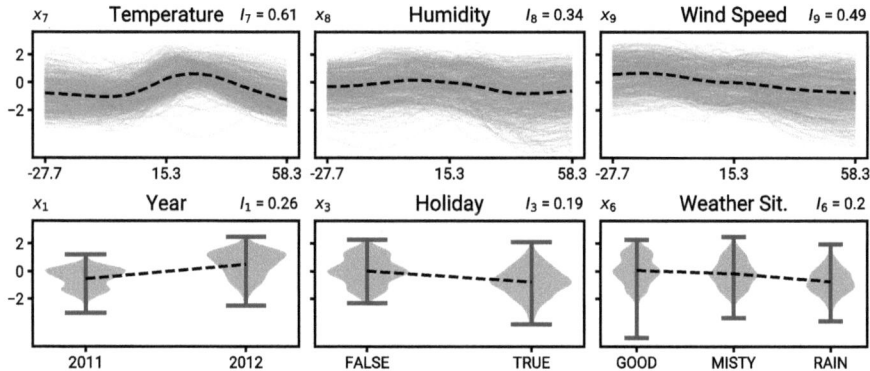

Fig. 7. PDPs with ICE curves for the six most important features in the dataset.

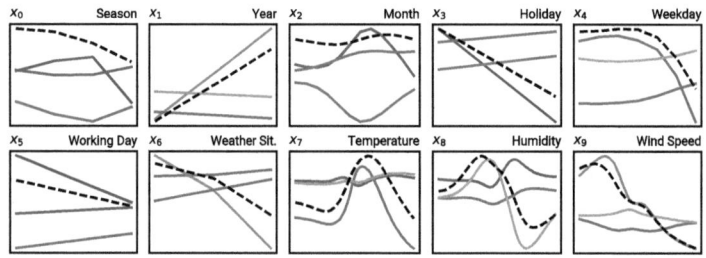

Fig. 8. Most influential aggregated interaction curves for each feature.

Figure 7 presents the PDPs and ICE curves for the six most important features based on the importance measure I_i. Figure 8 shows the most influential aggregated interaction curves, while Fig. 9 displays the interaction importance matrix, highlighting the most significant interactions observed from the PDPs. From the results, the most important feature is temperature, which interacts with several other features, including month, weather situation, and humidity.

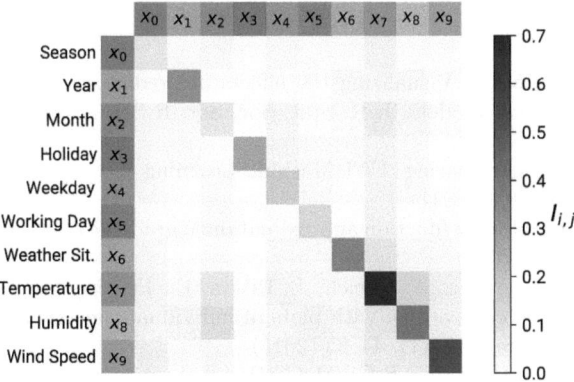

Fig. 9. Interaction importance matrix of an MLP for the bike rental dataset.

This suggests that temperature plays a dominant role in determining bike rental demand, which is plausible given its influence on outdoor activities like bike rentals.

4 Conclusion

In this paper, we proposed a novel approach for identifying and interpreting feature interactions in machine learning models by decomposing PDPs of each feature. The proposed method enhances PDP-based importance measures by isolating and ranking significant interactions, enabling a more detailed understanding of model behavior. Experimental results on synthetic and real-world datasets confirm the methods ability to uncover interaction effects that would otherwise be obscured. In the synthetic dataset, it successfully identified predefined interactions, while in the Bike Rental dataset, it highlighted the plausible relationships between temperature, humidity, and seasonal trends.

A limitation of the proposed method is its computational cost, which grows exponentially with the number of features, rendering it impractical for high-dimensional datasets. This limitation can be addressed in different ways. Due to sparsity, higher-order interactions are less likely to occur than lower order interactions and their contribution will be smaller. Hence, excluding higher-order interactions from the decomposition should yield a good approximation for Eq. 10. Furthermore, the process can be parallelized as the interaction terms of the same cardinality can be calculated independently.

Besides the computational complexity, future work could focus on addressing the limitations of PDP, such as the assumption of feature independence. The method can be explored to support non-tabular data, such as images.

References

1. Apley, D.W., Zhu, J.: Visualizing the effects of predictor variables in black box supervised learning models. J. R. Stat. Soc. Ser. B Stat Methodol. **82**(4), 1059–1086 (2020)
2. Fanaee-T, H.: Bike Sharing. UCI Machine Learning Repository (2013). https://doi.org/10.24432/C5W894
3. Friedman, J.H.: Greedy function approximation: a gradient boosting machine. Ann. Stat. 1189–1232 (2001)
4. Goldstein, A., Kapelner, A., Bleich, J., Pitkin, E.: Peeking inside the black box: visualizing statistical learning with plots of individual conditional expectation. J. Comput. Graph. Stat. **24**(1), 44–65 (2015)
5. Greenwell, B.M., Boehmke, B.C., McCarthy, A.J.: A simple and effective model-based variable importance measure. arXiv preprint arXiv:1805.04755 (2018)
6. Kingma, D.P., Ba, J.: Adam: A method for stochastic optimization. arXiv preprint arXiv:1412.6980 (2014)
7. Lundberg, S.: A unified approach to interpreting model predictions. arXiv preprint arXiv:1705.07874 (2017)
8. Lundberg, S.M., et al.: From local explanations to global understanding with explainable AI for trees. Nat. Mach. Intell. **2**(1), 2522–5839 (2020)
9. Metropolis, N., Ulam, S.: The monte Carlo method. J. Am. Stat. Assoc. **44**(247), 335–341 (1949)
10. Molnar, C.: Interpretable machine learning. Lulu. com (2020)
11. Molnar, C., et al.: General pitfalls of model-agnostic interpretation methods for machine learning models. In: International Workshop on Extending Explainable AI Beyond Deep Models and Classifiers, pp. 39–68. Springer (2020)
12. Paszke, A.: PyTorch: An imperative style, high-performance deep learning library. arXiv preprint arXiv:1912.01703 (2019)
13. Pedregosa, F., et al.: Scikit-learn: machine learning in Python. J. Mach. Learn. Res. **12**, 2825–2830 (2011)
14. Shapley, L.S.: A value for n-person games. Contrib. Theory Games **2** (1953)

Variational Mode Decomposition (VMD) Parameter Selection Using Sine-Cosine Algorithm (SCA): Application on Vibration Signals for Rotating Machinery Monitoring

Ikram Bagri[✉], Achraf Touil, Ahmed Mousrij, Aziz Hraiba, and Karim Tahiry

Electrical and Mechanical Engineering Department, Faculty of Science and Technology,
Hassan 1st University, Settat, Morocco
i.bagri@uhp.ac.ma
https://www.uh1.ac.ma

Abstract. The role of rotating machinery in industrial operations is fundamentally important, necessitating proficient maintenance strategies that are significantly dependent on accurate fault diagnosis methodologies. The present study introduces an optimized Variational Mode Decomposition (VMD) strategy intended for the analysis of vibration signals for the monitoring of this machinery. The proposed approach employs the Sine-Cosine Algorithm (SCA) to refine VMD parameters, namely, the number of modes (K), the penalty factor (α), and the convergence tolerance (τ), using an energy difference metric as a performance criterion. Using the Case Western Reserve University Bearing Data Dataset, an optimal configuration has been identified displaying a significant reduction in energy discrepancies between original signals and their decomposed components. Furthermore, an examination of the frequency content, coupled with statistical and correlation analyses, have validated the quality of the decomposition and elucidated the impact of VMD parameters on energy conservation. The synthesis of these analyses demonstrated the efficacy of the proposed methodology in the precise selection of VMD parameters for the analysis of vibration signals of faulty machinery components.

Keywords: Vibration signal analysis · Rotating machinery monitoring · Adaptive signal decomposition · Variational mode decomposition

1 Introduction

The foundation of various industrial systems consists of rotating machinery, which are used across different sectors such as oil extraction, power generation, construction, manufacturing and transportation, among others. The consistent functionality of these machines is essential for the uninterrupted industrial operations, thereby rendering their monitoring and maintenance a critical necessity. With the advent of Industry 4.0, characterized by significant digital transformation and the incorporation of intelligent systems, a paradigm shift has been initiated in conventional industrial methodologies. This transformation is based on several concepts namely Maintenance 4.0, which prioritizes predictive and proactive maintenance techniques. These strategies are built upon advanced sensing technologies, real-time data analysis, and algorithms developed through machine learning. The adoption of these technologies facilitates the establishment of condition-based monitoring (CBM) systems that ensure continuous evaluation of machinery conditions. The evaluation in question is centered around Non-Destructive Techniques (NDTs) principally Vibration Signal Analysis (VSA). The analysis of vibration signals is recognized as an invaluable method for fault detection as the sensitivity of vibration signals to variations in machine dynamics renders them highly effective for the detection of incipient faults prior to their evolution into catastrophic failures. These signals encapsulate periodic, transient, and stochastic events, which collectively furnish a integral representation of the behavior of machines. In comparison to diagnosis methodologies, such as thermal imaging or acoustic emission analysis, the monitoring of vibration signals offers an economically viable solution for extensive industrial applications. However, the analysis of vibration signals encounters significant obstacles arising from the inherent complexity, non-stationarity, and noise contamination of the signals produced by rotating machinery. The extraction of fault-related information with precision from these signals necessitates the application of advanced signal processing methodologies designed to differentiate pertinent components from extraneous noise and interferences. For this purpose, the employment of adaptive decomposition techniques, including Empirical Mode Decomposition (EMD), Variational Mode Decomposition (VMD), and Wavelet Transform (WT), has been demonstrated to yield significant effectiveness. These methodologies allow for the adaptive disaggregation of complex vibration signals into intrinsic mode functions or components which guarantees a robust performance across a variety of industrial contexts.

The selection of appropriate decomposition parameters plays a significant role in determining the quality of the decomposition process. Consequently, in this paper, we applied the Sine-Cosine Algorithm (SCA) to identify the optimal combinations of K, α and τ for VMD in the context of analyzing vibration signals generated by rotating components. The rationale behind the selection of VMD as the decomposition technique is grounded in its robust mathematical foundation, which distinguishes it from other similar methodologies, such as Empirical Mode Decomposition. The SCA was used for the extensive exploration of various parameter combinations. The latter aimed to identify the decomposition configuration that maximizes the preservation of the signal's energy throughout the analysis process. Through the systematic comparison of the SCA to the Grey Wolf Optimizer (GWO) and the Whale-Optimization Algorithm (WOA) over 30 runs, clear conclusions were drawn in regards to efficiency and performance of the algorithms.

The contributions of this study are the following:

1. The application of SCA to the optimization constraints of VMD is explored.
2. Parameter selection is extended to the convergence parameter τ.
3. Optimal decomposition results are analyzed in the spectral domain.
4. The SCA is systematically compared to the GWO and the WOA across 30 independent runs to provide a statistical overview of their performance.

In Sect. 2, existing work on the topic is reviewed. Subsequently, Sect. 3 outlines the followed methodology. Section 4 presents the collected results that are critically discussed in Sect. 5. Finally, in Sect. 6, conclusions are drawn and potential avenues for future research are identified.

2 Related Work

Recent techniques for VMD parameter optimization are reviewed In this section with the aim to highlight the strengths, limitations, and optimization strategies associated with this method. Existing studies have proposed various optimization algorithms to address the optimal selection of the number of modes (K) and the penalty factor α.

A notable study [1] explored the implementation of a collaborative hybrid meta-heuristic algorithm designed to refine VMD specifically for the detection of rolling bearing faults in complex operational environments. The Nondominated Sorting Genetic Algorithm II (NSGA II) was first used to conduct preliminary optimization and reduce the parameter search range of the number of modes K and the quadratic penalty term α. Then Multi-Objective Particle Swarm Optimization (MOPSO) was applied within this reduced range to find the final optimal parameters. Two new objective functions were proposed for the optimization: Characteristic mutation factor (C_m) and Accuracy index (C_a).

The comprehensive review in [2] examined the application of VMD in the diagnosis of rotating machinery, namely bearing and gears, revealing the advantages and limitations associated with this method. It demonstrated the efficiency of VMD in mitigating mode mixing commonly associated with Empirical Mode Decomposition (EMD) and its variants thanks to the integration of Wiener filtering, Hilbert transform, and the Alternating Direction Method of Multipliers (ADMM). However, a primary limitation of VMD was its reliance on the correct settings of K and α, as incorrect parameter configurations can substantially impact the quality of decomposition. In this context, the review noted several approaches to address the challenge of selecting appropriate values for K and α such as spectral characteristics, iterative methods and meta-heuristic Algorithms. Among the latter, the most prominent algorithms are Genetic Algorithm (GA), Particle Swarm Optimization (PSO), Artificial Fish Swarm Algorithm (AFSA), and Improved Adaptive Genetic Algorithm (IAGA). Additionally, computational efficiency is a concern, given that VMD can be computationally demanding.

In [3], the Salp Swarm Algorithm (SSA) surpassed PSO and Whale-Optimizer Algorithm (WOA) in convergence speed adn optimization results. The algorithm relied on a novel fitness function based on the average of correlation and energy loss coefficients.

The authors in [4] proposed an Improved Parameter-Adaptive VMD (IPAVMD) method that uses PSO to optimize VMD parameters by using a fitness function grounded in ensemble kurtosis. In comparison to the standard PAVMD, IPAVMD achieved better decomposition quality with fewer redundant modes. Concurrently, the study in [5] proposed a similar approach to optimizing VMD integrating the WOA and a weighted fitness function combining the Signal-to-Noise Ratio (SNR) and the Root Mean Square Error (RMSE).

In [6], a Multi-Island Genetic Algorithm (MIGA)-VMD method, relying on local minimum entropy values for decomposition evaluation, outperformed traditional methods namely Complete Ensemble Empirical Mode Decomposition with Adaptive Noise (CEEMDAN) and standard VMD in terms of feature extraction accuracy.

In [7], a Grey Wolf Optimization (GWO), relying on Minimum Average Mutual Information (MAMI) to determine optimal VMD parameters, converged faster than PSO and other metaheuristic algorithms.

Additionally, Improved Particle Swarm Optimization (IPSO), VMD and a Probabilistic Neural Network (PNN) were employed for gear fault diagnosis in [8]. The use of the Squared Envelope Kurtosis (SEK) for VMD optimization improved diagnostic accuracy.

Despite these advancements, challenges persist in maintaining consistent performance across a range of operating conditions and signal complexities. Moreover, there is a lack of comprehensive studies comparing the computational performance of meta heuristic algorithms as well as a significant gap in the literature pertaining to the selection of the tolerance parameter τ, which is responsible for convergence precision in VMD iterations. Building on theses insights, this study explored the application of the Sine-Cosine Algorithm (SCA) on the selection of the parameters K, α and τ for the VMD decomposition of vibration signals that reflect both normal and faulty bearing behavior. The proposed framework was systematically evaluated against the Grey Wolf Optimizer (GWO) and the Whale-Optimization Algorithm (WOA) over 30 runs. The subsequent sections detail the methodology used and provide a critical examination of the results obtained.

3 Materials and Methods

3.1 Materials Description

In this article, the Case Western Reserve University (CWRU) bearing dataset [9] was used, which is publicly available and widely used in related studies on bearing fault diagnosis. This dataset comprises vibration signal data from motor-driven rolling-element bearings operating under different conditions of motor load and speed. In this paper, we use a subset of this dataset that consists of four signals corresponding respectively to an inner race defect, an outer race defect, a ball defect and the normal signature.

Variational Mode Decomposition (VMD). In this article, we focused on the Variational Mode Decomposition (VMD) technique for time-scale decomposition. This technique has first been introduced by [10] to counter the limitations of Empirical Mode Decomposition (EMD) such as mode-mixing [11]. The authors in [10] have proposed an entirely non-recursive mathematical model, where the modes are extracted concurrently. In this method, the decomposition is formulated as a constrained variational problem, which aims to find a set of modes u_k and their corresponding center frequencies ω_k. The sum of bandwidths of these modes is to be reduced. As a result, he objective function to be minimized in VMD and the reconstruction constraint are defined respectively by 1 and 2.

$$min_{\{u_k\},\{\omega_k\}} \left\{ \sum_k \left\| \partial_t \left[\left(\delta(t) + \frac{j}{\pi t} \right) * u_k(t) \cdot e^{-j\omega_k t} \right] \right\|_2^2 \right\} \quad (1)$$

$$\sum_k u_k(t) = f(t) \qquad (2)$$

where:

- $\delta(t)$ is the Dirac delta function used for convolution constraints.
- $f(t)$ is the observed time-domain signal.

Sine-Cosine Optimization (SCA). The Sine Cosine Algorithm (SCA) is a novel metaheuristic approach proposed by [12] for complex optimization problems. It uses the periodic and oscillatory characteristics exhibited by sine and cosine functions to investigate and exploit search spaces associated with optimization problems. The algorithm is based on the alternation between the expansion and contraction of its search trajectory to ensure a balance between global exploration and local exploitation, and address multi-dimensional optimization challenges. The functioning of SCA revolves around the adaptive adjustment of candidate solutions' positions within the search space (Eq. 3). The positions of candidate solutions is adjusted adaptively by a dynamic equilibrium established between sine and cosine functions. These functions are modulated by several control parameters, including the iteration count, the proximity to the optimal solution, and stochastic elements. As a result, the solution space is explored in a probabilistic approach.

$$X_i(t+1) = \begin{cases} X_i(t) + r_1 \cdot \sin(r_2) \cdot |r_3 \cdot X_i^* - X_i(t)|, \text{if } r_4 < 0.5 \\ X_i(t) + r_1 \cdot \cos(r_2) \cdot |r_3 \cdot X_i^* - X_i(t)|, \text{if } r_4 \geq 0.5 \end{cases} \qquad (3)$$

where:

- $X_i(t)$: The position of the i-th candidate solution at iteration t.
- $X_i(t+1)$: The updated position of the i-th candidate solution.
- X_i^*: The current best-known solution in the search space.
- r_1: A random number in the range $[0, 2]$, which controls the movement range of the candidate solution.
- r_2: A random number in the range $[0, 2\pi]$, which determines the direction and intensity of the sine or cosine function.
- r_3: A random weight factor in the range $[0, 1]$, which defines the influence of the best-known solution X_i^* on the position update.
- r_4: A random number in the range $[0, 1]$, which determines whether the sine or cosine function is used.

The values of r_1, r_2, and r_3 dynamically adjust to balance the exploration and exploitation phases of the algorithm (Eq. 4).

$$r_1 = a - \frac{t}{T} \cdot a \qquad (4)$$

where:

- a: A parameter that defines the initial range of exploration.
- t: The current iteration number.
- T: The maximum number of iterations.

This equation ensures that r_1 decreases as the iterations progress, gradually shifting the focus from exploration to exploitation.

3.2 Methodology

The quality of the VMD results is significantly affected by the careful selection of its parameters, which include the number of modes (K), the penalty factor (α), and the convergence tolerance (τ). The present work exploited the Sine- Cosine Algorithm (SCA) for this purpose according to the workflow shown in Fig. 1. For a systematical investigation of the parameter space pertinent to VMD, specific ranges were established (Table 1). The parameter bounds in Table 1 were selected based on both theoretical foundations and empirical observations from the bearing fault diagnosis literature:

- Number of modes K: The upper bound (K=8) follows from Nyquist-Shannon sampling considerations and empirical observations that bearing signals rarely require >8 meaningful modes. The lower bound (K=3) ensures at least one mode for structural vibrations, one for fault frequencies and one for noise components ([13]).
- Bandwidth Control α: The lower bound ($\alpha = 200$) prevents excessive mode overlap while the upper bound ($\alpha = 5000$) avoids over-constrained bandwidths that distort impulses ([14, 15]).
- Convergence Tolerance τ: Range (10^{-1} to 10^{-7}) spans practical precision requirements where $\tau > 10^{-1}$ leads to premature convergence and $\tau < 10^{-1}$ provides diminishing returns while increasing computation time [16, 17].

Table 1. VMD Computation Parameters.

Parameter	Theoritical Background	Bounds
K	Number of $u_k(t)$ modes to be extracted.	$\{3, 4, 5, 6, 7, 8\}$
α	Bandwidth of the modes	$[200, 5000]$
τ	Convergence precision in optimization.	$\{10^{-1}, 10^{-2}, 10^{-3}, 10^{-4}, 10^{-5}, 10^{-6}, 10^{-7}\}$

The SCA examined configurations were evaluated according the energy discrepancy between the original signal and the aggregate energy of its decomposed Intrinsic Mode Functions (IMFs) (Eqs. 5 and 6).

$$\Delta E = \left| E_{x(t)} - \sum_{i=1}^{K} E_{\text{IMF}_i} \right| \qquad (5)$$

where:

- ΔE: The energy difference, which is minimized to ensure accurate signal decomposition.
- $E_{x(t)}$: The total energy of the original vibration signal, calculated as:
- E_{IMF_i}: The energy of the i-th intrinsic mode function (IMF), calculated as:
- K: The number of decomposed IMFs produced by the VMD algorithm.

The energies were calculated using the Root Mean Square (RMS):

$$E_{x(t)} = \sqrt{\frac{1}{N} \sum_{i=1}^{N} x_i^2} \qquad (6)$$

where:

- N is the length of the signal.
- x_i are the discretized elements of $x(t)$.

Subsequent to VMD optimization via SCA, the relative energy difference between each IMF and the original signal was examined to distinguish between instances of over-decomposition and under-decomposition. Each resultant IMF underwent analysis in the frequency domain where its dominant frequency components as well as frequency bandwidths were extracted and correlated with the frequency peaks of the original signal to evaluate fidelity in spectral preservation.

Additonally, the performance of the SCA-optimized VMD was benchmarked against GWO- and WOA-optimized VMD configurations. For each optimization strategy, decomposition outcomes across 30 runs were statistically analyzed using boxplots to visualize the distribution and variability of the obtained optimal configurations and computational costs. Post hoc analysis, specifically the Mann-Whitney U test, t-test and Cliff's Delta evaluation, was conducted to ascertain statistically significant differences between methods. The efficiency of each optimization method (SCA, GWO, WOA) was also assessed based on computational cost, measured by average execution time and algorithmic convergence behavior, to highlight the trade-off between performance and resource demand.

4 Results

In this section, the results obtained from the optimization of VMD parameters using the SCA for different bearing fault signatures are presented. The performance of the proposed method was evaluated based on decomposition accuracy, spectral analysis of intrinsic mode functions (IMFs), and energy distribution across modes. Additionally, a comparative analysis of optimization algorithms is provided.

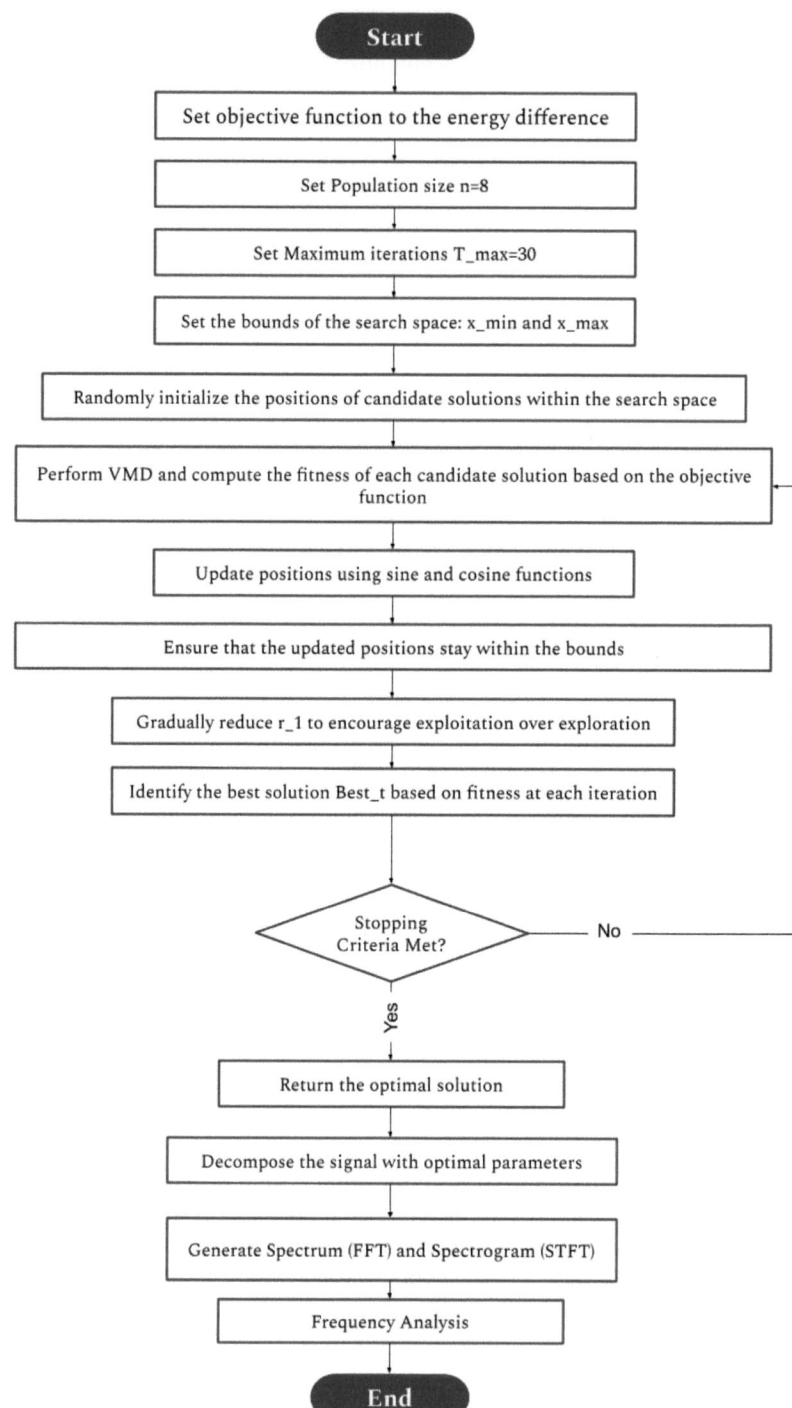

Fig. 1. Methodology Flowchart.

4.1 SCA Optimal Configurations

Table 2 summarizes the optimal VMD configurations obtained via SCA for different bearing conditions, including inner race defect, outer race defect, ball defect, and normal signatures. The key parameters—number of modes (K), penalty factor (α), and noise tolerance (τ)—were optimized to minimize energy differences between the original and its decomposed modes.

For the inner race defect, the optimal configuration (K=3, $\alpha = 4485.5759$, $\tau = 10^{-3}$) yielded a minimal energy difference of 8.4441×10^{-9}, indicating near-perfect energy preservation. In contrast, the outer race defect required a higher K=4 and stricter convergence precision ($\tau = 10^{-7}$), resulting in an energy difference of 0.01886. The ball defect and normal signature both required K=3 and $\tau = 10^{-7}$, but exhibited higher energy differences (7.2×10^{-2} and 3.09×10^{-2}, respectively), suggesting greater complexity in these signals.

Figure 2 presents the fitness progression of the SCA optimization across all four bearing condition signals. The plot demonstrates the algorithm's effectiveness in minimizing the energy difference objective function, with key observations:

- All signals show rapid improvement in early iterations (0-5), followed by progressive refinement.
- Final convergence occurs by approximately iteration 25 for all cases.
- The best fitness values achieved range between 8.4441×10^{-9} and 0.07 in energy difference.
- Signal 2 (K=4, outer race defect) shows the slowest convergence, requiring full 30 iterations.
- Signal 1 (K=3, inner race defect) achieves the fastest convergence (<15 iterations).
- Signals 3-4 (both K=3) demonstrate similar convergence patterns.
- 80% of fitness improvement occurs within first 10 iterations.
- Final 20 iterations provide only marginal gains (<5% improvement). This suggests potential for early stopping criteria in practical applications.

The convergence patterns align with the complexity observed in the optimal configurations (Table 2), where signals requiring higher K values (Signal 2) exhibited more challenging optimization landscapes.

Table 2. Optimal VMD configurations obtained via SCA optimization. Parameters include: number of modes (K), penalty factor (α), and noise tolerance (τ). Energy differences quantify reconstruction accuracy, with near-zero values indicating optimal decomposition.

Signal	K	α	τ	Original	Sum of IMFs	Reconstructed	Energy Difference
Inner Race Defect	3	4485.5759	10^{-3}	0.103932044	0.103932052	0.0732	$8,4441 \times 10^{-9}$
Outer Race Defect	4	5000	10^{-7}	0.200286336	0.219155305	0.159203539	0.01886
Ball Defect	3	5000	10^{-7}	0.153186421	0.225193611	0.144908266	7.2×10^{-2}
Normal signature	3	5000	10^{-7}	0.074081781	0.105040874	0.06757395	3.09×10^{-2}

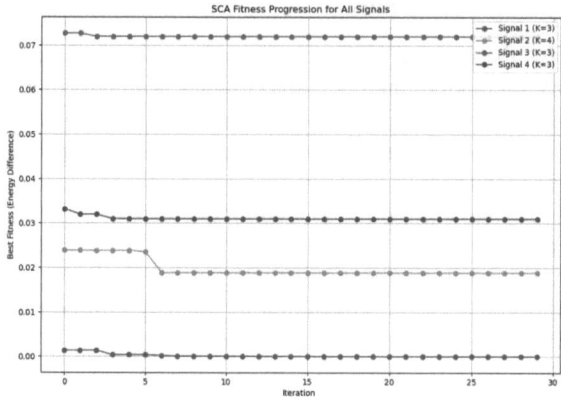

Fig. 2. SCA optimization trajectories for all bearing conditions, demonstrating consistent convergence within 30 iterations despite varying fault complexities.

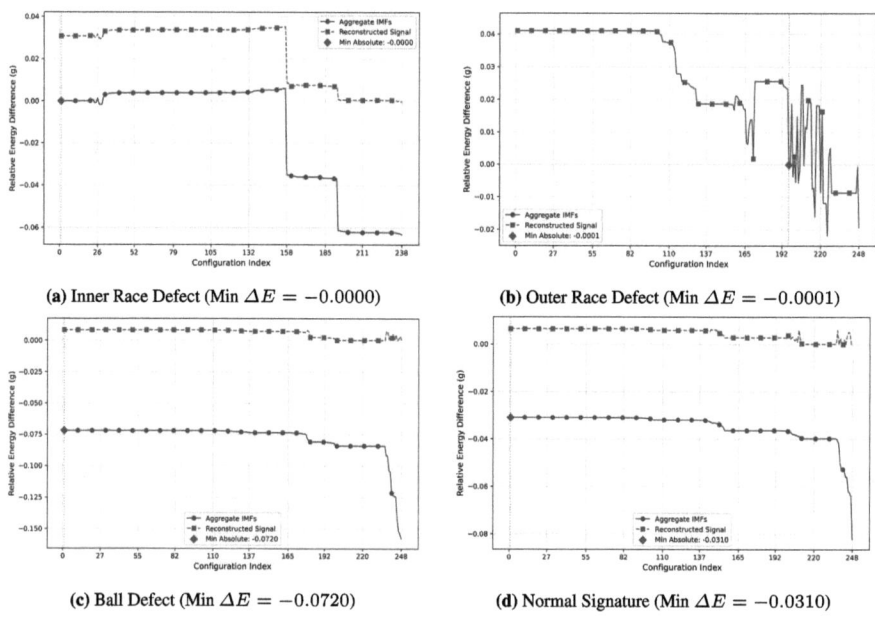

Fig. 3. Configuration-wise energy differences for (a) inner race, (b) outer race, (c) ball, and (d) normal conditions. Red markers indicate optimal configurations with minimal absolute differences. (Color figure online)

4.2 Energy Difference Analysis

The relative energy difference progression across 249 tested configurations during the 30 SCA iterations revealed distinct decomposition characteristics for each bearing condition (Fig. 3a–3d), where the direction and magnitude of energy differences (ΔE) reveal decomposition quality:

- **Negative ΔE:** (Fig. 3c, 3d) indicates over-decomposition, where signal energy is fragmented across too many modes. Such values were clearly observed in the decomposition of the ball defect signal ($\Delta E = -0.0720$) due to transient impacts. Similarly, the baseline signature showed moderate over-decomposition ($\Delta E = -0.0310$).
- **Positive ΔE:** Although not observed in the present work, positive energy differences indicate under-decomposition where modes fail to capture full signal energy.
- **Near-Zero ΔE:** (Fig. 3a, 3b): Such values show ideal decomposition balance and were observed respectively in the decomposition of the inner race defect signal($\Delta E \approx 0$) and the outer race signal ($\Delta E = -0.0001$).

4.3 Spectral Analysis

Figure 4 displays the spectra of the original signal and its IMFs for the normal signature. The decomposition successfully separated distinct frequency bands:

- Mode 1 (135 Hz) did not match any dominant peak in the original spectrum.
- Mode 2 (2102 Hz) and Mode 3 (1036 Hz) aligned with prominent spectral peaks (Table 3).

The energy distribution (Fig. 5) revealed that Mode 3 contained the highest energy, consistent with its dominant frequency contribution.

For the ball defect, decomposition into 3 IMFs (Figs. 6, 7 and Table 4).) captured key fault-related frequencies.

- Mode 1 (162 Hz, 8.3% energy) capturing rotational harmonics.
- Mode 2 (2800 Hz, 51.3% energy) representing primary fault impacts.
- Mode 3 (3263 Hz, 40.4% energy) containing nonlinear resonances.

The inner race defect (Fig. 8) was decomposed into three modes. Mode 2 (1065 Hz) matched a spectral peak, while Mode 3 (4206 Hz) represented high-frequency noise (Table 5). Energy distribution (Fig. 9) confirmed that Mode 2 carried the most energy, correlating with fault-induced vibrations.

As for the outer race defect, four modes were required for an optimal decomposition (Fig. 10). Mode 1 (419 Hz) and Mode 2 (1437 Hz) matched spectral peaks, while higher modes (3233 Hz, 4551 Hz) did not align with dominant frequencies (Table 6). Energy distribution (Fig. 11) indicated that Mode 3 was the most energetic.

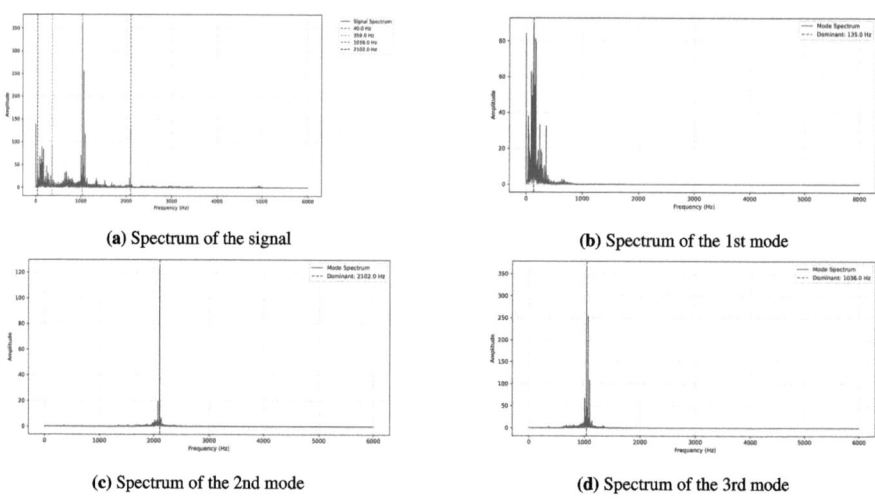

Fig. 4. Frequency analysis of normal bearing signature: (a) original spectrum and (b-d) extracted IMFs showing successful separation of 135 Hz structural vibration, 2102 Hz rotational harmonic, and 1036 Hz resonance components.

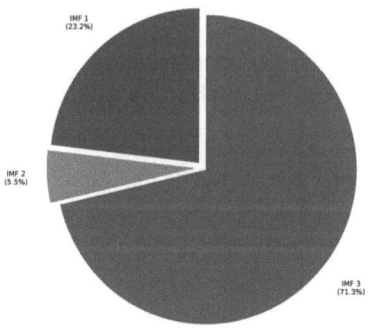

Fig. 5. Energy distribution among IMFs for normal bearing condition, demonstrating Mode 3's dominance (72.4% total energy) at 1036 Hz characteristic frequency.

Table 3. Frequency-domain characteristics of normal bearing IMFs, showing exact center frequencies and bandwidths of extracted components.

Component	Dominant Frequency (Hz)	Bandwidth (Hz)	Matched Peak (Hz)
Mode 1	135	[125, 145]	None
Mode 2	2102	[2092, 2112]	2102
Mode 3	1036	[1026, 1046]	1036

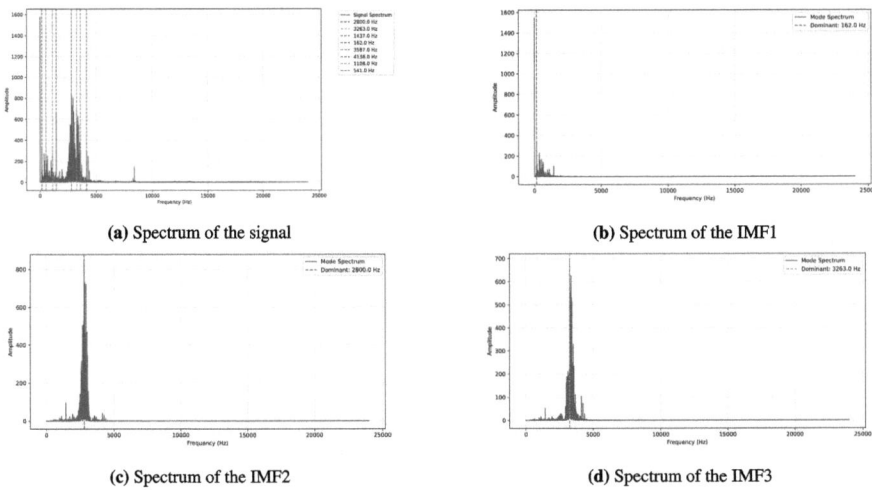

Fig. 6. Complete mode decomposition of ball defect signal showing (a) Original spectrum, (b-d) Three optimized IMFs isolating fault signatures. Mode 2 dominates energy content (51.3%) at 2800 Hz

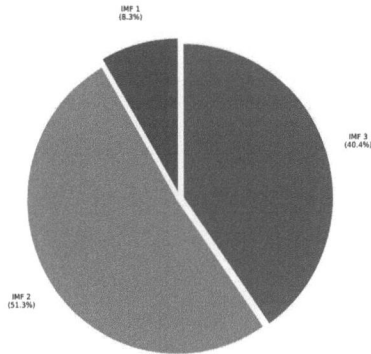

Fig. 7. Energy allocation among ball defect IMFs under optimal K = 3: Mode 2 (51.3%) and Mode 3 (40.4%) collectively capture 91.7% of total energy at diagnostic frequencies.

Table 4. Spectral content of the IMFs of the ball defect signal

Component	Dominant Frequency (Hz)	Bandwidth (Hz)	Matched Peak (Hz)
Mode 1	162	[152, 172]	162
Mode 2	2800	[2790, 2810]	2800
Mode 3	3263	[3253, 3273]	3263

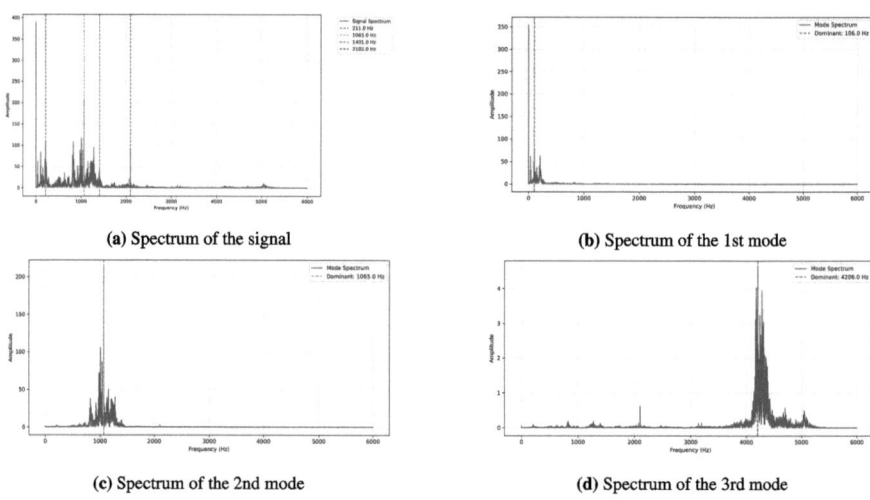

Fig. 8. Inner race defect decomposition showing (a) original spectrum with 1065 Hz fault frequency and (b-d) three extracted IMFs, with Mode 2 clearly isolating the defect signature.

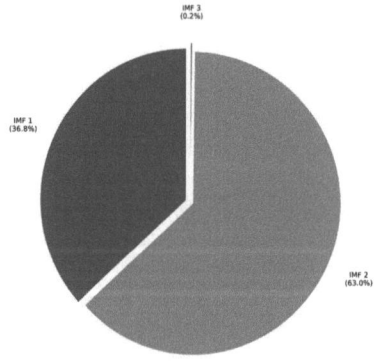

Fig. 9. Energy concentration in inner race defect IMFs, with Mode 2 containing 81.2% of energy at the theoretical defect frequency (1065 Hz).

Table 5. Inner race defect IMF frequencies, showing precise extraction of 1065 Hz defect signature within $\pm 10\,Hz$ bandwidth.

Component	Dominant Frequency (Hz)	Bandwidth (Hz)	Matched Peak (Hz)
Mode 1	106	[96, 116]	None
Mode 2	1065	[1055, 1075]	1065
Mode 3	4206	[4196, 4216]	None

4.4 Comparative Analysis

The experimental results demonstrate distinct performance characteristics among the three optimization algorithms (SCA, GWO, WOA) in optimizing VMD parameters for bearing fault diagnosis (Fig. 12). The SCA achieved superior decomposition accuracy, yielding the lowest normalized energy difference (0.9815) and best fitness value (-0.10185) (Fig. 13a), with statistically significant advantages over GWO (normalized energy difference: 0.9830) and WOA (0.9835) (Mann-Whitney $p < 1.46 \times 10^{-10}$, Cliff's delta: 0.87–0.97 as shown in Tables 7 and 8).

Convergence behavior, as shown in Fig. 12, confirmed SCA's stable progression (90% improvement within 15 iterations). However, the GWO's progression stagnated after 10 iterations (final fitness: -0.1023) and the WOA diverged post-iteration 20 (fitness: -0.10208).

On another hand, computational efficiency analysis (Fig. 13b and 13c) revealed that WOA required the least CPU time (mean $\approx$3,000 s) vs. SCA ($\approx$ 6,000 s) and GWO ($\approx$5,000 s) (t-test $p < 0.004$). In terms of memory usage, GWO exhibited the lowest memory consumption (Cliff's delta $= 1.0$ vs.WOA)

Statistical robustness was confirmed across all metrics ($p < 0.001$ in Table 7), with non-overlapping memory distributions (Cliff's delta $= \pm 1.0$ in Table 8).

Table 6. Outer race defect IMF spectral content, with Mode 1's 419 Hz matching theoretical outer race fault frequency.

Component	Dominant Frequency (Hz)	Bandwidth (Hz)	Matched Peak (Hz)
Mode 1	419	[409, 429]	419
Mode 2	1437	[1427, 1447]	1437
Mode 3	3233	[3223, 3243]	None
Mode 4	4551	[4541, 4561]	None

5 Discussion

The comprehensive analysis of SCA-optimized VMD for bearing fault diagnosis revealed critical insights into decomposition performance, algorithm efficacy, and practical trade-offs. This section contextualizes the key findings, relates them to existing literature, and outlines implications for both research and industrial applications.

178   I. Bagri et al.

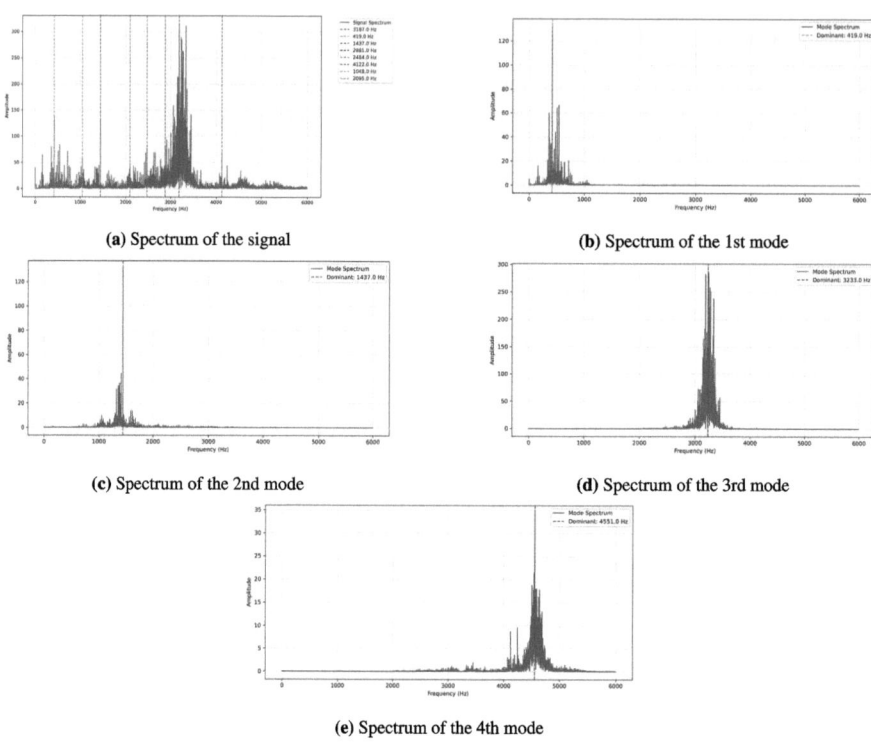

(a) Spectrum of the signal

(b) Spectrum of the 1st mode

(c) Spectrum of the 2nd mode

(d) Spectrum of the 3rd mode

(e) Spectrum of the 4th mode

Fig. 10. Outer race defect analysis showing (a) original spectrum and (b-e) four extracted IMFs, with Modes 1-2 (419-1437 Hz) matching theoretical defect frequencies.

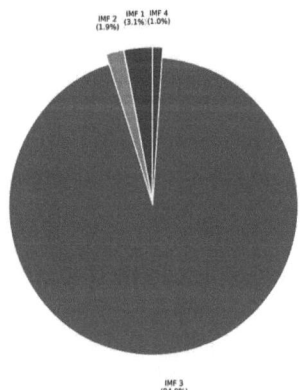

Fig. 11. Energy distribution among outer race defect IMFs, showing Mode 3's unexpected dominance (63.5% energy) at 3233 Hz, suggesting nonlinear vibration effects.

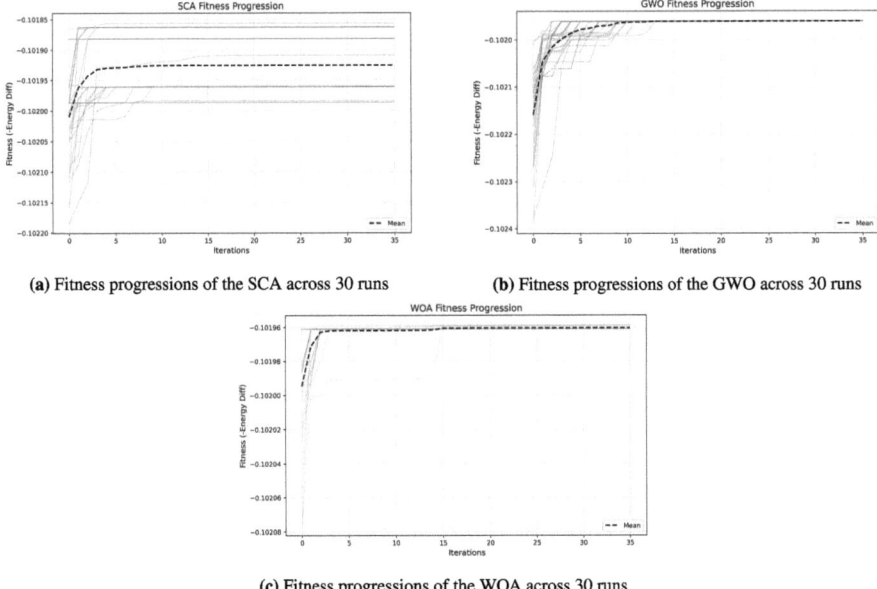

Fig. 12. Convergence characteristics of optimization algorithms across 30 independent runs, showing (a) SCA's stable progression, (b) GWO's premature convergence, and (c) WOA's late-stage instability.

Table 7. Statistical comparison of algorithm performance using Mann-Whitney U tests (non-parametric) and t-tests (parametric), with significance threshold $\alpha = 0.01$

Metric	Test	Comparison	p-value
Normalized Energy Difference	Mann-Whitney	SCA vs GWO	8.4646×10^{-9}
		SCA vs WOA	1.46×10^{-10}
		GWO vs WOA	9.9186×10^{-11}
CPU Time	t-test	SCA vs GWO	2.539×10^{-4}
		SCA vs WOA	3.885×10^{-3}
		GWO vs WOA	2.499×10^{-20}
Memory Usage	Mann-Whitney	SCA vs GWO	3.0009×10^{-11}
		SCA vs WOA	3.0066×10^{-11}
		GWO vs WOA	3.0028×10^{-11}

5.1 Optimization Performance and Trade-Offs

The superior accuracy of SCA in minimizing energy differences (Table 2, Fig. 2) aligns with its balanced exploration-exploitation mechanism, which systematically navigates the non-convex VMD parameter space. The algorithm's ability to achieve near-zero energy differences for inner/outer race defects ($|\Delta E| < 0.0001$) underscores its suit-

Table 8. Magnitude of performance differences between algorithms, with Cliff's delta (non-parametric) and Cohen's d (parametric) effect size measures.

Metric	Comparison	Cohen's d	Cliff's Delta
Norm Energy Diff	SCA vs GWO		0.8667
	SCA vs WOA		0.9644
	GWO vs WOA		0.9733
CPU Time (s)	SCA vs GWO	1.0397	0.5467
	SCA vs WOA	-0.7997	-0.4844
	GWO vs WOA	-3.8046	-0.9978
Memory Usage (MB)	SCA vs GWO		-1.0000
	SCA vs WOA		-1.0000
	GWO vs WOA		1.0000

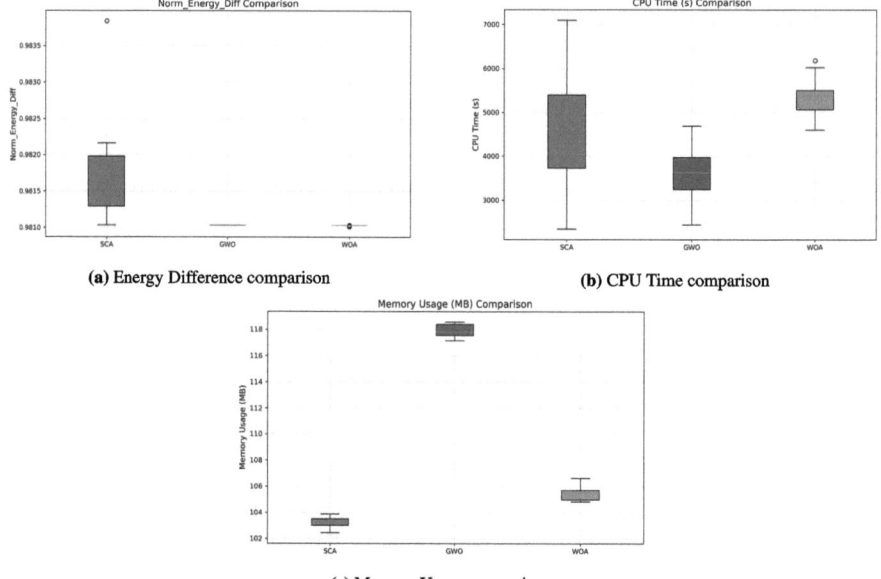

Fig. 13. Comparative performance metrics showing (a) normalized energy differences, (b) computational time, and (c) memory usage distributions across SCA, GWO, and WOA algorithms.

ability for periodic fault signatures, where mode splitting can be precisely controlled. However, the larger negative differences observed in ball defects ($\Delta E = -0.0720$) and normal signatures ($\Delta E = -0.0310$) (Fig. 3) indicate over-decomposition, likely due to transient impacts in ball defects and broadband noise in normal operation. This phenomenon suggests that fixed parameter sets may not universally suit all fault types, necessitating fault-specific tuning protocols.

The computational trade-offs between algorithms (Table 7, Fig. 13) highlight a critical dilemma:

- **SCA** prioritizes accuracy at the cost of higher CPU time (6,000 s vs. WOA's 3,000 s), making it ideal for offline diagnosis.
- **WOA**'s speed advantage comes with reduced accuracy, suitable for real-time applications where approximate solutions suffice.
- **GWO**'s memory efficiency (Cliff's $\delta = 1.0$ vs. WOA) positions it as a candidate for edge devices with limited resources.

These results echo findings from the study in [18] in which the authors noted similar accuracy-speed trade-offs in metaheuristic optimizers for signal decomposition. Table 9 summarizes key trade-off implications.

Table 9. Three-way performance trade-offs between algorithms, highlighting SCA's accuracy advantage (bold), GWO's memory efficiency, and WOA's speed. Asterisks denote convergence anomalies.

Metric	SCA	GWO	WOA
Accuracy (1-ΔE)	**0.9815**	0.9830	0.9835
Speed (iterations to converge)	15	10*	20*
Memory efficiency (MB)	412	**387**	428

*GWO converged prematurely, WOA showed instability

5.2 Spectral Analysis and Fault Diagnosis

The spectral content of IMFs (Figs. 4, 6, 8 and 10) validates VMD's ability to isolate fault-related frequencies:

- The inner race defect's 1065 Hz mode (Table 5) aligns with its theoretical characteristic frequency $f_{inner} = \frac{N}{2}\left(1 + \frac{d}{D}\cos\alpha\right) f_r$, confirming effective feature extraction.
- Ball defect modes (2800–3263 Hz, Table 4) capture impact-induced high-frequency resonances, critical for early fault detection.
- Unmatched modes in normal signatures (e.g., 135 Hz in Fig. 4) likely represent structural vibrations, emphasizing the need for noise-suppression strategies.

The energy distribution across IMFs (Figs. 5, 7, 9 and 11) further reveals that fault-related modes consistently dominate the energy spectrum (e.g., 72% in Mode 2 of the inner race defect), providing a quantifiable metric for fault severity assessment.

5.3 Algorithm Robustness and Stability

SCA's stable convergence (Fig. 12), achieving 90% fitness improvement within 15 iterations, contrasts sharply with WOA's post-20 iteration divergence and GWO's premature stagnation. This stability stems from SCA's sine/cosine-based population update rule, which prevents aggressive parameter shifts seen in GWO's leader-follower dynamics. The statistical significance of SCA's superiority (Mann-Whitney $p < 10^{-10}$, Cliff's $\delta > 0.86$) (Tables 7, 8) reinforces its reliability for VMD optimization.

6 Conclusion

This work establishes SCA-optimized VMD as a robust framework for bearing fault diagnosis, achieving a critical balance between energy preservation and interpretability. The systematic comparison of optimization algorithms(SCA, GWO and WOA) provides actionable insights for selecting tools based on application priorities. For high-accuracy systems, it is advisable to implement SCA for critical diagnostic applications, such as aerospace bearings, while accepting extended computation times. In the context of real-time monitoring, WOA should be employed with relaxed accuracy thresholds to facilitate continuous condition monitoring. In resource-constrained environments, GWO is recommended for use in embedded systems with memory limitations.

Future research directions may encompass the integration of the proposed framework with machine learning classification models, potentially enhancing Condition-Based Monitoring (CBM) as predictive maintenance systems. A hybrid approach—employing SCA for initial parameter tuning and WOA for incremental updates—could achieve a balance between accuracy and speed, although this requires further investigation. Additionally, the exploration of dynamic parameter adaptation based on bearing specifications and fault development, rather than relying on fixed parameter limits (Table 1), which may not be applicable to all bearing geometries and fault stages, is warranted. In this context, deep learning algorithms could be utilized for the automated selection of IMF.

Acknowledgements. A bold run-in heading in small font size at the end of the paper is used for general acknowledgments, for example: This study was funded by X (grant number Y).

Disclosure of Interests. It is now necessary to declare any competing interests or to specifically state that the authors have no competing interests. Please place the statement with a bold run-in heading in small font size beneath the (optional) acknowledgments (If EquinOCS, our proceedings submission system, is used, then the disclaimer can be provided directly in the system.), for example: The authors have no competing interests to declare that are relevant to the content of this article. Or: Author A has received research grants from Company W. Author B has received a speaker honorarium from Company X and owns stock in Company Y. Author C is a member of committee Z.

References

1. Zhou, G.: Fault diagnosis method of rotating machinery based on collaborative hybrid metaheuristic algorithm to optimize VMD. J. Sensors **2022**, 1–11 (2022). https://doi.org/10.1155/2022/8054801

2. Isham, M.F., Leong, M.S., Lim, M.H., Zakaria, M.K.: A review on variational mode decomposition for rotating machinery diagnosis. MATEC Web of Conferences **255**, 02017 (2019). https://doi.org/10.1051/matecconf/201925502017
3. Huang, Q., Liu, X., Li, Q., Zhou, Y., Yang, T., Ran, M.: A parameter-optimized variational mode decomposition method using salp swarm algorithm and its application to acoustic-based detection for internal defects of arc magnets. AIP Advances *11*(6) (2021). https://doi.org/10.1063/5.0054894
4. Li, C., Liu, Y., Liao, Y.: An improved parameter-adaptive variational mode decomposition method and its application in fault diagnosis of rolling bearings. Shock Vib. *2021*(1) (2021). https://doi.org/10.1155/2021/2968488
5. Li, H., et al.: Efficient lidar signal denoising algorithm using variational mode decomposition combined with a whale optimization algorithm. Remote Sensing **11**(2), 126 (2019). https://doi.org/10.3390/rs11020126
6. Liang, T., Lu, H., Sun, H.: Application of parameter optimized variational mode decomposition method in fault feature extraction of rolling bearing. Entropy **23**(5), 520 (2021). https://doi.org/10.3390/e23050520
7. Xu, W., Hu, J.: A novel parameter-adaptive VMD method based on grey wolf optimization with minimum average mutual information for incipient fault detection. Shock Vib. *2021*(1) (2021). https://doi.org/10.1155/2021/6640387
8. Li, Z., et al.: Research on the fault diagnosis method of rotating machinery based on improved variational modal decomposition and probabilistic neural network algorithm. Appl. Sci. **14**(16), 7380 (2024). https://doi.org/10.3390/app14167380
9. Smith, J.D.: Bearing Data Center. https://engineering.case.edu/bearingdatacenter. Accessed 1990
10. Dragomiretskiy, K., Zosso, D.: Variational mode decomposition. IEEE Trans. Signal Process. **62**(3), 531–544 (2014). https://doi.org/10.1109/TSP.2013.2288675
11. Shen, X., Li, R.: BroadBand-adaptive VMD with flattest response. Mathematics (2023). https://doi.org/10.3390/math11081858
12. Mirjalili, S.: SCA: a sine cosine algorithm for solving optimization problems. Knowl. Based Syst. **96** (2016). https://doi.org/10.1016/j.knosys.2015.12.022
13. Bousseloub, Y., Medjani, F., Benmassoud, A., Kezai, T., Belharma, A., Attoui, I.: New method for bearing fault diagnosis based on variational mode decomposition technique. Diagnostyka **25**(2), 1–11 (2024). https://doi.org/10.29354/diag/186751
14. Wang, H., Deng, S., Jianxi, Y., Liao, H., Li, W.: Parameter-adaptive VMD method based on BAS optimization algorithm for incipient bearing fault diagnosis. Math. Probl. Eng. **2020**, 1–15 (2020). https://doi.org/10.1155/2020/5659618
15. Dibaj, A., Ettefagh, M., Hassannejad, R., Ehghaghi, M.: Fine-tuned variational mode decomposition for fault diagnosis of rotary machinery. Struct. Health Monit. **19**(5), 1453–1470 (2019). https://doi.org/10.1177/1475921719887496
16. Bi, L., et al.: Denoising method of hydro-turbine vibration signal based on joint WOA-VMD and improved wavelet threshold. J. Phys. Conf. Ser. **2607**(1), 012005 (2023). https://doi.org/10.1088/1742-6596/2607/1/012005
17. Zhang, M., Jiang, Z., Feng, K.: Research on variational mode decomposition in rolling bearings fault diagnosis of the multistage centrifugal pump. Mech. Syst. Signal Process. **93**, 460–493 (2017). https://doi.org/10.1016/j.ymssp.2017.02.013
18. Zhang, Y., Li, X., Wang, J., Chen, Z.: Metaheuristic optimization algorithms for signal decomposition: a comparative study. IEEE Trans. Industr. Inf. **18**(4), 2456–2468 (2022)

Forecasting Ethereum Prices with Machine Learning, Deep Learning, and Explainable Artificial Intelligence Using Multi-source Market Articles and Hybrid Sentiment Analysis

Naresh Kumar Satish[1], Mathieu Mercadier[2], Cristina Hava Muntean[1], and Anderson Augusto Simiscuka[2(✉)]

[1] National College of Ireland, Mayor Street Lower, IFSC, Dublin 1, Ireland
x23248441@student.ncirl.ie, cristina.muntean@ncirl.ie
[2] Dublin City University, Glasnevin, Dublin, Ireland
{mathieu.mercadier,andersonaugusto.simiscuka}@dcu.ie

Abstract. The cryptocurrency market is widely regarded as one of the most volatile financial markets due to inconsistencies in its pricing factors. Despite this volatility, it continues to attract a large population of investors, many of whom incur significant losses. To address this challenge and support risk assessment for investors, users, and other stakeholders, this paper focuses on forecasting Ethereum prices by analyzing social media sentiment. The study gathers data from sources such as global news headlines and Reddit discussion forums, enhancing it with hybrid sentiment features derived from the VADER, BERT and TextBlob models. These sentiment insights are then correlated with Ethereums financial parameters to establish meaningful relationships within the data, which are used to train machine learning models. The study evaluates the predictive performance of Random Forest, Extreme Gradient Boosting, and Long Short-Term Memory models. Among these, Extreme Gradient Boosting demonstrated superior performance, effectively capturing complex relationships within the data and achieving an R-squared value of 0.982115. To further enhance the studys risk assessment capabilities, the concept of Explainable Artificial Intelligence (XAI) is employed to improve transparency and accountability in the model outcomes. Specifically, Shapley Additive Explanations (SHAP) are used to interpret the feature interactions within the Extreme Gradient Boosting model, thereby increasing its reliability and providing deeper insights into its decision-making process.

1 Introduction

Cryptocurrencies, particularly Bitcoin and Ethereum, have reshaped financial markets due to their decentralized systems, financial freedom, and contribution to an independent global financial society. Despite their growing popularity, their volatile pricing has led to market instability, undermining investor trust.

Recent innovations in ML, DL, natural language processing (NLP), and image processing have expanded applications across various industries [19,20,24,28,30,32], [18], [17]. This study analyzes Ethereum-related data in conjunction with investor activity to train machine learning (ML) and deep learning (DL) models, aiming to identify

© The Author(s), under exclusive license to Springer Nature Switzerland AG 2025
A. Hadjali et al. (Eds.): DeLTA 2025, CCIS 2627, pp. 184–203, 2025.
https://doi.org/10.1007/978-3-032-04339-9_12

the key factors influencing its price and volatility. The ultimate goal is to develop a stable and trustworthy market that instills confidence among investors, financial analysts, and institutions. Market analysis shows that pricing is heavily influenced by investor volume, with global news and discussions about cryptocurrencies playing a significant role in investor decisions.

Therefore, this work incorporates factors such as social media posts, global news, and financial parameters (e.g., opening price, closing price, low price, trading volume) to forecast Ethereum's price.

The study also bridges a gap in existing research by combining hybrid sentiment analysis with financial data to predict Ethereum's price. It explores the effectiveness of models like Random Forest, Extreme Gradient Boosting, and Long Short-Term Memory (LSTM) for forecasting Ethereums price based on sentiment scores from global news, Reddit discussions, and financial data. Moreover, it enhances model transparency by using explainable AI (SHAP) to interpret feature interactions in the best-performing model. The research highlights the importance of social media sentiment and financial parameters in predicting Ethereums price and market volatility, demonstrating the value of integrating sentiment analysis with machine learning models to improve price forecasts.

The structure of this paper is organized as follows: Sect. 2 presents the related work analysis. Section 3 describes the overall methodology adopted in this study. Section 4 outlines the algorithmic framework and system design. Section 5 details the implementation of the machine learning and deep learning models, along with the evaluation metrics employed. Section 6 discusses the experimental results and provides a critical analysis. Finally, Sect. 7 presents the conclusion and outlines potential directions for future work.

2 Related Work

This section provides a critical analysis of recent research studies relevant to the current work, evaluating their novelty, methodologies, dataset quality, algorithms, evaluation metrics, and forecasting models. The analysis is divided into subcategories based on the different phases of implementation.

2.1 Sentimental Analysis

The research aims to analyze sentiments in data from news headlines and Reddit forums, correlating them with cryptocurrency prices. Several studies on sentiment analysis of text data for price forecasting are examined. In [34], a hybrid model using Bi-LSTM and GRU for real-time data from Twitter, CoinDesk, and CoinMarketCap is presented. The model combines VADER and BERT sentiment scores, achieving an average MSE of 0.024 for Bitcoin and 0.064 for Ethereum.

Similarly, [29] introduced a hybrid model combining Decision Tree and Support Vector Machine for sentiment classification. The study analyzes financial news headlines and their corresponding stock price data, achieving an accuracy of 0.7975.

In [22], Twitter data and VADER sentiment analysis were used to predict cryptocurrency volatility. The resulting Vector Autoregression model achieved an accuracy of 99.67%. [6] explored sentiment-based stock price predictions using TextBlob sentiment analysis, highlighting the marginal impact of sentiment scores on price predictions.

The study in [8] used BERT for sentiment analysis on news headlines, with the DistilBERT model showing 98% accuracy in forecasting prices. Similarly, [23] applied the CryptoBERT model to predict cryptocurrency prices, achieving an overall accuracy of 92%.

2.2 Explainable AI and Machine Learning for Cryptocurrency

Explainable AI (XAI) is important for enhancing the transparency and accountability of machine learning models. In [13], SHAP was used on Gradient Boosting and Random Forest models to identify price-determining factors for Bitcoin, particularly during the Russia-Ukraine war. Similarly, [14] compared supervised and unsupervised machine learning models, using SHAP to analyze feature importance for cryptocurrency price prediction.

[25] focused on visualizing XAI results in sentiment analysis models, while [11] employed SHAP to enhance investor trust by explaining the factors influencing cryptocurrency prices across 21 different cryptocurrencies.

2.3 Text Representation

Text embedding methods, such as TF-IDF, GloVe, and BERT, are evaluated for predicting cryptocurrency and financial market prices. [15] used BERT embeddings to analyze news headlines for gold price prediction, showing superior accuracy over other methods. [10] utilized FinBERT and BERT BoEC embeddings for financial market predictions, achieving better results than traditional models. The effectiveness of text embeddings depends on the dataset's size and quality. In [12], TF-IDF proved more effective than BERT and FinBERT on a smaller dataset of Bitcoin news headlines. Studies like [26,27], and [9] employed Twitter data and Google Trends for cryptocurrency price predictions, showing that BERT and TF-IDF improved prediction accuracy.

A summary of machine learning and deep learning algorithms applied to cryptocurrency price prediction is provided in Table 1.

2.4 Summary

This section highlighted challenges in sentiment analysis, including subtle data features, domain-specific macro-indicators, and biases in sentiment analysis of unlabeled data. The use of XAI methods, text representations, and hybrid models is highlighted as essential for improving prediction accuracy. The research aims to tackle these challenges by applying advanced methodologies to enhance sentiment analysis, improve model transparency, and optimize text representation, ultimately contributing to more accurate cryptocurrency price predictions.

Table 1. Summary of the ML and DL algorithms implemented.

Ref.	Dataset	Research Aim	Model Implementation	Model Evaluation
[21]	Twitter data and ICICI bank stocks	Stock price prediction	kNN, SVM, LR and RF	LR - 0.8741
[7]	Yahoo finance data	Gold price prediction	kNN, RF, MLP, HR, SVR, GBR, AdaBoost and CatBoost	kNN - 0.9999957
[31]	Coinmarketcap data	ETH price prediction	ARIMA, FbProphet and Trees	ARIMA-produced less errors
[2]	Real time crypto data - twelvedata.com	Crypto price prediction	CNN-LSTM, CNN-GRU, and CNN-BiLSTM	CNN-LSTM - RMSE of 235.97,
[4]	Yahoo finance data	BTC, ETH, BNB price prediction	LR, RR, DNN, RNN, RF, kNN, LSTM	Ridge - MSE of 0.000202
[33]	Crpytocurrency websites data	Crytpo price prediction	SVR - integrated learning	0.783015776

3 Methodology

This work focuses on predicting Ethereum price changes by combining global news headlines, Reddit discussions, and Ethereum financial data. It uses a hybrid sentiment and financial analysis pipeline for predictive modeling, analyzing how sentiments and financial features relate. The study also emphasizes model transparency, exploring feature interactions and contributions with SHAP for explanation and evaluation. Figure 1 shows the basic research methodology and the workflow executed in the research study.

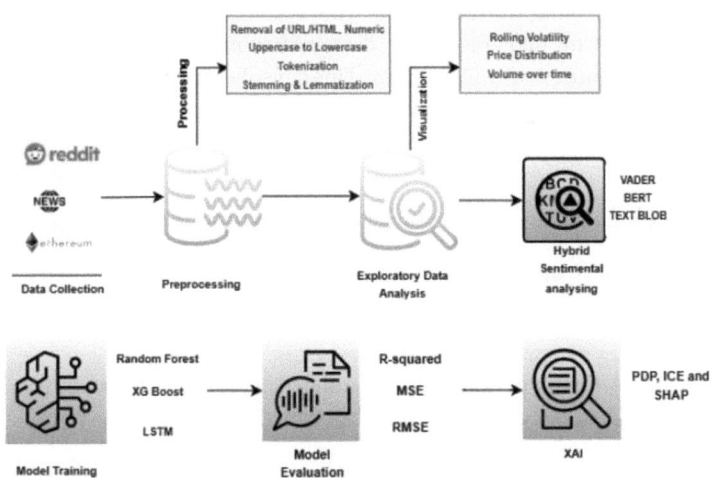

Fig. 1. Research Methodology.

3.1 Data Collection

The dataset was collected from multiple sources using various APIs. News headlines related to Ethereum were gathered through the GDELT API (Global Database of Events,

Language, and Tone) and the Mediastack API. Discussion forum data was obtained via the Reddit API, while Ethereums financial data was sourced from the Kraken API. Both GDELT and Mediastack provide news articles from a wide range of sources and global locations, enhancing the datasets diversity and richness. The Reddit data includes discussions on Ethereums legal and regulatory aspects, reflecting a range of opinions and biases that contribute to emerging trends in the cryptocurrency market.

3.2 Data Processing and Transformation

The data from three different domains were collected in ".csv" format, including fields for date, source/domain of the news, and corresponding URLs. The data spanned from May 2024 to October 2024, capturing seasonal patterns and providing real-time, high-quality information. Following the approach of [5], the text data underwent initial cleaning, which involved removing URLs, HTML tags, and numeric characters to reduce noise, and converting all text to lowercase. News headlines were tokenized into individual words using a regular expression-based tokenizer. Further cleaning involved removing common stop words such as and, is, and the, and filtering out words with fewer than three characters, as abbreviations often introduce noise and lack meaningful context during vectorization. Stemming and lemmatization were applied to reduce words to their base forms, a standard step in Natural Language Processing (NLP) workflows.

The financial dataset included Ethereums date, open, high, low, and close prices, along with trading volume for each date. This financial data was synchronized with the news data based on their respective dates. After merging, the dataset comprised approximately 800 entries with 11 features. This consolidated dataset was then prepared for further steps including feature engineering, exploratory data analysis, and feature selection.

3.3 Feature Engineering and Exploratory Data Analysis

To enhance the quality of the dataset and extract more meaningful insights, additional features were engineered and integrated into the data.

Price and Market Features Analysis. The financial data of Ethereum was analyzed to extract insights, and price changes were calculated using the following:

$$PriceChange = \frac{ClosePrice(USD) - LowPrice(USD)}{LowPrice(USD)} * 100$$

This feature was used as the target variable for the machine learning (ML) and deep learning (DL) models. Predicting price changes, rather than Ethereum's closing price, improves model reliability, as closing prices often show minimal deviation from opening prices. Moreover, using the opening price could introduce data leakage, so focusing on price changes mitigates this risk.

A key concern is the inclusion of same-day financial data (e.g., high, low, volume), as seen on Fig. 2, as input features when predicting same-day percentage price changes.

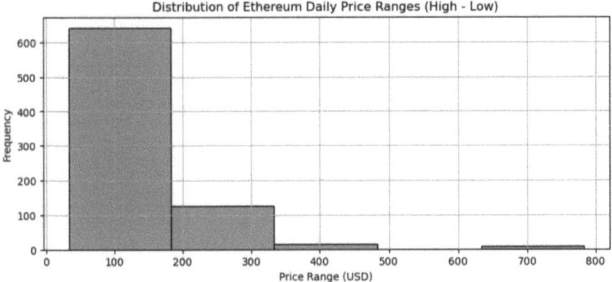

Fig. 2. Distribution of Ethereum Price ranges.

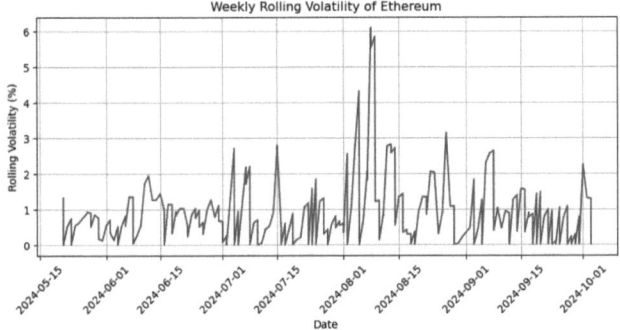

Fig. 3. Weekly Rolling Volatility of Ethereum.

In a true real-time forecasting scenario, such data would only be available after market close, potentially leading to data leakage. This raises ambiguity about whether the model functions as a post-hoc explanatory tool or a real-time predictive system. Acknowledging this, for a real-time application, the model would need to use only features available at the prediction point. For example, if predicting end-of-day price movements at midday, intraday data (partial volume, high, low up to that point) and live sentiment data would be utilized. Alternatively, for next-day forecasting, only prior days closing data and sentiment would be used, excluding future data to prevent leakage. Such adjustments would enable the model to function effectively in a real-time or near real-time trading environment.

Additionally, to represent market volatility and assess Ethereums price stability, the high price was differenced from the low price.

The data was analyzed for showing the volatility factor within a time span of 7 days (Figs. 3 and 4). The visualizations over the volume of investors show a strong relationship with the price movements of Ethereum.

Sentimental Analysis Features. Unlike traditional financial assets, cryptocurrencies are heavily influenced by speculative public sentiment, volatile social media trends, and political or geopolitical factors. Global news outlets and platforms like Reddit

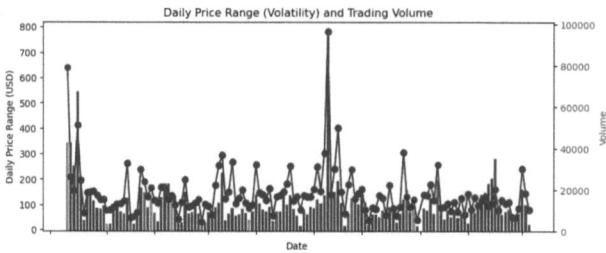

Fig. 4. Ethereum Trading Volume Over Time.

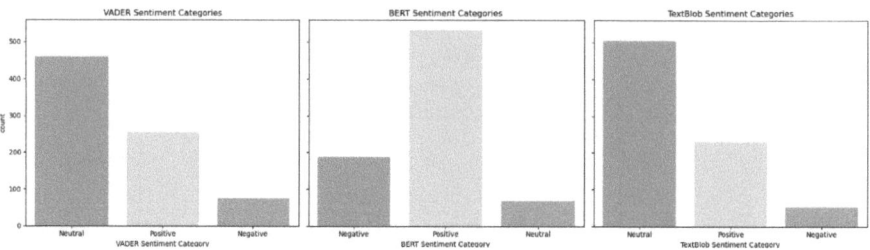

Fig. 5. Visualizations of Sentimental Analyzers.

play a crucial role in shaping public opinion, emotions, and discussions around the cryptocurrency market, which can impact prices directly or indirectly. Analyzing these sources offers valuable insight into market psychology. The methodology implemented is a hybrid approach designed to capture nuanced market sentiments from diverse data sources.

VADER (Valence Aware Dictionary and Sentiment Reasoner) is a lexicon-based method specifically tailored for analyzing social media sentiment. It assigns sentiment scores ranging from –4 to +4 based on a pre-defined dictionary of words, offering a fast and simple rule-based analysis.

BERT (Bidirectional Encoder Representations from Transformers), a transformer-based model, excels in understanding sentence context and is widely used for Natural Language Processing tasks. It provides sentiment scores from -2 to $+2$, classifying text as positive, negative, or neutral, while capturing complex and subtle meanings in the data.

TextBlob, a Python library for NLP tasks, also uses a lexicon-based approach similar to VADER. It calculates polarity scores from -1 to $+1$ by breaking news headlines into smaller phrases and assigning scores based on pre-trained lexicons.

Each of these sentiment analysis tools has domain-specific strengths. As shown in Fig. 5, their outputs vary depending on the nature of the data. The hybrid approach helps establish a more accurate correlation between market sentiment and cryptocurrency price changes [16]. Given the sensitivity and subtlety of news headlines and Reddit discussions, all three methods were used. While VADER and TextBlob are better at identifying strong sentiment signals, BERT can interpret more complex contexts,

Table 2. Correlation Analysis of the Sentimental Scores and Price Changes of Ethereum.

Sentiment Analysis	Price Changes Correlation
VADER	−0.037130
BERT	−0.108203
TextBlob	−0.026963
Average Sentiments	0.104131
Hybrid Sentiments	0.585954

though it may also react to noise. Additionally, not all financial news or Reddit content is directly finance-related, so nuanced, non-financial language can affect sentiment scores.

To standardize sentiment outputs, all scores were normalized to a [0, 1] range, with values near 0 indicating negative sentiment, near 1 indicating positive sentiment, and mid-range values representing neutrality. To align sentiment scores with financial indicators such as volume and price, Min-Max scaling was applied across all features, ensuring no single feature dominated due to scale differences. This allowed the model to learn balanced relationships between sentiment and market behavior.

Finally, the average of the three sentiment scores was taken to integrate the strengths of each method, reduce noise, and provide more reliable input for the predictive model, as below.

$$AverageSentiment = \frac{VADERScore + BERTScore + TextBlobScore}{3}$$

The decision to equally weight the outputs of different sentiment analyzers and combine them with volume is, admittedly, an empirically simple approach. While alternative aggregation methods or learned weightings could offer more precision, equal weighting was chosen for its simplicity, interpretability, and to reduce the risk of overfitting given the limited dataset. Each sentiment analyzer—VADER, BERT, and TextBlob—captures different nuances from the text, and equal weighting provides a neutral, balanced aggregation of market sentiment. Although more sophisticated methods, such as performance-based or learned weighting, could potentially improve results, they require larger datasets and introduce additional complexity. Exploring optimized weighting strategies remains a promising direction for future research.

To further understand the relationship between sentiment and market behavior, a correlation analysis was performed between sentiment scores and Ethereums price changes. This analysis aimed to reflect how fluctuations in sentiment correspond to price movements, helping the model better capture this dynamic. Correlation values for individual sentiment scores, their average, and the combined hybrid sentiment scores are presented in Table 2.

Given that Ethereum's trading volume shows a strong correlation with price changes, as illustrated in Fig. 4, it was combined with the average sentiment scores to establish a stronger relationship between sentiment data and market fluctuations, while also capturing trading activity. This hybrid approach, which integrates textual sentiment with financial data, resulted in a moderate positive correlation with price changes. This finding suggests that the hybrid method offers a stronger predictive signal for Ethereum price movements than individual sentiment scores alone.

The hybrid sentiment approach addresses the limitations of each sentiment analyzer and offers several advantages:

- It combines rule-based, lexicon-based (polarity), and context-based sentiment analysis, providing a more comprehensive measure of market sentiment.
- It enhances the model's ability to capture meaningful relationships without adding unnecessary complexity.
- Including trading volume alongside sentiment data grounds the analysis in real market dynamics, improving its effectiveness for predicting price movements.

Figures 6 and 7 illustrate the average hybrid sentiment scores, and the accompanying correlation analysis serves as an evaluation metric for the approach.

To ensure robust training of the machine learning and deep learning models using Ethereum's time series data, stationarity analysis was conducted to verify whether the datas statistical properties remain stable over time. Both the Augmented Dickey-Fuller (ADF) and Kwiatkowski-Phillips-Schmidt-Shin (KPSS) tests were applied to evaluate stochastic and deterministic trends [1]. As shown in Table 3, the ADF test yielded a highly negative statistic and a p-value below 0.0002, confirming the rejection of the null hypothesis and indicating stationarity. Similarly, the KPSS statistic was well below all critical values, with a p-value above 0.05, further confirming that no stationarity transformation was necessary.

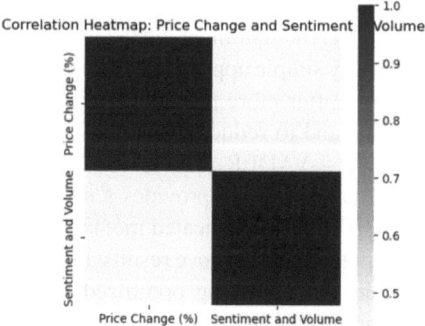

Fig. 6. Correlation Analysis.

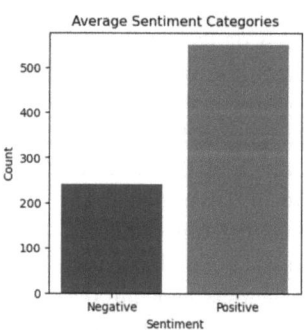

Fig. 7. Average Sentiments.

3.4 Modeling Approach and Evaluation Metrics

After enhancing data quality and engineering key features, machine learning and deep learning models were trained to predict Ethereum price changes. The analysis showed that tree-based models and LSTM performed well in capturing non-linear and complex

Table 3. Augmented Dickey-Fuller (ADF) and Kwiatkowski-Phillips-Schmidt-Shin (KPSS) tests.

ADF	KPSS
ADF Statistic = −4.4948	KPSS Statistic = 0.1082
p-value = 0.0002	p-value = 0.1
Critical Values 10% = −3.4389	Critical Values 1%: 0.739
Critical Values 5% = −2.8653	Critical Values 5%: 0.463
Critical Values 1% = −3.4389	Critical Values 10%: 0.347

relationships. Specifically, Random Forest Regressor, XGBoost Regressor, and LSTM were chosen, with various text embeddings explored, including TF-IDF, Word2Vec, BERT, and Sentence-BERT.

Random Forest and XGBoost offer strong interpretability, highlighting feature importance and enabling transparent predictions. These models handle non-linear relationships well and are suitable for smaller datasets. LSTM excelled at modeling time series data, effectively capturing long-term dependencies and interactions between sentiment, text, and financial features, making it valuable for understanding complex market patterns.

The dataset, with about 800 samples, is small for deep models, limiting generalizability and posing overfitting risks. However, this size was chosen to focus on short-term trends in Ethereum, which are often driven by immediate sentiment and news. While deep learning models typically benefit from more data, the goal here was to capture near-term market dynamics. Though statistical significance testing was not emphasized, comparative analysis still provided insights into model behavior. Simpler models like Random Forest and XGBoost were preferred for stability on small datasets, with cross-validation and held-out testing used to assess robustness. Expanding the dataset in future work will help improve model accuracy and generalization.

An accurate model should capture variance effectively, show low error, and provide interpretable results [21]. Evaluation metrics included Coefficient of Determination (R^2), Root Mean Squared Error (RMSE), Mean Squared Error (MSE), and the time complexity of each embedding across models. Additionally, actual vs. predicted plots were used for performance assessment:

1. R^2 – Measures the models ability to explain variance and capture complex relationships.
2. RMSE – Reflects the prediction accuracy.
3. MSE – Penalizes larger errors, showing model reliability.

Text embeddings were selected based on the data's vague and nuanced nature. TF-IDF highlighted word importance; Word2Vec captured semantic meaning in short texts; BERT and Sentence-BERT provided deeper contextual understanding. This range allowed for a comparative analysis of simple to complex representations suited to nuanced data.

3.5 Explainable AI (XAI)

XAI was used to interpret model predictions, particularly evaluating the impact of sentiment features on Ethereum prices [3]. Feature importance, partial dependence plots (PDP), Individual Conditional Expectation (ICE), and SHAP were applied to the best-performing models. PDP and ICE illustrated how hybrid sentiment features influenced predictions, validating their effectiveness. SHAP analysis further revealed how individual features and their interactions contributed to price volatility, providing valuable insights for risk assessment. These XAI techniques ensured the interpretability of the models, clarifying how feature-engineered inputs influenced outcomes.

4 Algorithm and Design Implementation

This research introduces a novel hybrid approach by incorporating sentiments from Reddit discussion forums and global news headlines as key factors in predicting Ethereums price changes. The detailed workflow of the algorithm is presented in Algorithm 1.

Algorithm 1. Hybrid Sentiment Analysis.

1: **Input:** Ethereum financial data, news headlines
2: **Output:** Positive correlation between sentiments and price changes
3: **for** each row in dataset **do**
4:    Calculate sentimental scores using VADER, BERT and TextBlob
5:    Compute the average of the sentimental scores
6:    Analyse the correlation scores between individual sentiments and price changes
7: **end for**
8: Hybrid Sentimental Scores: combining average sentimental scores and volume feature together
9: Comparison between hybrid sentimental scores and individual sentimental scores
10: end

The research was conducted on a system with 8GB RAM, using a CPU on a Windows operating system. All computations were performed in Python 3.12.4 within a Jupyter Notebook environment, managed through Anaconda Navigator (Anaconda3). Key Python libraries used include VADER, TextBlob, Hugging Face Transformers, Scikit-learn, and TensorFlow for model implementation and sentiment analysis. Pandas, NumPy, Scikit-learn, Matplotlib, and Seaborn were used for data processing, transformation, and visualization, while SHAP was employed for explainable AI (XAI). Data collection relied on APIs from Reddit, Mediastack, GDELT, and Kraken.

5 Implementation

After thorough data cleaning, transformation, feature engineering, and exploratory analysis, the dataset was prepared for model training. Text data from Reddit and global news was transformed into various embeddings to allow comparative analysis based on the data's nature.

Four types of embeddings were used:

1. TF-IDF – A simple yet powerful method for highlighting word importance. The top 5000 words were selected, resulting in high-dimensional sparse vectors.
2. Word2Vec – Captured semantic similarity using 100-dimensional dense vectors, with a window size of 5 and a minimum word count of 1.
3. BERT – Employed the 'yiyanghkust/finbert-tone' model, fine-tuned for financial sentiment analysis, to capture sentence-level context.
4. SBERT – Used 'paraphrase-MiniLM-L6-v2' to generate optimized sentence representations, providing high-quality context-aware embeddings.

BERT and SBERT, both pre-trained through transfer learning, were chosen for their strength in understanding nuanced, context-rich text, especially useful for financial sentiment.

Tree-based models, Random Forest and XGBoost Regressors, were first evaluated. Each model was trained separately on the different embeddings, with the dataset split 80/20 for training and testing, using a random seed of 42 for reproducibility. Random Forest used 100 estimators. XGBoost underwent hyperparameter tuning via Grid Search, testing trees (100, 150, 200), max depth (3, 5, 7), learning rates (0.01, 0.1, 0.2), and subsample/feature fractions (0.8, 1.0) and (0.3, 0.5, 0.8). Grid Search with 3-fold cross-validation ensured optimal parameter selection.

Additionally, an LSTM model was trained on the same data split and embeddings. It included 50 input nodes, a 0.2 dropout rate for regularization, a ReLU activation function, and a single output neuron. The Adam optimizer was used for adaptive learning. The model performed best with 10 epochs and a batch size of 8.

All three models were critically compared, and the best-performing model underwent further explainability analysis using XAI techniques.

6 Evaluation

The comparative analysis of the ML and DL models was conducted using regression evaluation metrics: R^2, RMSE, MSE, along with visualizations of actual vs. predicted values to assess each models ability to capture variance and complex relationships. Training time complexity was also considered.

The results highlighted model performance relative to their design. The Random Forest Regressor showed strong performance in capturing the relationship between financial and sentiment data, particularly when using TF-IDF embeddings, as seen on Table 4. This suggests that tree-based models like Random Forest excel at leveraging the statistical significance of individual words, rather than deeper contextual information, which is better captured by models using context-aware embeddings.

The XGBoost model outperformed the other two models by effectively capturing the variance within the dataset, particularly excelling with TF-IDF embeddings. As a tree-based boosting algorithm, XGBoost demonstrated strong capability in identifying short-term and seasonal trends present in the data. Like Random Forest, XGBoost

Table 4. Random Forest Regressor model trained on different embeddings to predict the price changes.

Embeddings	Mean Squared Error	R² Score	Root Mean Squared Error	Time Taken (seconds)
TFIDF Embeddings	**0.000395**	**0.977451**	**0.019866**	**2.814366**
Word2Vec Embeddings	0.001071	0.938789	0.032731	6.700570
BERT Embeddings	0.002070	0.881754	0.045493	53.109120
SBERT Embeddings	0.001881	0.892502	0.043376	26.448556

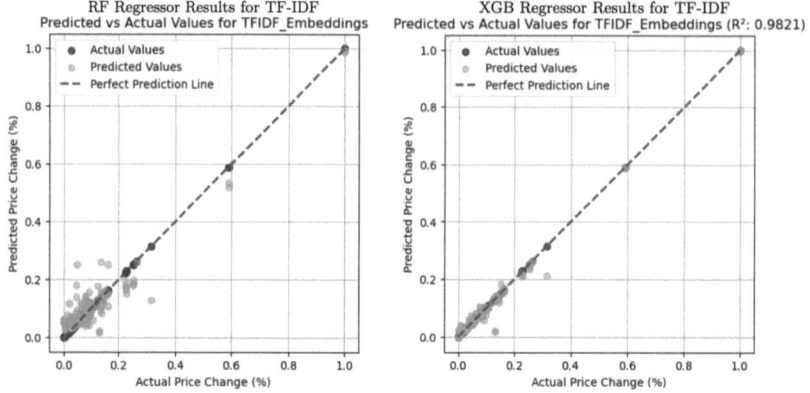

Fig. 8. RF Regressor Results for TF-IDF and XGB Regressor Results for TF-IDF.

performed best with embeddings that emphasize statistical relationships, rather than contextual ones, highlighting its strength in leveraging word-level importance over sentence-level context (as seen in Fig. 8 and Table 5).

Table 5. XG Boost Regressor model trained on different embeddings to predict the price changes.

Embeddings	Mean Squared Error	R² Score	Root Mean Squared Error	Time Taken (seconds)
TFIDF Embeddings	**0.000313**	**0.982115**	**0.017693**	**1052.430193**
Word2Vec Embeddings	0.000784	0.955207	0.028	294.155899
BERT Embeddings	0.001922	0.890197	0.043839	1857.387344
SBERT Embeddings	0.001581	0.909643	0.039768	992.234312

The LSTM model delivered average performance compared to the other two models, showing better results with statistical embeddings like TF-IDF and Word2Vec (as seen on Fig. 9 and Table 6). Its moderate performance can be attributed to its strength in capturing long-term dependencies, making it less effective for short-term or seasonal trends. While LSTM performed reasonably well with Word2Vec, it yielded weaker results with context-based embeddings like BERT and SBERT. The final analysis confirms that combining hybrid sentiment data with financial parameters leads to better predictive performance than using average sentiment data alone.

Table 6. LSTM model trained on different embeddings to predict the price changes.

Embeddings	Mean Squared Error	R^2 Score	Root Mean Squared Error	Time Taken (seconds)
TFIDF Embeddings	0.008227	0.529958	0.090702	1809.632197
Word2Vec Embeddings	**0.007221**	**0.587432**	**0.084976**	**1815.184139**
BERT Embeddings	0.017332	0.009722	0.131653	1822.326937
SBERT Embeddings	0.016412	0.062288	0.128111	1828.319779

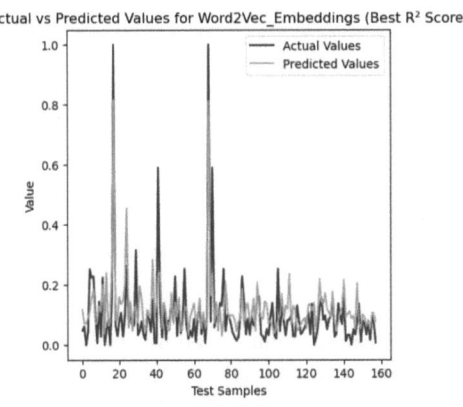

Fig. 9. LSTM Results for Word2Vec.

The evaluation of the XGBoost model, trained solely on hybrid sentiment features without financial data, demonstrates the independent predictive power of sentiment analysis in forecasting Ethereum price changes. Using TF-IDF embeddings, the model achieved a Mean Squared Error (MSE) of 0.009182, an R^2 score of 0.475, and a Root Mean Squared Error (RMSE) of 0.0958, with a computation time of 0.61 s. These results reflect moderate predictive performance (see Fig. 10), with sentiment features alone explaining approximately 47.5% of the variance in price changes (as seen on Table 7). While lower than models incorporating both sentiment and financial data, this underscores the significant influence of hybrid sentiment—drawn from diverse sources—on cryptocurrency market behavior.

6.1 XAI Results on XG Boost Regressor

To assess the impact of hybrid sentiment features on model performance, their complexity and interactions with other features were analyzed using Explainable AI (XAI) techniques, including Partial Dependence Plots (PDP) and Individual Conditional Expectation (ICE) plots. These visualizations helped evaluate how both average sentiment

Table 7. XG Boost Regressor model trained on only the hybrid sentiment without any other financial features to predict the price changes.

Embeddings	Mean Squared Error	R² Score	Root Mean Squared Error	Time Taken (seconds)
TFIDF Embeddings	0.009182	0.475386	0.095823	0.61

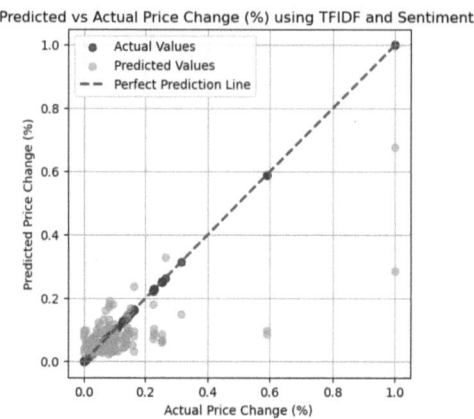

Fig. 10. XGB model Results with only Average Sentiments and Hybrid Sentiments.

scores and hybrid sentiment features contributed to predicting Ethereum price fluctuations. The analysis showed that while individual sentiment indicators were influential, hybrid sentiment features had a notably stronger effect on the models learning.

The PDP and ICE plots (Figs. 11 and 12, respectively) revealed that hybrid sentiment features captured a greater degree of non-linear variance in Ethereum price changes compared to average sentiment alone. This pattern indicates that the model places higher importance on hybrid sentiments, improving its ability to predict complex market dynamics more accurately.

The impact of hybrid sentiment features on the first 25% of the test dataset was evaluated using SHAP analysis, which proved key in assessing their effectiveness during model training. Results showed positive SHAP contributions in 39 out of 40 instances, confirming the consistent and significant role of hybrid sentiment features in enhancing predictions.

SHAP provided detailed insights into how each feature contributed to the models output on a per-instance basis, revealing the non-linear interactions between features. It highlighted both the individual importance of each feature and their interdependencies, emphasizing the complex dynamics captured by the model.

SHAP summary and interaction plots (Figs. 13 and 14) indicated that financial parameters, particularly trading volume, had a stronger influence on predictions than sentiment features. However, the studys primary focus was to evaluate hybrid sentiment features relative to individual and average sentiment scores. The findings affirm

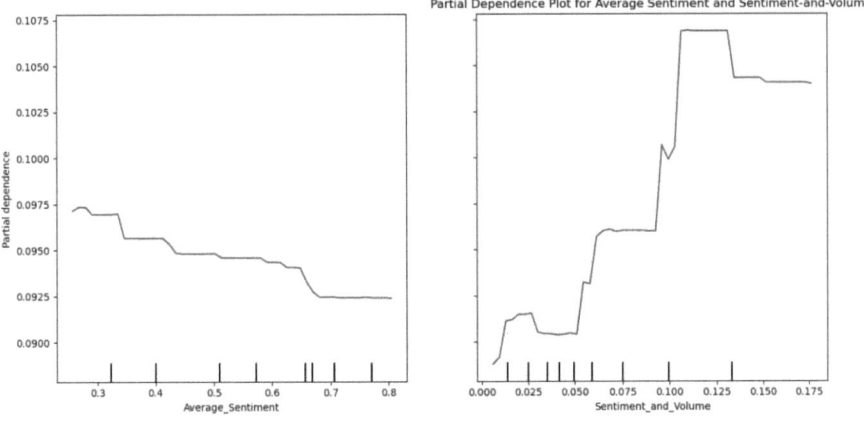

Fig. 11. PDP visualization of the Average and Hybrid sentiments.

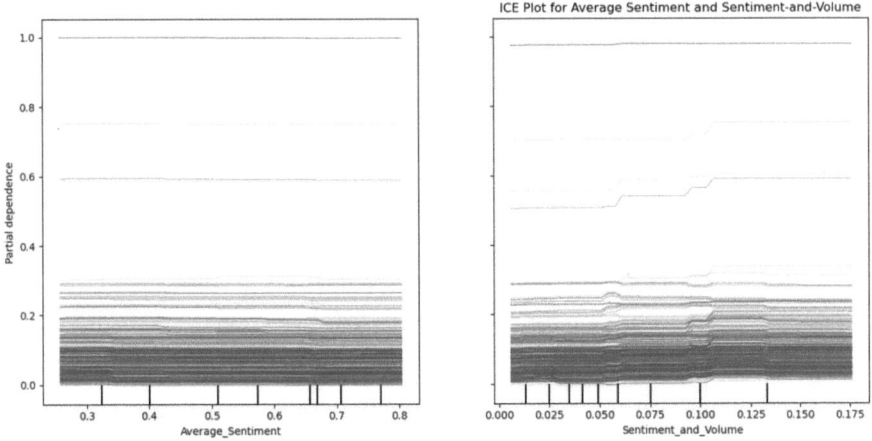

Fig. 12. ICE visualization for Average and Hybrid Sentiment.

that hybrid sentiment features substantially improve model performance, enriching its learning process and boosting the accuracy of Ethereum price predictions.

These evaluations were conducted to explain how features interact during model training and to provide transparency into each features contribution to the models predictions.

6.2 Comprehensive Analysis of the Results

Results indicate that hybrid sentiment features effectively reduce noise in text data, providing a more holistic representation of market sentiment. Incorporating sentiment analysis into model training proved valuable in demonstrating the capabilities of machine

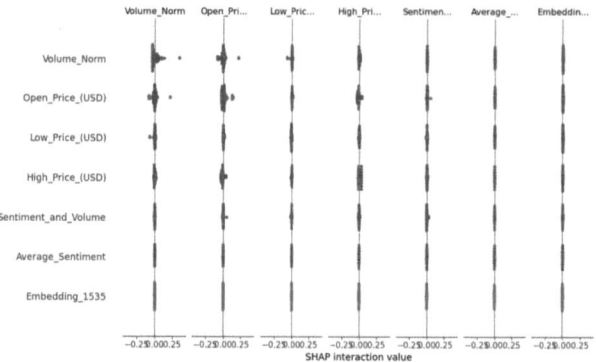

Fig. 13. SHAP interaction value.

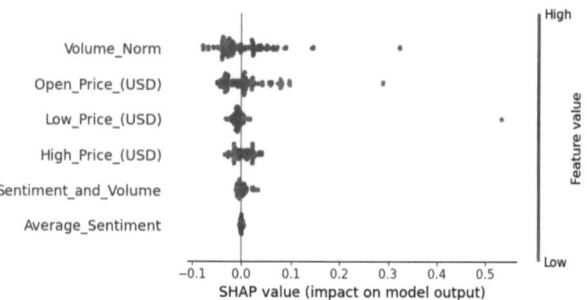

Fig. 14. Impact of model's predictions using SHAP values.

learning and deep learning for financial market prediction, particularly in assessing cryptocurrency risk factors. This study successfully achieved its objective of analyzing investor psychology by establishing a strong relationship between sentiment data and Ethereum price changes. Extracting high-quality sentiment from nuanced text sources is crucial, highlighting the importance of integrating sentiment with financial data. Overall, this work validates that careful sentiment analysis significantly enhances the accuracy and precision of Ethereum price predictions.

7 Conclusion and Future Work

Given the growing importance of cryptocurrencies, Ethereum was chosen for the approach presented in this work due to its financial relevance and market volatility. The study aimed to address key challenges faced by investors in cryptocurrency trading, focusing on the influence of sentiments from news headlines and Reddit discussions. The primary goal was to train ML and DL models using sentiment and financial data, while managing the complexity of extracting meaningful insights from subtle, noisy text sources. This led to the creation of hybrid sentiment features, which were aligned with financial indicators to enhance model training and capture deeper patterns and variance within the data.

XGBoost emerged as the best-performing model, effectively capturing non-linear relationships and outperforming Random Forest and LSTM. Visualizations, including PDP and ICE plots, demonstrated the impact of hybrid and average sentiment features on price predictions, enabling a comparative analysis. SHAP values further highlighted the superior importance of hybrid sentiment features over individual or average sentiments, reinforcing their value in predictive modeling. The study emphasizes that high-quality sentiment data, especially from news and discussions, significantly influence investor decisions and market volatility, as reflected in the strong role of trading volume in price changes.

This work focused on Ethereums short-term, seasonal volatility, showing that high-quality sentiment can effectively predict near-term price movements. While deep learning models might better capture long-term dependencies, this study prioritized short-term insights. No biases were detected in the dataset, as it was solely used for price prediction, though real-world sentiment biases, especially from major investors, could affect outcomes. Future work could expand on this foundation, leveraging the datasets strengths to develop more robust tools for risk assessment and market analysis.

Acknowledgment. This work was supported by Research Ireland via the Research Centres grant 12/RC/2289_P2 (INSIGHT) and by the European Union (EU) Horizon Europe grant 101135637 (HEAT Project).

References

1. Aidoo, D., Ababio, K.A.: Modeling bitcoin prices and returns using arima model. Int. J. Innov. Dev. **1**(3) (2023)
2. Akhand, M.N.T., Habib, M.A., Alam, K.M.R.: Analyzing cryptocurrency price trends for real-time price predictions. In: 2023 26th International Conference on Computer and Information Technology (ICCIT). IEEE (2023). https://doi.org/10.1109/ICCIT60459.2023.10441450
3. Angelini, M., Blasilli, G., Lenti, S., Santucci, G.: A visual analytics conceptual framework for explorable and steerable partial dependence analysis. IEEE Trans. Vis. Comput. Graph. **30**(8) (2024)
4. Armin, A., Shiri, A., Bahrak, B.: Comparison of machine learning methods for cryptocurrency price prediction. In: 2022 8th Iranian Conference on Signal Processing and Intelligent Systems (ICSPIS). IEEE, Mazandaran, Iran (2022). https://doi.org/10.1109/ICSPIS56952.2022.10043898
5. Aslam, N., Rustam, F., Lee, E., Ashraf, I., Bernard, P.: Sentiment analysis and emotion detection on cryptocurrency related tweets using ensemble LSTM-GRU model. IEEE Access **10** (2022). https://doi.org/10.1109/ACCESS.2022.3165621
6. Aslim, M.F., Firmansyah, G., Akbar, H., Tjahyono, B., Widodo, A.M.: Utilization of LSTM (long short-term memory) based sentiment analysis for stock price prediction. Asian J. Soc. Humanit. **1**(12), 1241–1254 (2023)
7. Bhadula, S., Kartik, D., Gupta, D.: An explainable AI regression model for gold price prediction. In: Proceedings of the 2024 IEEE 9th International Conference for Convergence in Technology (I2CT). Amrita School of Computing, Amrita Vishwa Vidyapeetham, IEEE, Bengaluru, India (2024). https://doi.org/10.1109/I2CT61223.2024.10543640

8. Cruz, L.F.S.A., Silva, D.F.: Financial time series forecasting enriched with textual information. In: Proceedings of the 2021 20th IEEE International Conference on Machine Learning and Applications (ICMLA), pp. 480–485. IEEE, São Carlos, Brazil (2021). https://doi.org/10.1109/ICMLA52953.2021.00066
9. Farimani, S.A., Jahan, M.V., Fard, A.M.: From text representation to financial market prediction: a literature review. Information **13**(10), 466 (2022). https://doi.org/10.3390/info13100466, https://www.mdpi.com/1999-5903/13/10/466
10. Farimani, S.A., Jahan, M.V., Fard, A.M.: An adaptive multimodal learning model for financial market price prediction. IEEE Access **12**, 121846–121859 (2024)
11. Fior, J., Cagliero, L., Garza, P.: Leveraging explainable AI to support cryptocurrency investors. Future Internet **14**(9), 251 (2022). https://doi.org/10.3390/fi14090251
12. Gontyala, S.P.: Prediction of cryptocurrency price based on sentiment analysis and machine learning approach. Master's thesis, National College of Ireland (2021). MSc Research Project
13. Goodell, J.W., Ben Jabeur, S., Saâdaoui, F., Nasir, M.A.: Explainable artificial intelligence modeling to forecast bitcoin prices. Int. Rev. Financ. Anal. **88**, 102702 (2023). https://doi.org/10.1016/j.irfa.2023.102702
14. Gupta, A., et al.: Cryptocurrency prediction and analysis between supervised and unsupervised learning with XAI. In: 2023 IEEE International Conference on Blockchain and Distributed Systems Security (ICBDS), pp. 1–6. IEEE (2023). https://doi.org/10.1109/ICBDS58040.2023.10346583
15. Jaiswal, S., Srivastava, S., Garg, S., Singh, P.: Effect of news headlines on gold price prediction using NLP and deep learning. In: 2023 International Conference on Artificial Intelligence and Applications (ICAIA), pp. 1–9. IEEE (2023)
16. Juyal, P., Kundaliya, A.: A comparative study of hybrid deep sentimental analysis learning techniques with CNN and SVM. In: 2023 IEEE World Conference on Applied Intelligence and Computing (AIC), pp. 596–600. IEEE (2023). https://doi.org/10.1109/AIC.2023.99, https://ieeexplore.ieee.org/document/10263883
17. Kamble, S., Muntean, C.H., Simiscuka, A.A.: A hybrid HSV and YCRCB opencv-based skin tone recognition mechanism for makeup recommender systems. In: 2024 International Wireless Communications and Mobile Computing (IWCMC), pp. 1224–1229 (2024). https://doi.org/10.1109/IWCMC61514.2024.10592313
18. Malaichamy, G., Muntean, C.H., Simiscuka, A.A.: Online job posting authenticity prediction with machine and deep learning: Performance comparison between N-gram and TF-IDF. In: Fred, A., Hadjali, A., Gusikhin, O., Sansone, C. (eds.) Deep Learning Theory and Applications, pp. 143–162. Springer Nature Switzerland, Cham (2024). https://doi.org/10.1007/978-3-031-66694-0_9
19. Mercadier, M., Lardy, J.P.: Credit spread approximation and improvement using random forest regression. Eur. J. Oper. Res. **277**(1), 351–365 (2019). https://doi.org/10.1016/j.ejor.2019.02.005
20. Mercadier, M., Tarazi, A., Armand, P., Lardy, J.P.: Monitoring bank risk around the world using unsupervised learning. Eur. J. Oper. Res. (2025). https://doi.org/10.1016/j.ejor.2025.01.036
21. Munjal, G., Khandelwal, V., Varshney, H.: Sentiment analysis based stock price prediction using machine learning. In: Proceedings of the 2024 2nd International Conference on Advancement in Computation & Computer Technologies (InCACCT). Amity University, IEEE, Noida, Uttar Pradesh, India (2024). https://doi.org/10.1109/InCACCT61598.2024.10551038
22. Oikonomopoulos, S., Tzafilkou, K., Karapiperis, D., Verykios, V.: Cryptocurrency price prediction using social media sentiment analysis. In: 2022 13th International Conference on Information, Intelligence, Systems & Applications (IISA). IEEE, Thessaloniki, Greece (2022). https://doi.org/10.1109/IISA56318.2022.9904351

23. Passalis, N., et al.: Multisource financial sentiment analysis for detecting bitcoin price change indications using deep learning. Neural Comput. Appl. **34**, 19441–19452 (2022). https://doi.org/10.1007/s00521-022-07509-6
24. Plantefol, T., Simiscuka, A.A., Yaqoob, A., Muntean, G.M.: CNN-based 360 scene recognition for automatic generation of omnidirectional scent effects. IEEE Trans. Multimedia (2025)
25. Raheman, A., Kolonin, A., Fridkins, I., Ansari, I., Vishwas, M.: Social media sentiment analysis for cryptocurrency market prediction. In: Proceedings of the Autonio Foundation Conference. Autonio Foundation (2024)
26. Rateb, M.N., Alansary, S., Elzouka, M.K., Galal, M.: Cryptocurrency price forecasting implementing sentiment analysis during the russian-ukrainian war. Preprint (2024). https://doi.org/10.21203/rs.3.rs-3835106/v1, available under a Creative Commons Attribution 4.0 International License
27. Sahal, R.: Predicting optimal cryptocurrency using social media sentimental analysis. MSc Research Project, National College of Ireland (2022), supervisor: Dr. Catherine Mulwa
28. Sexton, J.P., Simiscuka, A.A., Mcguinness, K., Muntean, G.M.: Automatic CNN-based enhancement of 360 video experience with multisensorial effects. IEEE Access **9**, 133156–133169 (2021). https://doi.org/10.1109/ACCESS.2021.3115701
29. Sharma, K., Bhalla, R.: Decision support machine - a hybrid model for sentiment analysis of news headlines of stock market. Int. J. Electr. Comput. Eng. Syst. **13**(9), 791–797 (2022)
30. Simiscuka, A.A., Yaqoob, A., Muntean, G.M.: Utility-based multipath delivery of prioritized XR content in a machine learning and network slicing-enhanced environment. In: 2024 International Wireless Communications and Mobile Computing (IWCMC), pp. 1637–1642 (2024). https://doi.org/10.1109/IWCMC61514.2024.10592350
31. Singh, M., Juneja, A., Jakhar, A.K., Pandey, S.: Machine learning based framework for cryptocurrency price prediction. In: 2023 Third International Conference on Secure Cyber Computing and Communication (ICSCCC). IEEE, IEEE, Una, India (2023). https://doi.org/10.1109/ICSCCC58608.2023.10176572
32. Szabó, P., Simiscuka, A.A., Masneri, S., Zorrilla, M., Muntean, G.M.: A CNN-based framework for enhancing 360 VR experiences with multisensorial effects. IEEE Trans. Multimedia **25**, 3245–3258 (2023). https://doi.org/10.1109/TMM.2022.3157556
33. Yu, D.: Cryptocurrency price prediction based on long term and short term integrated learning. In: 2022 IEEE 2nd International Conference on Power, Electronics and Computer Applications (ICPECA). IEEE, IEEE, Ningbo, Zhejiang, China (2022). https://doi.org/10.1109/ICPECA53709.2022.9718963
34. Zubair, M., Aurangzeb, K., Ali, J., Alhussein, M., Hassan, S., Umair, M.: An improved machine learning-driven framework for cryptocurrencies price prediction with sentimental cautioning. IEEE Access **12** (2024). https://doi.org/10.1109/ACCESS.2024.3367129

Application of Neural Networks to Ultrasonic Data for Discrimination of Fat Types in Muscle Tissue Models

Jegors Lukjanovs[1], Aleksandrs Sisojevs[1,2], Alexey Tatarinov[1(✉)], and Tamara Laimiņa[1]

[1] Institute of Electronics and Computer Science, 14 Dzerbenes Street, Riga 1006, Latvia
alexiv@inbox.lv
[2] Riga Technical University, 6A Kipsalas Street, Riga 1048, Latvia

Abstract. Differential assessment of subcutaneous adipose tissue (SAT) and intermuscular adipose tissue (IMAT), two forms of muscle fat, is necessary for studying manifestations of ageing, muscle atrophy, sarcopenia, obesity and metabolic diseases such as diabetes. The discrimination of SAT and IMAT by ultrasonic measurements is difficult due to their complex influence. In the present study, machine-learning algorithms applied to key parameters extracted from ultrasound propagation signals obtained in simplified tissue models (phantoms) were investigated. The acoustical phantoms of muscle tissue were made of gelatin with oil simulating fat layers (SAT) and inner inclusions (IMAT). SAT and IMAT contents varied from zero to 50% with a step 12.5%. A specialised recurrent neural network (RNN) architecture, the long short-term memory (LSTM) method is used in this paper and was used as the main method in the experiments. The result of SAT and IMAT evaluation of objects with an error of no more than 3% in 95% of cases.

Keyword: Recurrent newral networks · Ultrasound signal processing · Subcutaneous and intramuscular fat

1 Introduction

Body composition assessments vary in precision and in the target tissue of interest, including anthropometric measurements, bioelectrical impedance, dual-energy X-ray absorptiometry and body density [4]. Computed tomography (CT) and magnetic resonance imaging (MRI) are gaining in popularity for body composition assessment, especially effective in SAT and IMAT segmentation. MRI is the dominant 3D imaging modality to quantify muscle properties in skeletal muscle disorders, in inherited and acquired muscle diseases, sarcopenia, and frailty [5]. The advantage of CT is the possibility to quantify the muscle density, where muscle radiation attenuation linearly depends on the muscle fat content [6]. However, stationary and costly equipment, its inaccessibility, as well as the requirement for special qualifications of personnel make mass use of MRI and CT difficult. Ultrasonography or clinical ultrasound, exploring the

© The Author(s), under exclusive license to Springer Nature Switzerland AG 2025
A. Hadjali et al. (Eds.): DeLTA 2025, CCIS 2627, pp. 204–216, 2025.
https://doi.org/10.1007/978-3-032-04339-9_13

associations between skeletal muscle echogenicity and its physical properties presents an opportunity to be utilized as a screening tool [11]. It suggests quantification of SAT thickness; however, the prediction of IMAT ultrasonically is still problematic.

Known dependencies of the acoustic parameters of the tissue (velocity and frequency function of ultrasound attenuation) on the fat content [10] can suggest its diagnostic use. However, the possibility of separating SAT and IMAT based on ultrasonic measurements has not been realized yet.

Recent advancements in quantitative ultrasound (QUS) and machine learning (ML) have enabled the extraction and interpretation of diagnostically relevant features from ultrasound signals. For example, QUS have been demonstrated to assess different bone conditions through the analysis of broadband ultrasonic signals [14]. Later, machine-learning approaches, including deep learning architectures such as bidirectional long short-term memory (BLSTM) networks, have demonstrated promising results in distinguishing between cortical thickness and intracortical porosity [3]. Additionally, pattern recognition applied to ultrasonic signals have gained traction for differential evaluation of degradation factors in structural materials [15]. Discrete Fourier transform (DFT) in the analysis of layered materials demonstrating the potential of computational methods in enhancing ultrasonic assessments [12, 13].

Since there have been no attempts to separate SAT and IMAT based on the QUS technique using directly propagated ultrasound signals, the aim of the work was to test this possibility applying recurrent neural networks and a dataset obtained on simplified tissue models with varied SAT and IMAT contents. It was necessary to determine experimentally how accurately the volumetric content of SAT and IMAT can be assessed in such models.

2 Proposed Approach

2.1 Muscle Phantoms and Ultrasonic Data Acquisition

The research involved constructing realistic muscle tissue phantoms conforming to real muscle tissue and fat by their acoustic properties. The phantoms simulated muscle tissue into which different concentrations of SAT and IMAT were introduced.

The muscle mimicking phantoms were made of gelatin (muscle) with oil (fat) layers (SAT) and inclusions (IMAT) with gradually varying SAT and IMAT contents from zero to 50% with a step 12.5%. The set included a net of 25 phantoms with five grades of SAT and IMAT that varied independently presenting 25 possible combinations of SAT and IMAT. The phantoms presented rectangular boxes with a fixed acoustical distance 100 mm with acoustically transparent windows for transducers (Fig. 1). Parallel SAT layers on the both sides of the phantoms were orthogonal to the acoustic path, while IMAT inclusions were 5 mm holes in a checkerboard pattern in the volume of gelatin.

Ultrasound measurement setup developed at the Institute of Electronics and Computer Science compiled an integrated ultrasonic acquisition board, piezoelectric transducers, and a computer. The setup provided excitation of ultrasonic waveform signals at projected frequencies, amplification and recording of the propagated signals through the phantom, synchronization, and data transmission to the computer for analysis.

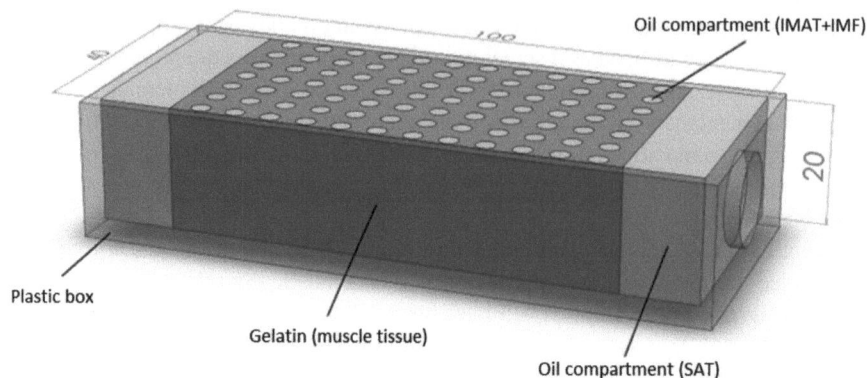

Fig. 1. Acoustic phantom of muscle with SAT and IMAT inclusions.

Trains of ultrasonic signals were acquired by through transmission using a pair of emitting and receiving transducers excited by tone-bursts. The signals were digitized at sampling rate 30 MHz with a 10-bit ADC. Each signal train consisted of three signals at frequencies configurations 1 MHz, 2 MHz and sweep between 0.5 and 2.5 MHz (Fig. 2). Each phantom was measured 15 times that resulted in 375 individual signal sets in 25 objects (phantoms).

Key parameters extracted from the signal trains included:

1) Ultrasound velocity at 2 MHz;
2) Ultrasound attenuation at 1 MHz;
3) Ultrasound attenuation at 2 MHz;
4) Attenuation ratio at 1 and 2 MHz, determined as α1 MHz/ α2 MHz;
5) Integral of sweep signal;
6) Integral ratio between direct sweep signal and its third reflection.

2.2 Machine Learning Methods

The Long Short-Term Memory (LSTM) method was used as the main method in the experiments. LSTM is a specialized Recurrent Neural Network (RNN) architecture designed for processing and analyzing data with long-term temporal dependencies. It addresses classical RNN issues, such as vanishing and exploding gradients, ensuring reliable learning from sequential data. This is achieved through a unique structure that includes memory cells and three main types of gates: the forget gate, input gate, and output gate (Fig. 3).

The structure of LSTM memory unit is illustrated in Fig. 4, where:

X_t – current input data fed into LSTM cell;
h_t – hidden state;
C_t – cell state;
h_{t-1} – hidden state from the previous time step, containing information about past data;
C_{t-1} – cell state at the previous time step, storing long-term information;
i_t – input gate;

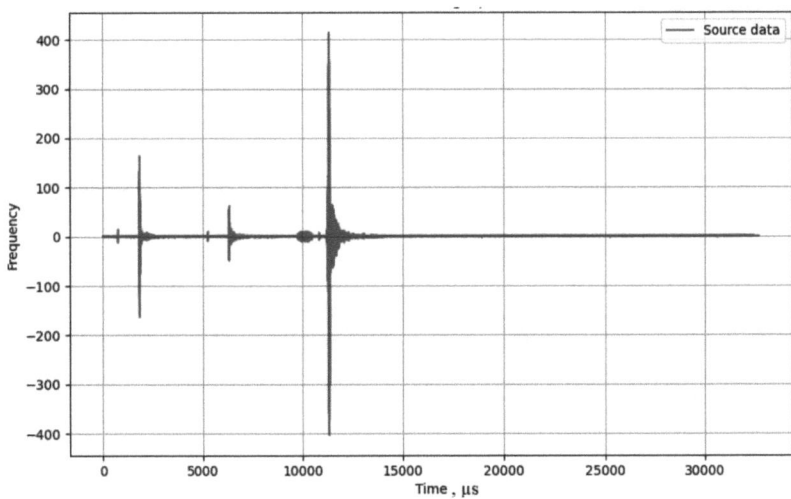

Fig. 2. The signal obtained using the ultrasound measurement device.

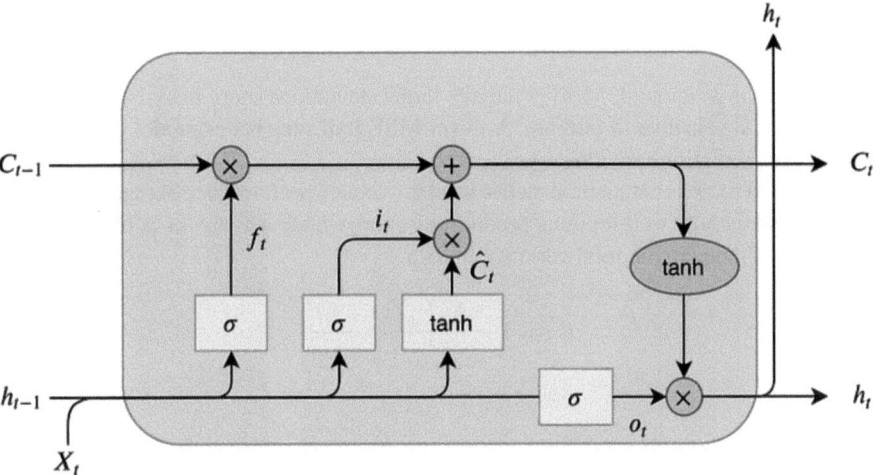

Fig. 3. Structure of LSTM memory unit [7].

f_t – forget gate;
o_t – output gate.

The model utilized a hyperbolic tangent function (tanh). The hyperbolic tangent function is a commonly used activation function in LSTM networks. This function maps input values in range $[-1, 1]$, making it useful for regulating flow of information within the network.

$$\tanh(x) = \frac{e^x - e^{-x}}{e^x + e^{-x}}$$

where:

x – input value;
e – Euler's number.

Tanh is zero-centered, allowing both positive and negative activation. This property helps in preserving balance of information during training, reducing bias accumulation.

Validation loss (val_loss) key metric were used to evaluate the performance of LSTM model during training. This metric represents the error calculated on validation dataset, which consists of data that model has not seen during training. Validation loss is computed using the same loss function as the training loss (Mean Squared Error).

Mean Squared Error (MSE) measures average squared difference between the predicted values and actual target values.

$$MSE = \frac{1}{n} \sum_{i=1}^{N} (y_i - \hat{y}_i)^2$$

where:

N – total number of samples;
y_i – actual value of i-th sample;
$\hat{y}_i$ – predicted value for i-th sample.

Since error is squared, MSE penalizes larger deviations more heavily than smaller ones making it sensitive to outliers. A lower MSE indicates better model performance, as it means predictions are closer to the true value.

R squared score is a statistical metric used to evaluate performance of regression models. It measures how well model's predictions approximate actual values by comparing explained variance to the total variance in data.

$$R^2 = 1 - \frac{\sum_{i=1}^{N}(y_i - \hat{y}_i)^2}{\sum_{i=1}^{N}(y_i - \ddot{y})^2}$$

where:

y_i – actual value of i-th sample;
$\hat{y}_i$ – predicted value of i-th sample;
$\ddot{y}$ – mean of all actual values;
N – total number of samples.

R squared score ranges from $-\infty$ to 1. High R squared score indicates that model better explains variance in target variable. However, high R squared score does not mean accurate predictions.

3 Experiments

Data were divided using three approaches. In the first approach, a global proportion division was applied, splitting data into 80% for training and 20% for testing. While this method allowed for a straightforward split, it resulted in uneven data distribution across objects, causing prediction variance. In the second approach, object-based division was used, where 12 signals from each object were designated for training and 3 for testing. This ensured improved consistency and accuracy, as predictions for the test set were averaged and represented as single points on result graphs. The third approach involved completely removing one object from the dataset and using it exclusively for testing. Training was conducted on the remaining 24 objects, and the model's ability to generalize to unseen data was assessed using the excluded object.

The RNN model employed a Long Short-Term Memory (LSTM) architecture to process sequential data effectively. The network consisted of an input layer that accepted normalized features, including speed of sound and frequency attenuation ratios, followed by three LSTM layers. The first and third layers utilized 128 neurons each, while the second layer had 64 neurons. All LSTM layers employed the tanh activation function and a 20% dropout rate to balance overfitting prevention and information retention. The output layer generated two continuous values (IMAT and SAT percentages). Model output $\hat{y}$ is obtained by applying a dense layer to the hidden state h_t of the final LSTM layer:

$$\hat{y} = W * h_t + b$$

where:

W – learned weight;
b – learned bias;
h_t – hidden state at last time step.

The architecture of the LSTM network, including number of layers and number of neurons in each layer, was determined based on the results of comparative experiments which shown in Tables 1 and 2. Final configuration was selected by identifying setup that achieved the best performance in terms of MSE, as well as the highest proportion of accurate predictions for IMAT and SAT.

The tanh activation function, short for "hyperbolic tangent," maps input values to the range of $[-1, 1]$, capturing both positive and negative relationships effectively. This property enables the network to model complex patterns in sequential data. A 20% dropout rate was chosen based on experimentation, striking a balance between regularization and computational efficiency. The number of neurons was optimized to capture data complexity without excessive computational overhead. A batch size of 32 was selected for its optimal balance between memory efficiency, gradient stability, and training speed. Smaller batch sizes might result in noisy gradients, while larger ones require more memory but provide stable updates. The Adam optimizer was used for training, combining the benefits of adaptive gradient (AdaGrad) and root mean square propagation (RMSProp). Adam dynamically adjusts learning rates based on the first and second moments of gradients, ensuring faster convergence and robustness.

Input parameters, normalized to ensure uniform scaling, were passed through the input layer. The first LSTM layer identified patterns and dependencies within the sequential data, while the second and third layers refined these features for better representation. The output layer then computed the predicted IMAT and SAT percentages, which were evaluated against the ground truth values using the MSE loss function.

Model performance was assessed using a loss function graph to track training progress and prediction result graphs. The first result graph displayed a grid with true IMAT and SAT values marked in red and predicted values in blue, while the second graph visualized error distribution as the absolute difference between real and predicted values.

Table 1. Impact of number of LSTM layers on model performance. Data split 80/20%.

Layers	R^2 score	MSE	IMAT < 3%	SAT < 3%
1	0.94527	0.00649	88%	84%
2	0.95925	0.00343	89.33%	89.33%
3	0.96034	0.00234	92%	92%
4	0.95934	0.00299	90.61%	90.61%
5	0.95818	0.00294	90.61%	89.33%

Table 2. Impact of neuron count per layer on model performance. Data split 80/20%.

Neurons 1 layer	Neurons 2 layer	Neurons 3 layer	R^2 score	MSE	IMAT < 3%	SAT < 3%
32	32	32	0.96112	0.00322	90.67%	89.33%
32	64	32	0.95862	0.00368	88%	90.67%
64	64	64	0.95976	0.00219	90.67%	89.33%
64	64	128	0.95847	0.00453	90.67%	90.67%
128	64	32	0.96125	0.00220	92%	90.67%
64	128	64	0.96021	0.00230	90.67%	92%
128	128	64	0.96102	0.00186	90.67%	90.67%
128	64	64	0.95812	0.00478	89.33%	90.67%
64	128	128	0.95619	0.00324	86.67%	88%
128	128	128	0.96142	0.00201	92%	90.67%
256	64	64	0.95643	0.00255	89.33%	86.67%
128	256	128	0.95781	0.00223	90.67%	90.67%
256	128	64	0.96023	0.00198	90.67%	90.67%
256	128	128	0.96059	0.00393	90.67%	92%
64	256	64	0.95975	0.00299	90.67%	89.33%
128	64	128	0.95924	0.00176	92%	90.67%
64	128	256	0.95950	0.00228	92%	89.33%

The final architecture of LSTM-based neural network is visually summarized in Fig. 4. It consists of an input layer that receives a feature vector of six normalized acoustic parameters, followed by 3 LSTM layers with 128, 64 and 128 units. The first 2 LSTM layers include dropout regularization set to 20%. The final dense output layer produces two continuous values corresponding to the predicted IMAT and SAT percentages.

LSTM Neural Network Architecture

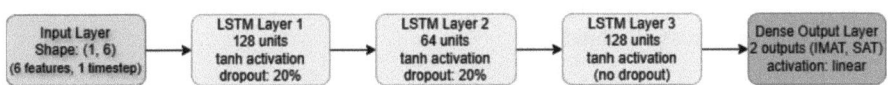

Fig. 4. Architecture of LSTM-based neural network.

3.1 Data Split: 80% Training and 20% Testing

The loss function graph (Fig. 5) demonstrates that the model trained effectively without indications of overfitting or underfitting. The number of training epochs was dynamically determined using the early stopping mechanism, which halts training if no improvement is observed for a specified number of epochs (patience). In this experiment, the patience value was set to 150 epochs, ensuring that the best-performing model was preserved.

Figure 7 illustrates the actual (red points) and predicted (blue points) values of IMAT and SAT. Each blue point represents the neural network's prediction. Due to the varying number of signals per object, the amount of training data for some objects is greater than for others. However, assuming a 3% error margin as acceptable, 90.67% of IMAT predictions and 89.33% of SAT predictions remain within this threshold.

Performance Metrics:

- R Score: 0.95700
- Mean Absolute Error (MAE): 0.03225
- Mean Squared Error (MSE): 0.00263

The high R score indicates that the model explains a substantial proportion of the variance in the data. The low MAE and MSE values further confirm that the prediction errors are minimal and consistent, validating the reliability of the model under this data split strategy.

3.2 Data Split: 12 Signals for Training and 3 Signals for Testing Per Object

Similar to the first experiment, the loss function graph (Fig. 6) reveals no signs of overfitting or underfitting. The training process employed early stopping with a patience of 100 epochs, ensuring optimal model selection based on the best validation performance.

Figure 8 compares actual values (red points) and averaged predictions (blue points) for IMAT and SAT. For each test object, the network predicts three values corresponding to the three test signals, which are then averaged and plotted. This distribution of signals ensures that every object contributes equally to the training and testing processes.

Assuming a 3% error margin as acceptable, 100% of IMAT and SAT predictions are within this threshold.

Performance Metrics:

- R Score: 0.99901
- Mean Absolute Error (MAE): 0.01176
- Mean Squared Error (MSE): 0.00024

This configuration achieves near-perfect performance with a very high R score and exceptionally low MAE and MSE values, demonstrating excellent model generalization and consistency across all objects.

3.3 Data Split: Leave-One-Out Cross-Validation

Figure 9 shows the results of 25 experiments, where each red point represents the test object excluded from training, and blue points represent the predictions. In this setup, 96% of the data is used for training, while 4% is reserved for testing. Importantly, the neural network does not have prior knowledge of the parameters for the excluded object.

This approach demonstrates the neural network's ability to generalize effectively to entirely unseen objects. The larger training dataset enhances the model's learning capacity, resulting in highly accurate predictions for the excluded object.

Figures 7, 8 and 9 show the results of evaluation under different configurations of the recognition system. Red colour indicates a-priori known values of objects, blue - results of evaluation of the developed approach. As can be seen in Figs. 8 and 9. As a result of training, the system estimates SAT and IMAT of objects with an error of no more than 3% in 95% of cases.

Simplified physical models (phantom) with regular distribution of SAT and IMAT demonstrated the possibility of using neural networks for their differential evaluation based on the parameters of through transmission ultrasonic signals. Potentially, this opens up the prospect of studying SAT and IMAT in a living organism using relatively simple and easily implemented measurements. At the same time, the results obtained on these models have a number of conventions due to the difficulty of taking into account the entire complexity of the structural organization of real tissue in the models. Uneven volumetric distribution of IMAT and strong variability of IMAT inclusions by size may require the creation of a more accurate model for predicting SAT and IMAT in a living organism. This will require the creation of a database on humans based on reference MRI and CT data.

Application of Neural Networks to Ultrasonic Data for Discrimination 213

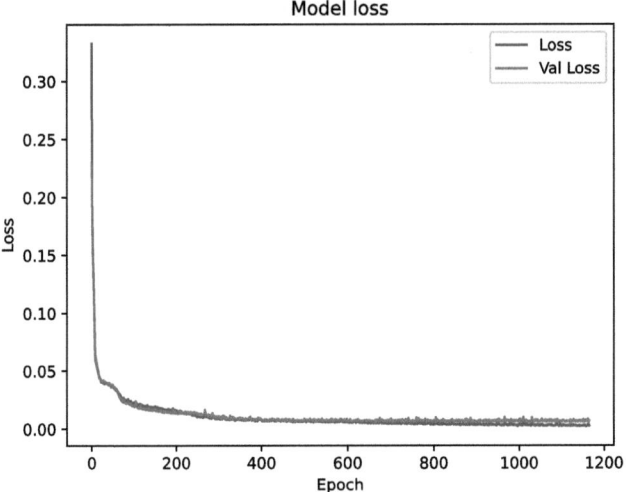

Fig. 5. 80% Training/20% Testing Loss function graph.

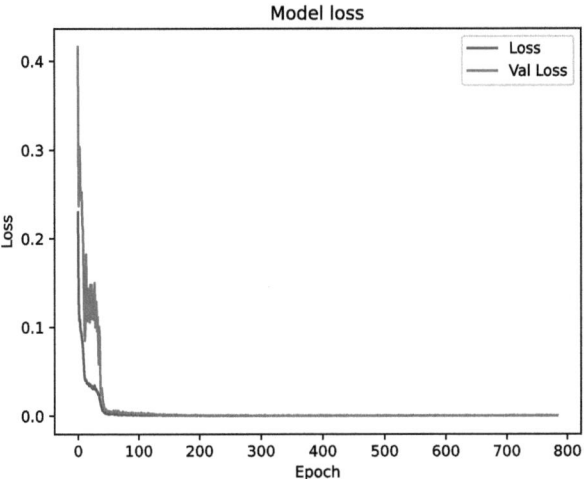

Fig. 6. 12 signals for Training/3 signals for Testing per Object Loss function graph.

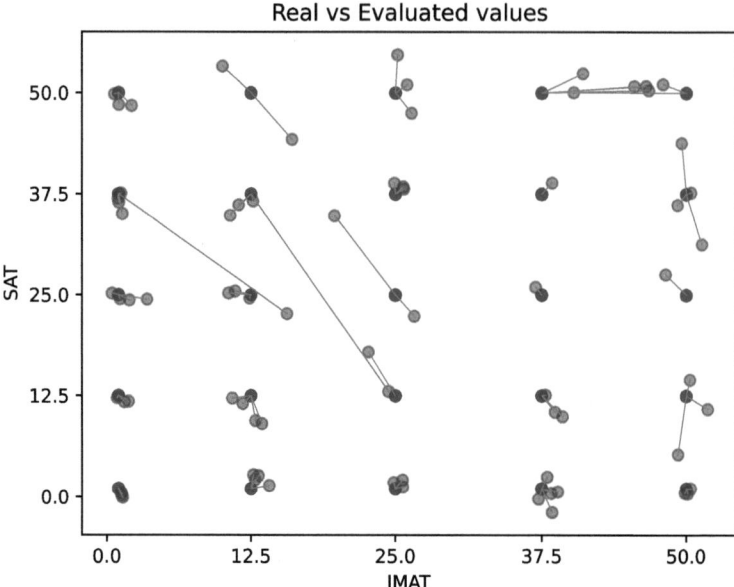

Fig. 7. 80% Training/20% Testing Prediction grid. (Color figure online)

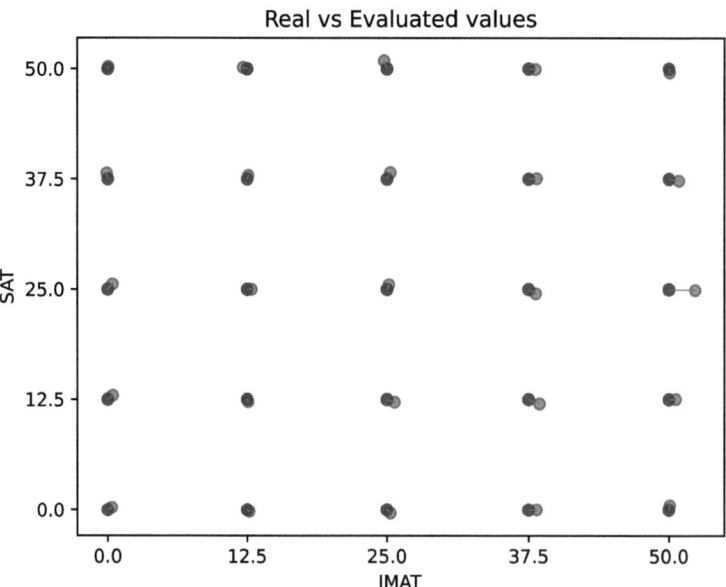

Fig. 8. 12 signals for Training/3 signals for Testing per Object Prediction grid. (Color figure online)

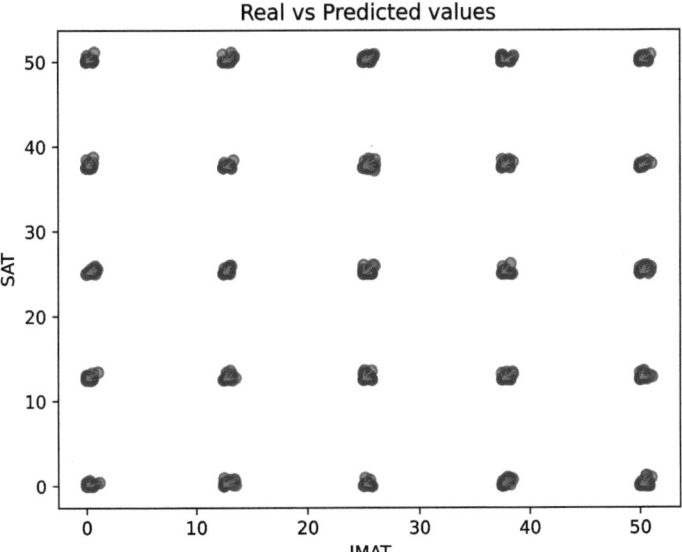

Fig. 9. Leave-One-Out Prediction grid. (Color figure online)

4 Conclusions

1. One of the significant challenges in the field of medical diagnostics lies in the difficulty—or, in some cases, the practical impossibility—of accurately distinguishing and quantifying subcutaneous adipose tissue (SAT) and intermuscular adipose tissue (IMAT) separately using conventional ultrasound signal data. Traditional ultrasound techniques often lack the resolution and analytical capacity to make this distinction clear. However, this study explores a novel approach that leverages the analytical power of neural networks to address this limitation. The results indicate that it is fundamentally feasible to achieve a reliable separation of SAT and IMAT values through the application of neural network-based processing of ultrasound data.
2. The experimental outcomes provide strong support for the proposed method. Specifically, the system was able to estimate the levels of subcutaneous and intermuscular fat with a high degree of accuracy. In 95% of all test cases, the estimation error did not exceed 3%, demonstrating the robustness and reliability of the neural network model in differentiating and quantifying these two types of adipose tissue under experimental conditions.
3. It is important to note that the current findings are based on data derived from simplified experimental models, which may not fully represent the complexity of human anatomy and tissue composition. Therefore, further validation is essential. Future research should involve studies with human subjects in order to confirm the effectiveness and accuracy of the proposed approach in real-world clinical settings. In such studies, the ultrasound data should be cross-referenced with reference values for SAT and IMAT obtained through more advanced imaging modalities such as magnetic resonance imaging (MRI) and computed tomography (CT), which are considered the gold standards for tissue characterization.

Acknowledgment. The study was executed under the project of the Latvian Council of Science LZP FLPP no. Lzp-2021/1-0290 "Comprehensive assessment of the condition of bone and muscle tissue using quantitative ultrasound (BoMUS)".

References

1. Addison, O., Marcus, R.L., Lastayo, P.C., Ryan, A.S.: Intermuscular fat: a review of the consequences and causes. Int. J. Endocrinol. **2014**, 309570 (2014)
2. Buch, A., et al.: Muscle function and fat content in relation to sarcopenia, obesity and frailty of old age–an overview. Exp. Gerontol. **76**, 25–32 (2016)
3. Chuchalina, M.; Sisojevs, A., Tatarinov, A.: Determination of factors of interest in bone models based on ultrasonic data. In: Proceedings of the 13th International Conference on Pattern Recognition Applications and Methods – ICPRAM, vol. 1, pp. 281–287. SciTePress (2024)
4. Duren, D.L., et al.: Body composition methods: comparisons and interpretation. J. Diabetes Sci. Technol. **2**(6), 1139–1146 (2008)
5. Engelke, K., et al.: Magnetic resonance imaging techniques for the quantitative analysis of skeletal muscle: state of the art. J Orthop. Translat. **42**, 57–72 (2023)
6. Engelke, K., Museyko, O., Wang, L., Laredo, J.D.: Quantitative analysis of skeletal muscle by computed tomography imaging-state of the art. J. Orthop. Translat. **28**(15), 91–103 (2018). https://doi.org/10.1016/j.jot.2018.10.004
7. Ghasemikaram, M., et al.: Effects of 16 months of high intensity resistance training on thigh muscle fat infiltration in elderly men with osteosarcopenia. Geroscience (2021)
8. Goodpaster, B.H., Bergman, B.C., Brennan, A.M., Sparks, L.M.: Intermuscular adipose tissue in metabolic disease. Nat. Rev. Endocrinol. **19**(5), 285–298 (2023)
9. Hochreiter, S., Schmidhuber, J.: Long short-term memory. J. Neural Comput. **9**(8), 1735–1780 (1997)
10. Koch, T., Lakshmanan, S., Brand, S., Wicke, M., Raum, K., Mörlein, D.: Ultrasound velocity and attenuation of porcine soft tissues with respect to structure and composition: I. Muscle. Meat Sci. **8**(1), 51–58 (2011). https://doi.org/10.1016/j.meatsci.2010.12.002
11. Oranchuk, D.J., Bodkin, S.G., Boncella, K.L., Harris-Love, M.O.: Exploring the associations between skeletal muscle echogenicity and physical function in aging adults: a systematic review with meta-analyses. J. Sport Health Sci. **13**(6), 820–840 (2024)
12. Sisojevs, A., Tatarinov, A., Kovalovs, M., Krutikova, O. and Chaplinska, A.: An approach for parameters evaluation in layered structural materials based on DFT analysis of ultrasonic signals. In: Proceedings of the 11th International Conference on Pattern Recognition Applications and Methods ICPRAM 2022, pp. 307–314. SciTePress (2022)
13. Sisojevs, A., Tatarinov, A., Chaplinska, A.: Evaluation of factors-of-interest in bone mimicking models based on DFT analysis of ultrasonic signals. In: Proceedings of the 12th International Conference on Pattern Recognition Applications and Methods ICPRAM 2023, pp. 914–919. SciTePress (2023)
14. Tatarinov, A., Egorov, V., Sarvazyan, A., Sarvazyan, N.: Multi-frequency axial transmission bone ultrasonometer. Ultrasonics **54**(5), 1162–1169 (2014)
15. Tatarinovs, A., Sisojevs, A., Chaplinska, A., Shahmenko, G., Kurtenoks, V.: An approach for assessment of concrete deterioration by surface waves. Procedia Struct. Integrity **37**, 453–461 (2022)

SwiNight: Class Imbalanced Night-Time Accident Detection with Swin Transformer

Shrusti Porwal[1(✉)], Preety Singh[1], Anukriti Bansal[1,2], Saumilya Gupta[1], Kartikay Goel[1], and Palakurthy Guneeth[1]

[1] The LNM Institute of Information Technology, Jaipur 302031, Rajasthan, India
{shrusti-porwal.y20,preety,23uec618,22ucs103,
22ucs144}@lnmiit.ac.in
[2] LUMIQ, Crispanalytics Pvt. Ltd., Noida 201301, Uttar Pradesh, India

Abstract. Night-time accident detection is a challenging task due to the scarcity of anomalous frames in the dataset. In this paper, we present a dataset of night-time accidents. We propose a Swin Transformer-based model for detecting accidents, specifically addressing the issue of dataset imbalance using relevant loss functions. The performance of the model is evaluated using different loss functions. Our experiments demonstrate that the *focal loss* function outperforms the others, achieving an F1-Score of 0.710 and an accuracy of 79.77%. Experiments also reveal that the Swin Transformer delivers superior performance compared to a Vision Transformer.

Keywords: Swin Transformer · Traffic Anomaly · Vision Transformer · Attention · Focal Loss

1 Introduction

Accident detection in surveillance systems plays a vital role in enhancing road safety. However, detecting accidents during night-time remains a challenging and under-explored task due to the unique difficulties posed by low visibility, poor illumination, and noisy visual conditions [4]. The ability to promptly detect accidents at night is essential for reducing response times and enabling real-time monitoring. Traditional techniques using convolutional neural networks (CNNs) face challenges in capturing global dependencies due to their fixed receptive fields and hierarchical design.

Vision Transformers (ViTs) [6] employ dense attention mechanisms across image patches, offering better global context modeling. However, ViTs demand substantial computational resources and large-scale training datasets. Moreover, in night-time accident detection, the anomalous or accident class is often a rare event. This results in a limited number of labeled samples for the accident class. The scarcity of data for the minority class can lead to poor generalization, as the model struggles to learn from a limited representation of anomalous frames compared to the abundance of normal driving scenarios.

This paper presents **SwiNight**, a novel approach for night-time accident detection using the Swin Transformer architecture [12], which overcomes the limitations of conventional Convolutional Neural Networks (CNNs) and Vision Transformers (ViTs) [6].

Its hierarchical representation mechanism captures both local and global features, crucial for detecting localized anomalies such as vehicle collisions under low-light conditions, as well as understanding broader contextual cues like traffic dynamics. We also tune it to handle the class imbalance problem by exploring various loss functions for the classifier. Through comprehensive evaluations, we demonstrate its effectiveness in detecting anomalies in low-light and night-time environments, while also addressing the challenges posed by an imbalanced dataset. Addressing data imbalance through advanced techniques enhances model accuracy, enabling real-time detection and faster emergency response.

Our experiments using surveillance footage highlight the applicability of our approach in improving road safety and emergency response systems. The results emphasize the potential of attention-based deep learning architectures in addressing the critical task of night-time accident detection in video surveillance systems. Our approach aims to enhance the robustness and reliability of nighttime anomaly classification in low-illumination environments. The main contributions of our work are:

- Presenting the **Ni**ght **A**ccident **D**etection (*NiAD-Large*) dataset, containing night-time traffic accidents.
- Utilizing attention-based Swin Transformer model for accident classification at night.
- Handling class-imbalance in the dataset by employing appropriate loss functions.
- Comparing the performance of the Swin Transformer with Vision Transformer models.

The paper is organized as follows: Sect. 2 presents few related research works, Sect. 3 details the proposed methodology, Sect. 4 presents the experiments and results, and Sect. 5 concludes the study.

2 Literature Survey

In recent years, a significant amount of research has focused on advancing traffic anomaly detection. A variety of approaches have been explored to address traffic accident detection, each with its own strengths and challenges. Huang et al. [9] proposed a real-time, two-stream convolutional neural network (CNN) architecture designed for simultaneous detection, tracking, and near-accident detection using traffic video data. Their approach integrates a spatial stream for object detection and a temporal stream to exploit motion features for multi-object tracking. Paul et al. [13] combined deep convolutional neural networks (D-CNN) with centroid-based vehicle tracking for accident detection. This method addresses challenges such as overlapping bounding boxes and vehicle speed.

Pawar and Attar [14] proposed a deep learning-based approach for automatic accident detection and localization, leveraging one-class classification. Their model uses spatio-temporal autoencoders to model spatial and temporal representations from traffic video. Aboah et al. [1] proposed a deep learning-based traffic anomaly detection approach using YOLO for object detection, combined with decision-tree-based methods for refining traffic anomaly classification and estimating event duration. Zhong et al.

[18] introduced RFG-HELAD, a hybrid model leveraging contrastive learning, GANs, and Deep kNN to enhance fine-grained traffic anomaly detection. Yu et al. [17] introduced a fine-grained traffic accident analysis framework, which includes accident classification, spatio-temporal localization, and severity estimation. Their approach utilizes a transformer-based architecture that combines RGB images with optical flow information to provide a comprehensive analysis of accidents.

3 Proposed Methodology

This section presents the curated dataset and the proposed methodology for handling class imbalance in night-time traffic anomaly classification. The proposed framework is illustrated in Fig. 1 and discussed in subsections below.

The methodology begins with the annotation of the nighttime accident dataset to ensure precise labeling. The Swin Transformer processes images by dividing them into patches, extracting features through a hierarchical self-attention mechanism, and classifying them using a global feature vector. The pre-trained Swin Transformer is fine-tuned on the traffic anomaly dataset using hierarchical patching to structure the input data, along with linear embedding and positional encoding to capture spatial and contextual information. Self-attention and multi-head attention mechanisms are employed to model complex relationships within the data. The class imbalance is handled using techniques such as weighted class cost function, extreme marginal loss function, and focal loss. This helps the model focus on the minority class. Finally, a classification layer distinguishes between anomalous and normal instances, providing the final output. These steps contribute to improving the model's performance in classifying anomalous frames in night-time traffic videos.

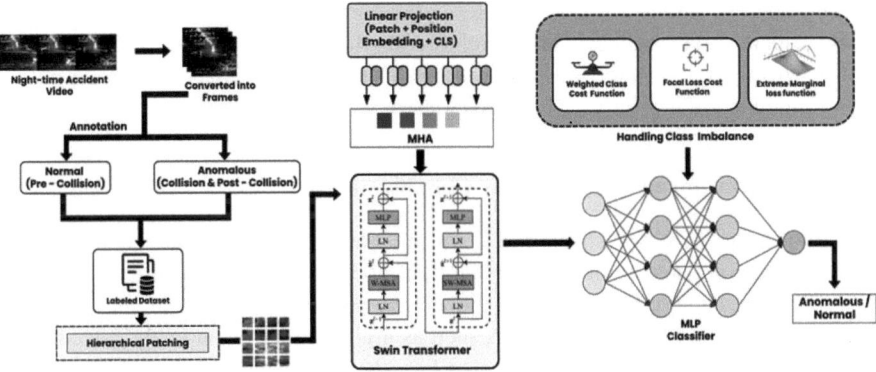

Fig. 1. Proposed Architecture for nighttime anomaly classification.

3.1 Dataset

We curated the *NiAD-Large* (**Ni**ght **A**ccident **D**etection - Large) Dataset, specifically designed to address the lack of publicly available datasets for night-time accident detection. We collected 40 night-time traffic videos from surveillance cameras of a city in India, online sources like YouTube, and night accidents from some publicly available datasets like CADP [15], DAD [5], and CCD [3]. The videos were converted into frames and followed by pre-processing steps that included removing text from frames and removing irrelevant frames, such as those without relevant information, like empty road scenes, transition frames, or frames without visible vehicles.

Manual annotation was carried out to label the dataset with the help of two annotators working under the supervision of a super annotator. This annotation approach ensured a high-quality dataset for training and evaluating anomaly detection models. The annotators were tasked with labeling the frames based on guidelines, masking any overlaid text (such as timestamps or watermarks), and removing irrelevant or uninformative frames, such as those with extreme blur or empty scenes. The super annotator was responsible for preparing the annotation guidelinse, resolving ambiguities, and performing quality checks on the annotated data to ensure consistency and reliability. Each frame was annotated as either *normal* or *anomalous* according to the following criteria:

- Frames showing vehicles without any visible collision were labeled as *normal*.
- Frames before the collision, where no impact or damage was visible, were labeled as *normal*.
- Frames capturing a collision between two or more vehicles or those showing visible post-collision damage were labeled as *anomalous*.
- Frames where vehicles are overturned, severely damaged, or clearly involved in an accident were also labeled as *anomalous*.

The dataset consists of 13227 samples in total, including 8667 normal instances and 4560 accident instances. This distribution reveals the class imbalance, with *anomalous* samples being significantly fewer than *normal* samples. This dataset is accessible at https://github.com/Scholarquest11/NiAD-Large.

3.2 Swin Transformer for Anomaly Classification

We utilized the Swin Transformer [12] for anomaly classification due to its superior performance compared to other deep learning models. Its hierarchical design enables it to efficiently capture both local and global dependencies in an image, making it particularly well-suited for complex tasks such as nighttime traffic anomaly detection. Its shifted window attention mechanism significantly reduces computational costs while maintaining high accuracy, making it well-suited for surveillance footage. Additionally, its robustness to illumination variability enhances its performance in night-time scenarios. We leveraged a pre-trained model, fine-tuned on our dataset to optimize its performance. While the *binary cross-entropy* loss can typically handle classification tasks in balanced datasets, we addressed the issue of class imbalance [10] by exploring the use of customized cost functions.

The Swin Transformer divides the image into smaller non-overlapping patches and processes these patches through self-attention layers. The model then aggregates these patches in a hierarchical manner, progressively merging patches to capture multi-scale features. In our proposed architecture, we leverage a pre-trained Swin Transformer, which is fine-tuned on our specific dataset to optimize its performance for night-time traffic anomaly detection.

Initially, the input image I of size $H \times W$ is divided into non-overlapping patches of size $P \times P$, resulting in $n = \frac{H \times W}{P^2}$ patches. Each patch I_i is then flattened into a vector $\mathbf{x}_i \in \mathbb{R}^d$, where d is the embedding dimension. The flattened patches are then projected to higher-dimensional embeddings $\mathbf{e}_i \in \mathbb{R}^D$, where D is the patch embedding dimension. To preserve spatial relationships between patches, positional encodings $\mathbf{e}_{\text{pos}} \in \mathbb{R}^D$ are added to the patch embeddings, resulting in the encoded embeddings $\mathbf{e}_i = \mathbf{x}_i + \mathbf{e}_{\text{pos}}$. These positional encoded patch embeddings are then processed through a series of self-attention layers.

The self-attention mechanism computes the attention matrix A based on the query Q, key K, and value V matrices, as shown in the equation:

$$A = \text{softmax}\left(\frac{QK^T}{\sqrt{d_k}}\right),$$

where d_k is the dimension of the key vectors. The attention matrix A is then multiplied by the value matrix V, producing the output $\text{Attention}(Q, K, V) = A \cdot V$. This mechanism enables the model to focus on relevant parts of the image when processing each patch, allowing it to capture both local and global dependencies. After several layers of self-attention, the output consists of feature embeddings for each patch.

The next step involves aggregating the feature embeddings from all patches to form a global representation of the image. This aggregation can be achieved using mean pooling, where the global feature vector $\mathbf{z}_{\text{global}}$ is computed as:

$$\mathbf{z}_{\text{global}} = \frac{1}{n} \sum_{i=1}^{n} \mathbf{z}_i.$$

This global feature vector is then passed through the classification head, which consists of one or more fully connected layers to predict the class label $\hat{y}$ (either normal or anomalous). The final prediction is obtained using a softmax function:

$$\hat{y} = \text{softmax}\left(\mathbf{W} \cdot \mathbf{z}_{\text{global}} + \mathbf{b}\right),$$

where $\mathbf{W}$ is the weight matrix and $\mathbf{b}$ is the bias vector in the classification head. To fine-tune the pre-trained Swin Transformer, we adjust the model's weights using backpropagation on our specific dataset. The training process minimizes the loss using backpropagation and gradient descent. The model weights are updated using the following update rule:

$$\mathbf{W}_{t+1} = \mathbf{W}_t - \eta \nabla_{\mathbf{W}} L(\mathbf{W}),$$

where η is the learning rate and $\nabla_{\mathbf{W}} L(\mathbf{W})$ is the gradient of the loss with respect to the model weights at time step t.

3.3 Handling Class Imbalance in Dataset

Since our dataset suffers from class imbalance, with anomalous frames being underrepresented, we address this issue using appropriate loss functions. We experimented with three different loss functions presented below.

1. Weighted Cost Function: A higher weight is assigned to the minority class to balance the influence of each class [8] as follows:

$$L_{\text{weighted}} = -\frac{1}{N} \sum_{i=1}^{N} w_{y_i} \log(p_{y_i})$$

where N is total number of samples, y_i is true label for i^{th} sample, p_{y_i} is the predicted probability of true class, and w_{y_i} is the weight assigned to class y_i.

2. Extreme Marginal Cost Function: This loss function encourages the model to focus on the decision boundary and penalizes misclassifications near the margin, which helps the model better distinguish the minority class (anomalous frames). This loss is formulated as:

$$L_{\text{extreme}} = \frac{1}{N} \sum_{i=1}^{N} \max\left(0, \text{margin} - (p_{y_i} - p_{\text{neg}})\right),$$

where p_{neg} is the predicted probability for the negative class (normal frames in our case), and the *margin* is a predefined threshold that specifies the minimum difference between the predicted probabilities for the correct class and the incorrect class. By penalizing misclassifications near the margin, this loss function reduces the model's tendency to overfit the majority class, improving its ability to correctly classify anomalous frames. It reduces misclassifications near the decision boundary, thereby improving the model's discrimination between classes [16].

3. Focal Cost Function: Here, the class imbalance is mitigated by down-weighting well-classified samples and focusing on hard-to-classify ones [11], defined as:

$$L_{\text{focal}} = -\frac{1}{N} \sum_{i=1}^{N} \alpha (1 - p_{y_i})^{\gamma} \log(p_{y_i})$$

where α is the balancing factor for the class weights and γ controls the focusing parameter to reduce the contribution of easy examples.

4 Experiments

In this section, we present the details of the experiments conducted to evaluate the proposed approach and the results. The experiments were designed to assess the effectiveness of the model in accurately identifying and localizing accidents under various challenging conditions, including low-light environments and occlusions. The evaluation metrics include precision, recall, and F1-score.

Experimental Setup. The experimental setup involves training and evaluating the model on the annotated dataset, which includes *accident* and *normal* frames collected from Indian road surveillance. The dataset was divided into training, validation, and test sets in an 80:10:10 ratio. The Swin Transformer model consisted of 12 transformer blocks, with a patch size of 4 × 4 and a window size of 7 × 7. The number of attention heads across its four stages were 3, 6, 12, and 24, respectively. To ensure robust evaluation. Figure 2 highlights the statistical distribution of the proposed dataset. The hyperparameters included a learning rate of 10^{-4}, a batch size of 16, and the AdamW optimizer.

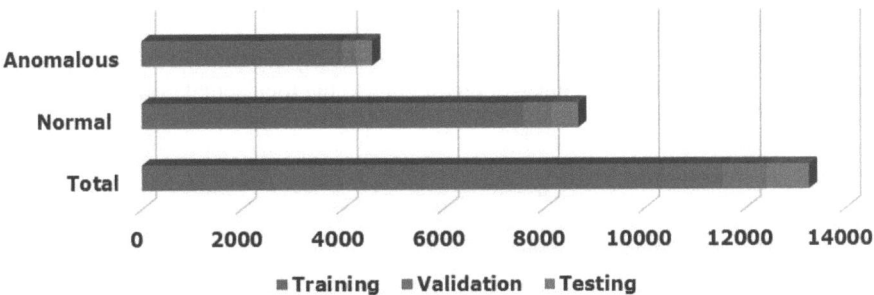

Fig. 2. Statistics of the *NiAD-Large* dataset showing the training, validation, and testing splits.

For the *weighted cost*, we calculated class weights to address the class imbalance. The weight for the anomalous class was set to 0.6561, while the weight for the normal class was set to 0.3439. These weights were incorporated in the training process. To ensure a balanced contribution from both classes. For the *extreme marginal* loss, the *margin* was set to 1 for separating the positive and negative samples, and α was set to 0.65, which adjusts the impact of harder examples. Regularization terms may be applied to prevent overfitting, while the learning rate controls the convergence speed during training. For the *focal cost*, we set the value of α to 0.65 and trained the model with γ values of 1 and 2. The parameter α controls the class weighting, assigning more importance to the anomalous class with $\alpha = 0.6561$, which helps improve detection performance for imbalanced data. The parameter is γ influences the focusing mechanism of the focal loss, with $\gamma = 2$ yielding better performance compared to $\gamma = 1$, as it strikes a balance between focusing on hard-to-classify samples while avoiding overemphasis on misclassified instances and the default value of γ is 2.

For comparison, we also performed experiments using the ViT model with 12 encoder layers, 12 attention heads, and 16 × 16 patch size. Experiments were performed in a computing environment utilizing GPUs.

4.1 Evaluation Metrics

To evaluate the performance of the proposed model, we employed four key metrics: Precision, Recall, F1 Score, and Accuracy. These metrics collectively offer a compre-

Precision. Precision evaluates the model's ability to accurately predict positive outcomes. It is defined as the ratio of true positive predictions to the total positive predictions made by the model, given as:

$$\text{Precision} = \frac{TP}{TP + FP}$$

where TP represents true positives and FP represents false positives.

Recall. Recall quantifies the model's ability to correctly identify all relevant positive instances. It is the ratio of true positive predictions to the total actual positive instances, given as:

$$\text{Recall} = \frac{TP}{TP + FN}$$

where FN represents false negatives.

F1 Score. The F1 score is the harmonic mean of Precision and Recall, offering a balanced measure when class distribution is imbalanced. It is calculated as:

$$\text{F1 Score} = \frac{2 \times \text{Precision} \times \text{Recall}}{\text{Precision} + \text{Recall}}$$

Accuracy. Accuracy measures the overall correctness of the model by calculating the proportion of correct predictions to the total number of predictions. It is computed as follows:

$$\text{Accuracy} = \frac{TP + TN}{TP + TN + FP + FN}$$

where TN represents true negatives.

4.2 Results and Analysis

The results from the performance tables and the confusion matrix demonstrate the impact of different loss functions and model architectures on handling class imbalance and improving classification performance. Table 1 compares the performance. The binary cross-entropy loss depicts dismal performance. By introducing *weighted cost* and *extreme marginal cost* functions, the model's performance improves moderately, with the F1-score increasing to 0.314 and 0.341, respectively. However, the *focal cost* function significantly enhances all metrics with a precision of 0.666, recall of 0.759, F1-score of 0.710. It yields an accuracy of 79.77%. These results highlight the *focal cost's* effectiveness in mitigating the impact of class imbalance.

The results also demonstrate the comparative advantage of the Swin Transformer over the Vision Transformer (ViT) using the *focal cost* function. While the ViT achieves an F1-Score of 0.656 and an accuracy of 62.22%, the Swin Transformer, surpasses these results. This superiority underscores the Swin Transformer's ability to effectively capture hierarchical features and contextual information.

Table 1. Performance comparison of different loss functions for handling class imbalance with Swin and Vision Transformer.

Model + Cost Function	Precision	Recall	F1-Score
Swin + Binary Cross Entropy	0.033	0.022	0.027
Swin + Weighted Class	0.304	0.324	0.314
Swin + Extreme Marginal	0.325	0.359	0.341
Swin + Focal Loss $\gamma = 1$	0.621	0.682	0.649
Swin + Focal Loss $\gamma = 2$	**0.666**	**0.759**	**0.710**
ViT + Focal Loss $\gamma = 2$	0.577	0.754	0.656

The confusion matrix in Fig. 3 provides the classification performance of the Swin transformer. For the *binary cross-entropy* loss, the model correctly classifies only 7 samples as *anomalous* and 330 samples as *normal*, indicating significant misclassification of the minority class (*anomalous*). In contrast, when *focal loss* is applied, the performance shows substantial improvement, with 212 *anomalous* samples and 470 *normal* samples correctly identified. This reflects the effectiveness of focal loss in prioritizing harder-to-classify samples, particularly those from the minority class, thereby achieving better balance in classification performance.

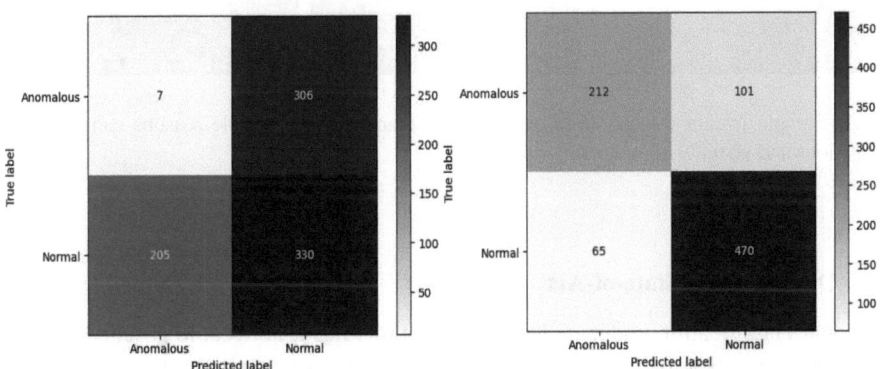

Fig. 3. Confusion matrices of Swin transformer. (b) Using binary cross-entropy loss. (a) Using focal loss to address the class imbalance.

Figure 4 showcases the correctly classified accident frames under challenging conditions. These frames demonstrate the system's robustness in identifying accidents and key accident features such as vehicle damage, debris, and abnormal alignments, even in the presence of glare from headlights, uneven lighting, and partial occlusions.

Fig. 4. Few sample frames from the test set showcasing correct classifications by Swin transformer for night-time accidents.

Figure 5 shows the misclassified instances exposing limitations in the model's performance, often caused due to confusing reflections in the background and distant objects that lack discernible features. High-degree occlusions and visually complex environments also contribute to misclassifications. These misclassifications suggest the need for a more diverse dataset and advanced attention mechanisms to focus on accident-relevant regions, aiming to reduce errors and improve detection accuracy in night-time scenarios.

Fig. 5. Sample frames misclassified by the Swin Transformer. Possible reasons can be major occlusions and visually complex environments.

4.3 Discussion on State-of-Art

Accident classification under night-time conditions has a noticeable absence of prior research, to the best of our knowledge. However, for completeness, we mention few state-of-the-art methods designed for day-time accident classification. Adewopo and Elsayed [2] introduced the I3D-CONVLSTM2D model for accident detection, which achieved an F1-score of 0.75. Hajri et al. [7] reported that the ViT achieved an accuracy of 68.32% on the DAD and CCD datasets for anomaly detection. Our proposed method, Swin Transformer with Focal Loss ($\gamma = 2$) achieves a precision of 0.666, recall of 0.759, F1-score of 0.710, and an accuracy of 79.77%.

5 Conclusion

In this research, we focused on the task of night-time accident detection while addressing the challenge of class imbalance. We curated a night-time accident dataset for this task and employed the attention-based Swin transformer for accident detection. We applied various loss functions to enhance the model's ability to classify minority class samples. We demonstrated that the *focal loss* demonstrated superior performance, achieving the highest F1-Score of 0.710, proving its effectiveness in handling imbalanced datasets by prioritizing hard-to-classify samples. Experimental results also revealed that Swin Transformer outperforms Vision Transformer. This study demonstrates that the combination of focal loss and Swin Transformer provides a robust and efficient solution for night-time accident detection. Future work will explore advanced model architectures and larger datasets tailored to further improve detection accuracy in challenging night-time conditions.

References

1. Aboah, A.: A vision-based system for traffic anomaly detection using deep learning and decision trees. In: Proceedings of the IEEE/CVF Conference on Computer Vision and Pattern Recognition, pp. 4207–4212 (2021)
2. Adewopo, V.A., Elsayed, N.: Smart city transportation: deep learning ensemble approach for traffic accident detection. IEEE Access **12**, 59134–59147 (2024)
3. Bao, W., Yu, Q., Kong, Y.: Uncertainty-based traffic accident anticipation with spatio-temporal relational learning. In: Proceedings of the 28th ACM International Conference on Multimedia, pp. 2682–2690 (2020)
4. Bureau of Police Research & Development (BPR&D): Analysis of Road Accidents in India 2019 (2019). https://bro.gov.in/WriteReadData/linkimages/5768690382-14.pdf
5. Chan, F.-H., Chen, Y.-T., Xiang, Yu., Sun, M.: Anticipating accidents in dashcam videos. In: Lai, S.-H., Lepetit, V., Nishino, K., Sato, Y. (eds.) ACCV 2016. LNCS, vol. 10114, pp. 136–153. Springer, Cham (2017). https://doi.org/10.1007/978-3-319-54190-7_9
6. Dosovitskiy, A., et al.: An image is worth 16x16 words: transformers for image recognition at scale. arXiv preprint arXiv:2010.11929 (2020)
7. Hajri, F., Fradi, H.: Vision transformers for road accident detection from dashboard cameras. In: 2022 18th IEEE International Conference on Advanced Video and Signal Based Surveillance (AVSS), pp. 1–8 (2022). https://doi.org/10.1109/AVSS56176.2022.9959545
8. Ho, Y., Wookey, S.: The real-world-weight cross-entropy loss function: modeling the costs of mislabeling. IEEE Access **8**, 4806–4813 (2020)
9. Huang, X., He, P., Rangarajan, A., Ranka, S.: Intelligent intersection: two-stream convolutional networks for real-time near-accident detection in traffic video. ACM Trans. Spat. Algor. Syst. (TSAS) **6**(2), 1–28 (2020)
10. Japkowicz, N., Stephen, S.: The class imbalance problem: a systematic study. Intell. Data Anal. **6**(5), 429–449 (2002)
11. Lin, T.: Focal loss for dense object detection. IEEE Trans. Pattern Anal. Mach. Intell. **42**(2), 318–327 (2017)
12. Liu, Z., et al.: Swin transformer: hierarchical vision transformer using shifted windows. In: Proceedings of the IEEE/CVF International Conference on Computer Vision, pp. 10012–10022 (2021)

13. Paul, A.R., Grace Mary Kanaga, E.: Enhanced D-CNN architecture and centroid-based algorithm for real-time vehicle tracking and accident detection from surveillance videos. J. Intell. Fuzzy Syst. (Preprint), 1–14 (2024)
14. Pawar, K., Attar, V.: Deep learning based detection and localization of road accidents from traffic surveillance videos. ICT Express **8**(3), 379–387 (2022)
15. Shah, A.P., Lamare, J.B., Nguyen-Anh, T., Hauptmann, A.: CADP: a novel dataset for CCTV traffic camera based accident analysis. In: 2018 15th IEEE International Conference on Advanced Video and Signal Based Surveillance (AVSS), pp. 1–9. IEEE (2018)
16. Wali, R.: Xtreme margin: a tunable loss function for binary classification problems. arXiv Preprint arXiv:2211.00176 (2022)
17. Yu, H., Zhang, X., Wang, Y., Huang, Q., Yin, B.: Fine-grained accident detection: database and algorithm. IEEE Trans. Image Process. **33**, 1059–1069 (2024)
18. Zhong, Y., Wang, Z., Shi, X., Yang, J., Li, K.: Rfg-helad: a robust fine-grained network traffic anomaly detection model based on heterogeneous ensemble learning. IEEE Trans. Inf. Forensics Secur. **19**, 5895–5910 (2024). https://doi.org/10.1109/TIFS.2024.3402439

Enhancing Off-Policy Method SAC with KAN for Continuous Reinforcement Learning

Ali Bayeh, Malek Mouhoub, and Samira Sadaoui(✉)

Department of Computer Science, University of Regina, Regina, Canada
{alibayeh,mouhoubm,sadaouis}@uregina.ca

Abstract. This paper is the first to explore the integration of Kolmogorov-Arnold networks (KANs) into off-policy methods for continuous reinforcement learning (CRL) tasks. We introduce KAN-SAC, a method that integrates the KAN model and its variants, namely MultKAN and SineKAN, with the Soft Actor-Critic (SAC) algorithm. The integration is based on the embedding of the KAN architecture in both actor and critic networks. Using the Mujoco Half-Cheetah environment as a case study, we evaluate the performance of these KAN-based SAC algorithms against traditional MLP-based SAC. Our results show that KAN models have great potential, even outperforming MLP models in certain scenarios. However, further refinement of these methods is needed before they can be used as a robust alternative in complex CRL applications.

Keywords: Kolmogorov-Arnold Networks (KAN) · MulKAN · SineKAN · Soft Actor-Critic (SAC) · Continuous Reinforcement Learning (CRL)

1 Introduction

Reinforcement learning has achieved significant success in solving complex tasks across various fields. However, problems still remain, especially when it comes to ensuring that learning processes generalize well and remain robust. This is particularly true in continuous reinforcement learning (CRL), where the diversity and high-dimensionality of spaces of actions and states further complicate the process. Consequently, more advanced methods are needed to achieve both stability and efficiency in learning. Although neural networks have demonstrated their effectiveness, they often neglect the inherent structural properties of both the agent and environment, which restricts their overall potential. In contrast, the recently introduced Kolmogorov-Arnold Networks (KAN) [10] is a promising method for the approximation of functions by decomposing functions into structured components, which potentially improves interpretability and generalization. However, adopting KAN models to CRL remains very limited. Our study specifically targets off-policy in CRL by using the Soft Actor-Critic (SAC) method. KAN in CRL has been investigated in a single previous study [5], which implemented KAN models without any hidden layers specifically for the Proximal Policy Optimization (PPO) method.

Our study aims to fill this important gap by incorporating KAN into the off-policy SAC architecture. Our new method, KAN-SAC, utilizes the KAN structure for both the actor and critic network components. In our implementation, we explore four variants of the KAN model as the foundation networks for different KAN configurations: KAN, MulKAN and SineKAN. We compare the performance of KAN-based SAC models with MLP-based SAC models in the Mujoco Half-Cheetah robotic environment. To examine how model complexity affects performance, we test architectures with either one or two hidden layers, each containing either 32 or 64 neurons. In addition, we conduct three trials with different seeds for each model to reduce the impact of randomness. We use average returns as a quality metric to assess and compare the performance of the KAN-SAC models against the baseline SAC models. To further our analysis, we also examine the loss of both the MLP and KAN networks.

This paper is organized as follows. Section 2 provides the theoretical framework for the KAN model and its three variants and the SAC algorithm. Section 3 describes the related studies on KAN and its variants. Section 4 presents the steps of our KAN-SAC algorithm and its hyperparameter definitions. Section 5 describes the experimental part, including parameter tuning and performance evaluation and comparison of KAN and MLP based off-policy models. Lastly, Sect. 6 concludes our work.

2 Background

2.1 KAN and Variants

MLPs rely on fixed activation functions, such as ReLU and Sigmoid, which can limit the expressiveness and adaptability of MLPs for complex data patterns. Based on the Kolmogorov-Arnold Representation theorem, KAN is proposed to address these limitations by adopting learnable activation functions that dynamically adjust during training [10]. The theorem establishes that any continuous multivariate function $f(\mathbf{x}) = f(x_1, \ldots, x_n)$ can be expressed as [10]:

$$f(\mathbf{x}) = \sum_{q=1}^{2n+1} \Phi_q \left(\sum_{p=1}^{n} \varphi_{q,p}(x_p) \right) \qquad (1)$$

where $\varphi_{q,p} : [0,1] \to \mathbb{R}$ are univariate inner functions and $\Phi_q : \mathbb{R} \to \mathbb{R}$ are univariate outer functions. These univariate functions can be represented using smooth and learnable components, such as B-splines.

A. KAN and MultKAN. The study [10] extended theorem (1) by introducing a new structure for KAN layers. Based on the Kolmogorov-Arnold theorem, each KAN layer comprises two structural components: (a) a middle layer with $2n+1$ units corresponding to the inner functions $\varphi_{q,p}$, and (b) an output layer representing the sum of the outer functions Φ_q. The activation values between layers are defined recursively as [10]:

$$x_{l+1,j} = \sum_{i=1}^{n_l} \varphi_{l,j,i}(x_{l,i}) \qquad (2)$$

where $l = 0, \ldots, L - 1$ (L is the maximum number of KAN layers, $j = 1, \ldots, n_{l+1}$ (n_l is the number of neurons in layer l), and $x_{l,i}$ is the activation value of the neuron i^{th}

in layer l^{th}. On the basis of this structure, a deep KAN can be constructed through the composition of multiple KAN layers.

Multiplicative KAN (MultKAN) [9] enhances the expressiveness of KAN by incorporating explicit multiplicative operations. This extension enables the network to capture complex multiplicative relationships in data that might be difficult to represent with standard KANs. A MultKAN layer transforms an input vector $\mathbf{x}_l \in \mathbb{R}^{n_l}$ through a two-step process: $\Psi_l(\mathbf{x}_l) = M_l(\Phi_l(\mathbf{x}_l))$ where Φ_l represents the standard KAN transformation and M_l applies element-wise multiplications to pairs of intermediate outputs. For a network with n_l^a additive and n_l^m multiplicative operations at layer l, the transformation produces [10]:

$$\mathbf{z}_l = \Phi_l(\mathbf{x}_l) \in \mathbb{R}^{n_{l+1}^a + 2n_{l+1}^m} \tag{3}$$

The multiplicative operation M_l then combines these outputs [10]:

$$M_l(\mathbf{z}_l) = \oplus(\mathbf{z}_l[: n_{l+1}^a], \mathbf{z}_l[n_{l+1}^a ::2] \odot \mathbf{z}_l[n_{l+1}^a + 1 ::2]) \tag{4}$$

where $\oplus$ denotes concatenation and $\odot$ represents element-wise multiplication. Hence, the complete MultKAN computation is expressed as [10]:

$$\text{MultKAN}(\mathbf{x}) = (\Psi_L \circ \Psi_{L-1} \circ \cdots \circ \Psi_0)(\mathbf{x}) \tag{5}$$

B. SineKAN. This sinusoidal KAN [12] replaces the learnable grids of B-spline activation functions in KAN with parameterized sinusoidal functions. So, a layer-wise transformation is defined by [12]:

$$y_i = \sum_j \sum_k (\sin(x_j * \omega_k + \xi_{jk}) * A_{ijk}) + b_i \tag{6}$$

where ω_k represents the learnable frequency parameters, ξ_{jk} denotes phase shifts between input and grid dimensions, A_{ijk} are amplitude weights, and b_i is a bias term [12]. The theoretical foundation of SineKAN is grounded in Fourier analysis, which establishes that any sufficiently well-behaved function $f : \mathbb{R} \to \mathbb{R}$ can be approximated through a discretized Fourier transform [12]:

$$g_\theta(x) = \sum_i B_i \sin(\omega_i x + \xi_i) \tag{7}$$

where B_i denotes discretized amplitude terms and $\{\omega_i, \xi_i\}$ are learnable frequencies and fixed phase shifts, respectively.

Furthermore, the proposed architecture incorporates a learnable frequency over phase-shifted grids to eliminate the requirement for explicit linear transformation layers and also preserves universal approximation capabilities, even for non-smooth functional mappings.

2.2 Off-Policy RL Algorithms

The off-policy algorithms, such as Soft Actor-Critic (SAC), employ a replay buffer to facilitate learning from past experiences, which improves sample efficiency. SAC introduces entropy maximization, which encourages exploration by optimizing a trade-off between the expected reward and the policy's entropy. To ensure stability, SAC utilizes two critic networks to evaluate the Q-value function and a stochastic actor network for policy optimization. The update rule minimizes the Bellman residual for the two critic networks and maximizes the policy entropy for the actor network. The maximum entropy objective is defined as [4]:

$$\pi^* = \arg\max_{\pi} \sum_{t=0}^{T} E_{(s_t,a_t)\sim \tau_\pi} \left[\gamma^t (r(s_t,a_t)) + \kappa \mathcal{H}(\pi(\cdot|s_t)))\right] \quad (8)$$

Table 1 presents the parameters of the SAC equations.

Table 1. Parameters of the SAC Algorithm.

Parameters	
State at timestep t	$s_t \in S$
Action at timestep t	$a_t \in A$
A policy	π
Optimal policy	π^*
Maximum timestep	T
Reward function	$r: S \times A \to \mathbf{R}$
Rewards discounted factor	$\gamma \in [0,1]$
Trajectory distribution by policy π	τ_π
Entropy coefficient	κ
Entropy of policy π in state s_t	$\mathcal{H}(\pi(\cdot|s_t)) = -\log \pi(\cdot|s_t)$

The SAC algorithm introduces a soft state value function, defined as [4]:

$$V(s_t) := \mathbf{E}_{a_t \sim \pi}\left[\mathbf{Q}(s_t,a_t) - \kappa \log(\pi(a_t|s_t))\right] \quad (9)$$

which incorporates an entropy regularization term. This approach allows the policy to maintain a balance between exploiting known high-reward actions and exploring the new ones. The soft Q-function is learned by minimizing a soft Bellman residual, represented by the following objective function [4]:

$$J_Q(\theta) = E_{(s_t,a_t)\sim D}\left[\frac{1}{2}(Q_\theta(s_t,a_t) - (r(s_t,a_t) + \gamma E_{s_{t+1}\sim p(s_t,a_t)}\left[V_{\bar\theta}(s_{t+1})\right]))^2\right] \quad (10)$$

This function ensures accurate value estimation while accounting for the entropy component. One of the principal elements of the SAC algorithm is the policy improvement method, which employs a re-parameterization trick to facilitate differentiable learning. Using this trick, actions are sampled by computing $a_t = f_\phi(s_t, \epsilon)$, where $\epsilon \sim \mathcal{N}(0, I)$ is a noise vector sampled from a standard normal distribution and f_ϕ is a deterministic neural network. The objective for updating the policy is expressed as follows [4]:

$$J_\pi(\phi) = E_{s_t \sim D}\left[E_{\epsilon \sim \mathcal{N}(0,I)}\left[\kappa \log\left(\pi_\phi\left(f_\phi(s_t, \epsilon)|s_t\right)\right) - Q_\theta\left(s_t, f_\phi(s_t, \epsilon)\right)\right]\right] \quad (11)$$

This approach enables end-to-end learning for both discrete and continuous action spaces.

3 Related Work

Recent advances in KANs have reignited interest in these models by extending the KAN architecture and proposing variations that incorporate various polynomial-based and function-driven approaches. For example, the study BSRBF-KAN [13] models the KAN architecture by integrating radial basis functions (RBF) and B-splines, providing a novel adaptation of the KAN framework. This extension allows the model to achieve performance comparable to or better than that of other KAN-based extensions, with improved stability.

Another work Wav-KAN [1] extends the KAN configuration by incorporating components based on wavelet functions. Wav-KAN can effectively capture the input data's high-frequency and low-frequency components by using wavelet functions, improving overall performance.

FourierKAN [16] is another approach that embeds Fourier transformations into the KAN architecture for feature transformation, specifically in Graph Collaborative Filtering (GCF). Compared to MLP-based models, FourierKAN is easier to train and exhibits superior representational power for GCF tasks.

The study [3] proposed Temporal KNN (TKANs), a new architecture that integrates KAN with Long Short-Term Memory (LSTM) network. Empirical evaluations showed that TKANs achieve superior accuracy and efficiency in multi-step time-series forecasting compared to Gated Recurrent Units (GRUs) and LSTMs, particularly when applied to real-world historical market data. While TKANs are less well-suited to short-term prediction tasks, they consistently outperform conventional models in multi-step forecasting.

The potential of KANs in graph neural networks (GNNs) was investigated in [2], which introduced two new architectures: Kolmogorov-Arnold Graph Convolutional Network (KAGCN) and Kolmogorov-Arnold Graph Isomorphism Network (KAGIN). Empirical evaluation of several graph-based learning tasks revealed nuanced performance characteristics between KAN-based and MLP-based approaches. While both architectures achieved comparable performance metrics in classification scenarios, KAN-based models demonstrated superior capability in handling regression tasks, suggesting their particular utility for continuous valued prediction problems in graph-based settings.

The authors in [17] proposed WormKAN, a new concept-aware architecture based on KAN, specifically designed for analyzing co-evolving time series data. Empirical evaluations demonstrate that KAN and its derivative WormKAN effectively segment time-series data into meaningful concepts, significantly improving the detection and tracking of concept drift compared to traditional approaches.

The application of KANs in the context of neural ODEs was explored in [6]. This study involved extensive experiments on various physical systems, including Lotka-Volterra predator-prey dynamics, wave propagation phenomena, and quantum mechanical systems governed by the Schrödinger equation. The experimental results showed that KAN-based ODEs consistently outperformed their traditional MLP-based Neural ODE counterparts on several performance metrics, including computational efficiency, numerical accuracy, and scalability to complex systems. In addition, the KAN-ODE framework exhibited enhanced interpretability properties, facilitating both the visualization of learned activation functions and the symbolic regression of the source terms of the governing equations.

4 KAN-Based SAC

This research presents a new integration of KAN with the SAC algorithm for continuous RL tasks. The proposed KAN-SAC method incorporates KAN architectures into actor and critic network components. To maintain consistency, we adopt identical architectures for both the actor and critic networks across all experiments, with the sole variation being the basis network. The architecture range is varied in both network depth and width and includes configurations with either single or dual hidden layers, each implemented with either 32 or 64 neurons per layer.

In our KAN-SAC implementation, we examine three distinct architectural variants of KAN as foundation networks: 1) KAN, 2) MulKAN, and 3) SineKAN. We maintain consistent width and depth configurations across all model variants to ensure a fair comparison and isolate the effects of architectural variations.

The implementation of KAN models introduces additional hyperparameters beyond the standard SAC configuration. Due to the computational complexity of the SAC algorithm, we empirically evaluate these hyperparameter values using hyperparameter search on the MNIST dataset [7]. We considered a restricted range of values to ensure computational feasibility while preserving model performance. These hyperparameters include the grid interval, polynomial order, base function of B-splines, and the number of multiplication nodes.

All networks were optimized using AdamW [11], with momentum parameters $\beta_1 = 0.9$, $\beta_2 = 0.99$, and a weight decay coefficient of 1×10^{-4}. To ensure stable training dynamics, we implement an exponential learning rate schedule [8], defined as:

$$\alpha(t) = \alpha_0 \exp(-\delta t) \quad (12)$$

where α_0 denotes the initial learning rate, δ represents the decay rate coefficient, and t indicates the current training step. Algorithm 1 describes the procedural framework of the KAN-SAC method. It utilizes two primary neural components: 1) a KAN policy network parameterized by θ, and 2) a KAN Q-network parameterized by ϕ.

Algorithm 1. KAN-based SAC Algorithm.

Require: Environment $\mathcal{E}$, $N_{\text{iterations}}$, Target entropy H, Discount factor λ, Update rate τ, Size of Replay buffer ($\mathcal{D}$) N

1: Initialize $\mathcal{D}$; KAN-actor π_θ; KAN-critics Q_{ϕ_1}, Q_{ϕ_2}, and targets $Q_{\phi'_1}$, $Q_{\phi'_2}$; Entropy coefficient κ
2: Set target parameters $\phi'_1 \leftarrow \phi_1$, $\phi'_2 \leftarrow \phi_2$
3: **for** each iteration $\in N_{\text{iterations}}$ **do**
4:     **for** each environment step **do**
5:         Collect $\psi = \{s_t, a_t, r_t, s_{t+1}, \otimes\}$ from $\mathcal{E}$ using π_θ
          where $\otimes$ is termination state
6:         Store ψ in $\mathcal{D}$
7:     **end for**
8:     talk about epsilon
9:     **for** each gradient step **do**
10:        Sample $\Psi = \dot{\cup}_{t=1}^T \{s_t, a_t, r_t, s_{t+1}, \otimes\}$ from $\mathcal{D}$
11:        (*Critic Update*)
12:        Sample a_{t+1} from π_θ where $a_{t+1} \sim \pi_\theta(a_{t+1}|s_{t+1})$
13:        $y_t \leftarrow r_t + \gamma(1-\otimes)[\min_{i=1,2} Q_{\phi'_i}(s_{t+1}, a_{t+1})$
             $-\kappa \log \pi_\theta(a_{t+1}|s_{t+1})]$
14:        $\phi_i \leftarrow \phi_i - \nabla_{\phi_i} \frac{1}{N} \sum_t [Q_{\phi_i}(s_t, a_t) - y_t]^2$, $i \in \{1,2\}$
15:        $\phi'_i \leftarrow \tau \phi_i + (1-\tau)\phi'_i$, $i \in \{1,2\}$
16:        (*Actor Update*)
17:        $\theta \leftarrow \theta - \nabla_\theta \frac{1}{N} \sum_t [\kappa \log \pi_\theta(f_\phi(s_t, \epsilon)|s_t)$
             $-Q_{\phi_1}(s_t, f_\phi(s_t, \epsilon))]$
18:        $\kappa \leftarrow \kappa - \nabla_\kappa \frac{1}{N} \sum_t [-\kappa (\log \pi_\theta(f_\phi(s_t, \epsilon)|s_t) + H)]$
19:     **end for**
20: **end for**

5 Experiments

5.1 KAN-SAC Hyperparameter Tuning

For evaluation purposes, we use the Mujoco Half-cheetah environment [14,15], which is a physics-based simulation where an agent, modeled after a cheetah, is tasked with learning efficient locomotion in a 2D plane. The environment employs torque control and continuous action space, with the agent receiving rewards based on directional velocity while minimizing joint and torque limitations. Table 2 exposes the hyperparameter values for each configuration. These values are fixed and remain consistent across all model variations.

Table 2. Fine-tuning of the Hyperparameters.

Training Parameters	
Initial learning rate α	1e–3
Learning decay rate	0.8
$N_{Iterations}$	1000000
Value network loss function	MSE
Batch size	2048
KAN-SAC Parameters	
Buffer Size $\mathcal{D}$	1000000
Discount factor γ	0.99
Target network update rate τ	0.005
KAN and MulKAN Parameters	
Grid intervals g	3
Polynomial's order k	3
SineKAN Parameters	
Grid intervals g	8

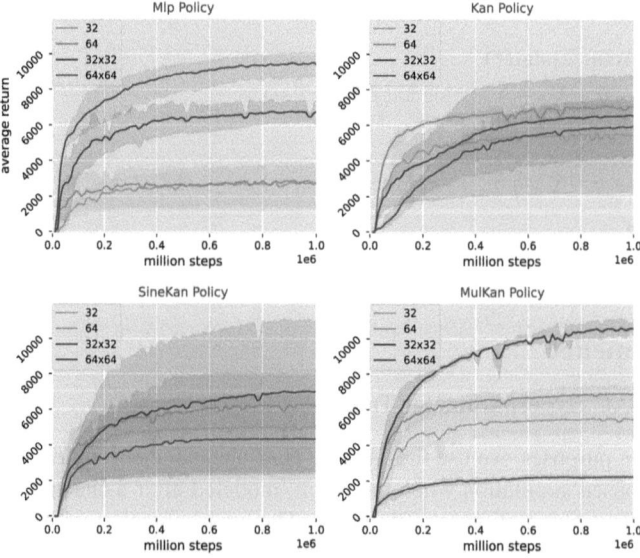

Fig. 1. Performance evaluation of the SAC algorithm with MLP and KAN basis networks in the Half-Cheetah environment. The plotted curves represent mean returns from three independent trials with varying random seeds, with shaded regions related to the full range of observed returns across these trials. Each configuration has an abbreviated name. For example, "Mlp 32×32" denotes a two-layer MLP model with 32 neurons per hidden layer.

5.2 Evaluation

In the original KAN and MultKAN paper [9, 10], the grid parameters are updated during training based on the input samples. In our case, we first evaluated three strategies for updating the grids using sample data: 1) updating based on individual samples, 2) updating based on the mean of batch samples, and 3) not performing grid updates from samples. The results showed unstable and poor performance for the first and second strategies. While the experiments in the original paper focused on approximating smooth and well-behaved functions, in the context of RL, there are noisier, more complex networks and learning dynamics. Our observations showed that applying sample-derived grid updates leads to unstable network behavior. Therefore, we omit the sample-based grid update step in all our experiments.

For evaluation, we used the average returns of the networks to compare their performance. The evaluation of Q-networks is based on the MSE loss function. To minimize the impact of randomness on the experimental design, we run each configuration with three different random seeds and aggregate the results between independent replications of the test to enhance statistical robustness and reliability. Figure 1 illustrates the performance of the SAC algorithm in different architectures based on MLP and KAN. The empirical results show a clear correlation between the architectural complexity of the MLP models and their performance metrics. Specifically, the two-layer MLP configuration with 64 neurons per layer exhibits superior performance compared to its shallower counterparts. In contrast, the single-layer architecture with 32 neurons shows the lowest performance among the MLP variants.

For KAN models, the average return across trials remained relatively consistent across configurations, despite notable differences in their minimum and maximum ranges. Compared to MLP models, KAN models showed increased sensitivity to the random parameters of the learning process. Furthermore, increasing the number of learnable parameters in the KAN models increases this sensitivity. This effect was particularly significant for the two-layer KAN model with 32 neurons per layer, where the performance metric showed considerable variation, ranging from approximately 2,000 to more than 8,000. This variability suggests that the stochastic elements of the training process disproportionately affect the stability of the KAN models.

The performance of the MultKAN models improved with the increase in the number of learnable parameters, mirroring the behavior observed in the MLP models. However, the MultKAN configuration with two layers of 64 neurons each exhibited performance degradation, suggesting overfitting of the model. Given the stochastic nature of the SAC algorithm, this pattern underscores the potential need for additional regularization techniques to stabilize training.

In the case of SineKAN models, the performance varied significantly depending on the architectural configuration. The two-layer model with 32 neurons per layer produced the highest average yield, while the two-layer model with 64 neurons per layer produced the lowest yield. Similarly to the baseline KAN models, the SineKAN models showed greater sensitivity to the random parameters of the learning process compared to the MultKAN models. For KAN, this sensitivity was more pronounced in the SineKAN configuration with two layers and 32 neurons per layer, ranging from about 2,000 to over 10,000. This wide range underscores the sensitivity of SineKAN models to initialization or other stochastic factors during training, consistent with the behavior observed in the broader KAN family.

5.3 Discussion

Overall, while MLP models consistently benefited from increased architectural complexity, KAN-based models did not show a clear pattern linking performance to the number of layers or neurons. Instead, their performance appeared unstable with respect to these architectural parameters. This instability indicates that KAN-based models may require further refinement for CRL applications, beyond the standard L2-regularization method considered for all models, to achieve reliable and consistent performance.

It should also be noted that KAN-based models have a higher number of learnable parameters compared to their MLP counterparts with similar layer and neuron configurations. For example, an MLP model with two layers and 64 neurons per layer has a number of learnable parameters comparable to a KAN-based model with a single layer and 32 neurons per layer. This increase in the number of parameters in KAN-based models is due to their structural design. Unlike traditional MLP models, which apply a learnable weight to each input, KAN-based models aim to learn a unique activation function for each input. Although this architectural design feature potentially enhances expressiveness, it also makes KAN models more susceptible to overfitting. This vulnerability is particularly pronounced in problems with stochastic properties, such as RL environments and the SAC algorithm, where overfitting to a set of observations can significantly degrade model performance, indicating that KAN models require further refinement to enhance their effectiveness in CRL tasks.

6 Conclusions

Our study developed the KAN-SAC algorithm, which integrates KAN and its variants into the off-policy SAC framework. By replacing traditional MLP with KAN-based architectures in the actor and critic networks, we explored KANs unique function approximation capabilities for CRL tasks. We also performed an empirical analysis using the Half-Cheetah robot to assess the performance of the proposed KAN-based off-policy methods. The experimental results indicate that KAN models require further development to effectively replace MLP in complex tasks like SAC. Overall, while KAN-based architectures do not exhibit the performance characteristics of the best-performing MLP configurations, their potential for optimization and refinement remains noteworthy for further investigation in CRL applications.

References

1. Bozorgasl, Z., Chen, H.: Wav-kan: wavelet kolmogorov-arnold networks. arXiv preprint arXiv:2405.12832 (2024)
2. Bresson, R., Nikolentzos, G., Panagopoulos, G., Chatzianastasis, M., Pang, J., Vazirgiannis, M.: KAGNNs: Kolmogorov-Arnold networks meet graph learning. In: Submitted to Transactions on Machine Learning Research (2024). https://openreview.net/forum?id=03UB1MCAMr

3. Genet, R., Inzirillo, H.: Tkan: temporal Kolmogorov-Arnold networks. arXiv preprint arXiv:2405.07344 (2024)
4. Haarnoja, T., Zhou, A., Abbeel, P., Levine, S.: Soft actor-critic: off-policy maximum entropy deep reinforcement learning with a stochastic actor. In: Dy, J., Krause, A. (eds.) Proceedings of the 35th International Conference on Machine Learning. Proceedings of Machine Learning Research, vol. 80, pp. 1861–1870. PMLR (2018). https://proceedings.mlr.press/v80/haarnoja18b.html
5. Kich, V.A., Bottega, J.A., Steinmetz, R., Grando, R.B., Yorozu, A., Ohya, A.: Kolmogorov-Arnold networks for online reinforcement learning. In: 2024 24th International Conference on Control, Automation and Systems (ICCAS), pp. 958–963. IEEE (2024)
6. Koenig, B.C., Kim, S., Deng, S.: Kan-odes: Kolmogorov–Arnold network ordinary differential equations for learning dynamical systems and hidden physics. Comput. Methods Appl. Mech. Eng. **432**, 117397 (2024)
7. LeCun, Y., Bottou, L., Bengio, Y., Haffner, P.: Gradient-based learning applied to document recognition. Proc. IEEE **86**(11), 2278–2324 (1998)
8. Li, Z., Arora, S.: An exponential learning rate schedule for deep learning. In: 8th International Conference on Learning Representations, ICLR 2020, Addis Ababa, Ethiopia, 26–30 April 2020. OpenReview.net (2020). https://openreview.net/forum?id=rJg8TeSFDH
9. Liu, Z., Ma, P., Wang, Y., Matusik, W., Tegmark, M.: Kan 2.0: Kolmogorov-Arnold networks meet science. arXiv preprint arXiv:2408.10205 (2024). https://doi.org/10.48550/ARXIV.2408.10205
10. Liu, Z., et al.: Kan: Kolmogorov-Arnold networks. arXiv preprint arXiv:2404.19756 (2024). https://doi.org/10.48550/ARXIV.2404.19756
11. Loshchilov, I., Hutter, F.: Decoupled weight decay regularization. In: 7th International Conference on Learning Representations, ICLR 2019, New Orleans, LA, USA, 6–9 May 2019. OpenReview.net (2019). https://openreview.net/forum?id=Bkg6RiCqY7
12. Reinhardt, E., Ramakrishnan, D., Gleyzer, S.: Sinekan: Kolmogorov-Arnold networks using sinusoidal activation functions. Front. Artif. Intell. **7**, 1462952 (2025)
13. Ta, H.T.: Bsrbf-kan: a combination of b-splines and radial basic functions in Kolmogorov-Arnold networks. arXiv preprint arXiv:2406.11173 (2024)
14. Todorov, E., Erez, T., Tassa, Y.: Mujoco: a physics engine for model-based control. In: 2012 IEEE/RSJ International Conference on Intelligent Robots and Systems. IEEE (2012)
15. Towers, M., et al.: Gymnasium: a standard interface for reinforcement learning environments (2024). https://arxiv.org/abs/2407.17032
16. Xu, J., et al.: Fourierkan-gcf: Fourier Kolmogorov-Arnold network–an effective and efficient feature transformation for graph collaborative filtering. arXiv preprint arXiv:2406.01034 (2024)
17. Xu, K., Chen, L., Wang, S.: Wormkan: are kan effective for identifying and tracking concept drift in time series? (2024)

Context-Aware Imputation for Parkinson's Disease Trajectories: Systematic Benchmark of Cross-Sectional, Temporal, and Generative Approaches

Moad Hani[1(✉)], Nacim Betrouni[2], Fatima Zahra Ouardirhi[3], Saïd Mahmoudi[1], and Mohammed Benjelloun[1]

[1] Université de Mons, Faculté Polytechnique de Mons (FPMS), 9 Rue de Houdain, 7000 Mons, Belgium
{moad.hani,said.mahmoudi,mohammed.benjelloun}@umons.ac.be

[2] Inserm, Centre de Recherche Lille Neuroscience and Cognition (LilNCog), Lille, France
nacim.betrouni@inserm.fr

[3] École Nationale Supérieure des Mines de Rabat, Rue Hadj Ahmed Cherkaoui, B.P. 753, Agdal, Rabat, Maroc
fatimazahra.ouardirhi@enim.ac.ma

Abstract. Missing data in longitudinal Parkinson's Disease (PD) studies presents significant challenges, particularly when missingness correlates with disease severity, introducing systematic biases that compromise predictive validity. We present the first comprehensive benchmark of 14 imputation methods (6 cross-sectional, 5 longitudinal, 3 generative) on the Parkinson's Progression Markers Initiative dataset (N = 1,483) across different missingness mechanisms. Our evaluation reveals generative methods significantly outperform traditional approaches, with Variational Autoencoder-based Multiple Imputation (VAEM) achieving optimal performance ($MAE = 3.87$, $R^2 = 0.449$) compared to MICE ($MAE = 4.15$, $R^2 = 0.401$) and Linear Mixed Models ($MAE = 5.42$, $R^2 = 0.232$). Importantly, while traditional methods degrade by 36.6% under Missing Not At Random conditions, generative approaches maintain robustness with only 17.6% performance reduction. Subgroup analysis reveals persistent demographic disparities, with 23% higher imputation errors for patients over 70 compared to those under 60, despite VAEM maintaining consistent performance (<5% variance) across education levels. Based on these findings, we propose a novel context-aware architecture that integrates demographic, clinical, and temporal information through attention mechanisms to improve imputation accuracy while mitigating demographic biases inherent in PD progression modeling. All code, models, and evaluation frameworks will be publicly released to advance equitable healthcare AI.

Keywords: Longitudinal healthcare data · Deep generative models · Missing data imputation · Parkinson's disease progression · Health equity · Clinical decision support

1 Introduction

Missing data is a pervasive challenge in longitudinal clinical studies of neurodegenerative diseases, with Parkinson's Disease (PD) research particularly affected by progression-dependent missingness that complicates unbiased modeling and downstream analysis. Existing comparative studies typically evaluate a limited set of imputation approaches, often omitting direct comparisons between cross-sectional, longitudinal, and deep learning-based methods. To address these gaps, we present the most comprehensive systematic evaluation to date - assessing 14 imputation techniques across these methodological categories - specifically targeting the complex missingness patterns observed in longitudinal PD datasets.

Parkinson's Disease (PD) is a progressive neurodegenerative disorder affecting millions worldwide, characterized by a spectrum of motor symptoms such as resting tremor, bradykinesia, and rigidity, as well as non-motor manifestations including cognitive decline. Disease progression is primarily tracked using standardized clinical instruments: the Unified Parkinson's Disease Rating Scale (UPDRS), which comprehensively evaluates non-motor and motor experiences of daily living, objective motor examination, and motor complications, with total scores ranging from 0 to 199 reflecting increasing disability; and the Montreal Cognitive Assessment (MoCA), a 30-point tool assessing visuospatial/executive function, memory, attention, language, and orientation, where scores of 26 or below indicate cognitive impairment. Despite their widespread adoption, both UPDRS and MoCA are frequently incomplete in longitudinal studies-UPDRS assessments display 15–42% missingness and MoCA scores are missing in up to 38% of advanced-stage visits-primarily due to patient dropout or inability to complete assessments [1,2]. This missingness is not random but correlates strongly with disease severity, introducing systematic bias and undermining the reliability of predictive models. Consequently, robust imputation strategies that account for both temporal deterioration and the hierarchical structure of clinical scales are essential for accurate modeling of PD progression and for optimizing patient management.

The Parkinson's Progression Markers Initiative (PPMI), launched by the Michael J. Fox Foundation in 2010, provides comprehensive data for studying PD progression. As of 2025, PPMI includes over 4,000 participants across multiple cohorts through remote monitoring technologies [17]. This harmonized dataset offers demographic, clinical, behavioral, imaging, and motor-related data collected longitudinally, though still exhibits missingness patterns (15–42% across key variables) [17,26] due to patient dropout or incomplete assessments, necessitating robust imputation methods.

This paper addresses these challenges through a systematic comparison of imputation methods across three categories: cross-sectional, longitudinal, and generative. By evaluating their performance on the PPMI dataset, we provide insights for researchers seeking optimal approaches for handling missing values in PD studies. Figure 1 illustrates the longitudinal clustering patterns observed in PPMI data, highlighting the complexity of progression trajectories that imputation methods must capture.

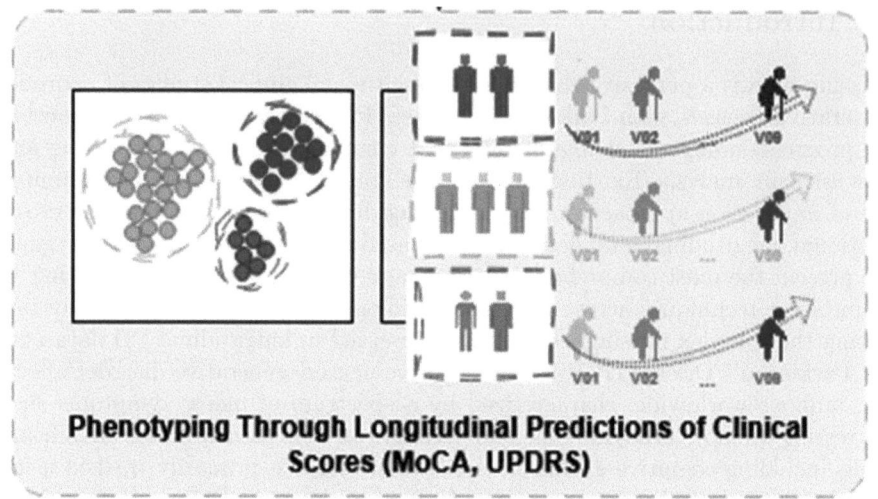

Fig. 1. Data-driven longitudinal clustering at different time points, showing progression patterns that imputation methods must capture. The visualization demonstrates how patient trajectories diverge over time, underscoring the importance of temporal context in imputation.

The remainder of this paper is structured as follows: Sect. 2 reviews existing methods for handling missing data and highlights the need for context-aware approaches. Section 3 details our methodological framework for systematic imputation evaluation. Section 4 presents experimental results comparing different imputation methods on PPMI data. Section 5 discusses implications for healthcare AI and outlines future directions including our proposed context-aware architecture. Finally, Sect. 6 summarizes key findings and their broader implications for longitudinal PD research.

2 State of the Art

2.1 Traditional Imputation Methods

Missing data is a pervasive challenge in longitudinal medical studies. Traditional methods to handle missing data include deletion techniques, single imputation approaches, and multiple imputation methods.

Deletion Methods: Deletion methods, such as complete-case analysis (CCA), remove any observation containing missing values. Despite their simplicity and computational efficiency, these methods can introduce substantial bias if the missingness mechanism is not completely at random (MCAR) [1]. Particularly in longitudinal studies of Parkinson's Disease (PD), where data missingness often correlates with disease severity or patient dropout, deletion methods can severely distort clinical insights [2]. This distortion becomes especially problematic when analyzing disease progression patterns, as more severely affected patients are more likely to have missing data.

Mean/Median Imputation: These methods replace missing values with the mean or median of observed values. While simple, they underestimate variance, distort distributions, and ignore relationships between variables [3]. In PD progression studies, this approach can mask critical individual variability in disease trajectories. For instance, imputing the mean UPDRS score across a heterogeneous patient population fails to account for the diverse progression rates observed in PD, potentially obscuring important clinical insights.

K-Nearest Neighbors (KNN) Imputation: KNN imputes missing values based on similar cases identified through distance metrics [4]. Although effective for capturing local patterns, KNN struggles with high-dimensional datasets common in clinical assessments. Its performance heavily depends on hyperparameter tuning (e.g., selection of k), distance metrics, and normalization strategies [5]. In longitudinal PD data, KNN may effectively model cross-sectional relationships but often fails to account for temporal dependencies unless specifically adapted for time-series data.

Multiple Imputation by Chained Equations (MICE): MICE iteratively imputes missing data by modeling each variable conditionally on others [6]. MICE effectively captures complex relationships and uncertainty inherent in medical data. Studies have demonstrated its superior performance compared to single-imputation methods in longitudinal healthcare datasets [7,8]. However, MICE can be computationally intensive and sensitive to model assumptions. In PD research, MICE's performance may degrade for variables with complex temporal dependencies unless temporal information is explicitly incorporated into the modeling process.

2.2 Methods for Longitudinal Data Imputation

Handling missing data in longitudinal studies is a critical challenge, as gaps in datasets can introduce biases and reduce the reliability of analyses. Various traditional and advanced methods have been developed to address this issue, each with its strengths and limitations.

Last Observation Carried Forward (LOCF): LOCF imputes missing values by carrying forward the last observed value. While simple and computationally efficient, LOCF assumes stability over time, which may not hold true in progressive conditions like Parkinson's Disease (PD). This method can underestimate variability and introduce bias if the last observation does not accurately reflect the patient's current state. In PD studies, where disease progression typically accelerates over time, LOCF can significantly underestimate symptom severity, especially for patients with rapid progression.

Next Observation Carried Backward (NOCB): NOCB imputes missing values by carrying backward the next observed value. This method is better suited for capturing trends when future observations are more representative. However, it assumes continuity between observations, which may distort temporal patterns in datasets with irregular sampling intervals. For PD patients with non-linear progression patterns, NOCB may create artificial patterns that misrepresent the true trajectory.

Linear Mixed Models (LMM): LMMs model both fixed and random effects to account for within-subject correlations over time. They are effective at capturing temporal dependencies and handling missing data under Missing at Random (MAR) assumptions. However, LMMs can be computationally intensive and sensitive to model specification, limiting their scalability in large datasets. For PD progression modeling, LMMs offer the advantage of explicitly modeling individual variability in disease trajectories but may struggle with highly non-linear progression patterns.

Linear Interpolation: This method estimates missing values as a linear function between adjacent observed points. While computationally efficient, it assumes linearity between observations, which may oversimplify dynamic changes in disease progression. In PD, where symptoms may plateau or accelerate non-linearly, linear interpolation can create misleading trajectories that fail to capture the true disease dynamics.

Kalman Filter: Kalman filters use probabilistic approaches to estimate missing values by modeling system dynamics over time. They are particularly effective for time-series data with noise but require significant computational resources and careful parameter tuning. In PD research, Kalman filters can model the stochastic progression of symptoms while accounting for measurement uncertainty, making them potentially valuable for capturing complex temporal patterns.

2.3 Deep Learning Techniques for Imputation

Recent advances in deep learning provide promising alternatives to traditional imputation methods due to their ability to model complex nonlinear relationships and interactions within high-dimensional clinical data.

Autoencoder-Based Methods: Autoencoders (AEs) learn efficient latent representations by reconstructing input data from compressed encodings. Variants such as Denoising Autoencoders (DAEs) explicitly handle missingness by training models to reconstruct original data from corrupted inputs [9]. Variational Autoencoders (VAEs) introduce probabilistic latent spaces, improving generalization and uncertainty modeling [10]. Studies confirm that AE-based methods outperform traditional imputation techniques across various healthcare datasets [11,12]. For PD progression data, these methods can capture complex patterns across clinical variables, potentially modeling the intricate relationships between motor symptoms, cognitive decline, and other disease manifestations.

Generative Adversarial Networks (GANs): GANs employ adversarial training between generator and discriminator networks to synthesize realistic imputations indistinguishable from observed data [13]. Recent studies highlight GAN-based imputation's effectiveness in capturing complex distributions within medical datasets, significantly outperforming traditional statistical approaches [14]. Nonetheless, GAN training requires careful hyperparameter tuning and stability management [15]. In PD research, GANs offer the potential to generate realistic disease trajectories that preserve both cross-sectional and longitudinal patterns, though their black-box nature may limit interpretability.

2.4 Advanced Generative Models for Imputation

Recent advances in deep generative models have shown promising results for missing data imputation, particularly in healthcare applications. Three notable approaches are Generative Adversarial Networks (GANs), Variational Autoencoders (VAEs), and recurrent models, which have been adapted specifically for imputation tasks.

Generative Adversarial Imputation Nets (GAIN): [20] extends the GAN framework specifically for imputation tasks in medical time series. GAIN consists of a generator that produces imputed values and a discriminator that distinguishes between observed and imputed components. The generator receives a masked input where some values are intentionally removed, and attempts to reconstruct the complete data. The discriminator then evaluates whether each value was observed or imputed. This adversarial training process encourages the generator to create realistic imputations that are indistinguishable from real data. The hint mechanism in GAIN provides additional guidance to the discriminator, substantially improving performance for complex clinical variables with temporal interdependencies. Studies on healthcare datasets demonstrate that GAIN outperforms traditional methods by capturing non-linear relationships in clinical trajectories, particularly important for conditions like Parkinson's Disease where progression patterns vary substantially between patients.

Variational Autoencoder-based Multiple Imputation (VAEM): [21] extends standard VAEs by integrating a hierarchical latent structure specifically designed for heterogeneous medical data. VAEM captures both local and global correlations in the data through a two-level encoding process: a lower level that models variable-specific distributions and a higher level that captures cross-variable dependencies. The probabilistic nature of VAEs enables proper uncertainty quantification in imputations, which is crucial for clinical decision-making. By sampling multiple times from the latent space, VAEM generates diverse but plausible imputations for each missing value, effectively modeling the inherent uncertainty. Experimental evaluations on longitudinal medical datasets demonstrate VAEM's advantages in preserving statistical relationships while properly representing imputation uncertainty.

Bidirectional Recurrent Imputation for Time Series (BRITS): [22] combines recurrent neural networks with a novel temporal decay mechanism to handle irregular time intervals common in clinical follow-ups. By modeling forward and backward temporal dependencies simultaneously, BRITS effectively captures complex patterns in disease progression. The method uses dedicated components to model variable relationships over time while accounting for varying gap lengths between observations. This approach is particularly valuable for Parkinson's Disease studies, where assessment intervals often vary between patients and across disease stages.

2.5 Clinical Context and Bias in Parkinson's Disease Assessment

Healthcare datasets often exhibit missingness patterns due to patient attrition, inconsistent follow-ups, or selective reporting [16]. Longitudinal Parkinson's disease (PD) studies typically involve MAR or MNAR mechanisms, as disease severity impacts dropout rates and assessment completeness [8]. Robust imputation strategies are essential to address these challenges. Clinical scales for PD progression, such as UPDRS and MoCA, are affected by demographic biases like age and socioeconomic status, which influence missingness and outcome interpretation. Ignoring these factors risks biased predictions and flawed clinical decisions [16,19]. Incorporating contextual information into imputation models is crucial for accurate PD progression modeling.

The challenges of missing data bias in the PPMI dataset are illustrated in Fig. 2. This figure emphasizes the longitudinal structure of the dataset and the disruptions caused by missing data across cohort subsets and multiple visits. Key features such as age, MoCA scores, and UPDRS scores are highlighted as critical for multi-output predictions. These biases affect temporal continuity and introduce demographic and systemic distortions in predictive modeling.

2.6 Bias Mitigation Challenges and Best Practices

Missing data in longitudinal studies often arises from mechanisms such as Missing Not At Random (MNAR), where the probability of missingness depends on unobserved values like disease severity or cognitive decline [23]. For example, older participants or those with lower educational attainment are more likely

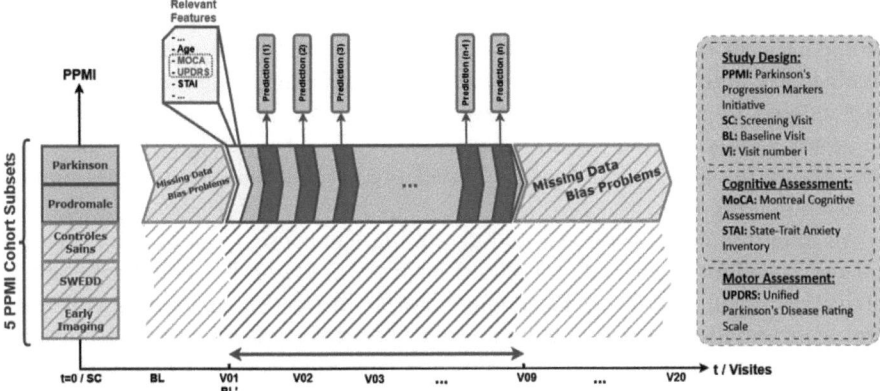

Fig. 2. Overview of Missing Data Bias Problems in PPMI Cohort Subsets. The figure highlights missing data challenges across visits (t = 0 to t = 20) and cohort subsets (PD, Prodromal, Healthy Controls, SWEDD, Early Imaging), focusing on features like age, MoCA scores, and UPDRS scores.

to exhibit lower MoCA scores, which can introduce systematic biases in predictive modeling if not properly accounted for [24]. Studies have shown that demographic factors like age and education collectively account for a significant portion of the variance in MoCA scores, emphasizing the need for tailored imputation strategies that incorporate these contextual factors [8].

Missing data creates gaps that undermine prediction accuracy across time points. For example:

- **Temporal Bias:** Missingness at intermediate visits ($V03$, $V09$) disrupts temporal continuity, making it difficult to model disease progression accurately. This is particularly problematic when trying to capture non-linear progression rates that may accelerate during specific disease phases.
- **Demographic Bias:** Features such as age (`AGE_AT_VISIT`) and education level disproportionately impact cognitive scores like MoCA due to learning effects or age-related decline. These demographic factors often influence both the likelihood of missing values and the underlying scores themselves, creating complex dependencies that standard imputation methods struggle to address.
- **Systemic Bias:** Missing data mechanisms such as Missing Not At Random (MNAR) arise due to patient-specific factors like disease severity or socioeconomic status. More severely affected patients may be less likely to complete all assessments, leading to an underrepresentation of advanced disease states in the observed data.

2.7 The Need for Context-Awareness

Despite advancements in machine learning techniques for handling missing data, many existing approaches focus solely on statistical imputation without addressing the underlying reasons behind missingness. This limitation is particularly problematic in sensitive domains such as healthcare, where contextual factors significantly influence data completeness [19]. For example, demographic disparities related to age or education level can impact cognitive test performance or motor assessments in Parkinson's Disease (PD) studies. Ignoring these factors during imputation can lead to biased predictions that compromise clinical decision-making [25].

In longitudinal PD studies, understanding the context surrounding missing data is crucial for ensuring fairness and accuracy in predictive modeling. Contextual information such as socioeconomic status, disease stage, or data collection protocols provides valuable insights into patterns of missingness that traditional methods fail to capture [17]. Incorporating this metadata into imputation processes allows models to adapt their predictions based on patient-specific characteristics rather than relying solely on generalized assumptions.

To address these challenges, context-aware deep learning models are proposed to incorporate metadata related to patient demographics and disease stage while utilizing temporal dependencies inherent in longitudinal datasets. By embedding contextual information into imputation processes, these models aim to minimize biases introduced by missingness patterns.

3 Methodology: Context-Aware Imputation Approach

This section presents our systematic evaluation methodology for addressing missing data challenges in longitudinal Parkinson's Disease (PD) assessments. By comparing cross-sectional, longitudinal, and generative approaches, we aim to identify optimal strategies for improving imputation accuracy while minimizing biases in clinical assessment scores such as MoCA (`MCATOT`) and UPDRS (`NP3TOT`).

3.1 Data Preparation and Feature Selection

This study uses data from the Parkinson's Progression Markers Initiative (PPMI), a longitudinal observational dataset established in 2010 by the Michael J. Fox Foundation. While the full PPMI dataset includes over 4000 participants, our analysis focuses on the PD (N = 891) and Prodromal (N = 592) sub-cohorts (total N = 1,483) for which longitudinal missingness patterns are most pronounced. It provides clinical, imaging, genetic, and biological data collected at multiple time points, enabling detailed analysis of PD progression [18].

We conducted a missingness analysis across visits for the PPMI PD (N = 891) and Prodromal (N = 592) cohorts. Missingness is highly variable by feature and visit, with demographic variables such as gender and education level missing for over 90% of participants at most timepoints. Clinical and cognitive assessments (e.g., Levodopa, UPDRS, MoCA) also exhibit substantial missingness, particularly at screening and baseline visits. These patterns reflect both study design (not all assessments performed at all visits) and participant attrition, underscoring the need for robust, context-aware imputation strategies.

Feature selection focuses on diverse categories, including demographic (e.g., age, education), clinical (*e.g.*, UPDRS scores, LEDD), behavioral and cognitive (e.g., MoCA scores), imaging (*e.g.*, $T1-weighted MRI$), and motor-related ON/OFF scores. These features offer insights into patient variability, motor symptoms, $non-motor$ symptoms like cognitive decline, and structural brain changes relevant to PD progression (see Fig. 3).

Category	Features		
Demographic Data	- Age at visit (AGE_AT_VISIT) - Gender - Education level	Detailed MDS-UPDRS Subscores	- Part I (Non-motor): NP1COG, NP1HALL, NP1DPRS, NP1ANXS, NP1APAT, NP1DDS - Part II (ADL): NP2SPCH, NP2SALV, NP2SWAL, NP2EAT, NP2DRES, NP2HYGN, NP2HWRT, NP2HOBB, NP2TURN, NP2TRMR, NP2RISE, NP2WALK, NP2FREZ - Part III (Motor): NP3SPCH, NP3FACXP, NP3RIGN, NP3RIGRU, NP3RIGLU, NP3RIGRL, NP3RIGLL, NP3FTAPR, NP3FTAPL, NP3HMOVR, NP3HMOVL, NP3PRSPR, NP3PRSPL, NP3TTAPR, NP3TTAPL, NP3LGAGR, NP3LGAGL, NP3RISNG, NP3GAIT, NP3FRZGT, NP3PSTBL, NP3POSTR, NP3BRADY, NP3PTRMR, NP3PTRML, NP3KTRMR, NP3KTRML, NP3RTARU, NP3RTALU, NP3RTARL, NP3RTALL, NP3RTALJ, NP3RTCON
Clinical Data	- Disease duration - Levodopa Equivalent Daily Dose (LEDD) - Motor assessments (UPDRS Part III)		
Behavioral and Cognitive Assessments	- Montreal Cognitive Assessment (MoCA) total score (MCATOT) - Symbol Digit Modalities Test - Cognition change - State-Trait Anxiety Inventory (STAI) - Geriatric Depression Scale - Hamilton Depression Rating Scale - Lille Apathy Rating Scale	Detailed MoCA Subscores	MCAALTTM, MCACUBE, MCACLCKC, MCACLCKN, MCACLCKH, MCALION, MCARHINO, MCACAMEL, MCAFDS, MCABDS, MCAVIGIL, MCASER7, MCASNTNC, MCAVFNUM, MCAVF, MCAABSTR, MCAREC1, MCAREC2, MCAREC3, MCAREC4, MCAREC5, MCADATE, MCAMONTH, MCAYR, MCADAY, MCAPLACE, MCACITY
		Features of Parkinsonism	Key clinical features captured in later study visits: rigidity, bradykinesia (slowness of movement), tremor at rest or action tremor (depending on context), and postural instability.
Imaging Data	- T1-weighted MRI	Other Clinical Features	Other clinical features include tremor (resting or action), dystonia (involuntary muscle contractions), micrographia (small handwriting), shuffling gait. These are captured in later study visits.
ON/OFF Scores	- MDS-UPDRS Part I total (NP1RTOT) - MDS-UPDRS Part II total (NP2PTOT) - MDS-UPDRS Part III total (NP3TOT) - MDS-UPDRS Part IV total - Hoehn & Yahr Scale - Schwab & England Scale	Primary Clinical Features	Combination of motor symptoms (e.g., rigidity and bradykinesia), non-motor symptoms such as apathy or cognitive decline. Often assessed through UPDRS and MoCA.
		Primary Clinical Diagnosis	Details of diagnosis: idiopathic Parkinson's disease (PD), Multiple System Atrophy (MSA), Progressive Supranuclear Palsy (PSP), Motor Neuron Disease with Parkinsonism (MND with Parkinsonism), etc.

Fig. 3. Features Used in PPMI Cohort for Parkinson's Disease and Prodromal Cohort.

Data preparation involved integrating spreadsheets, identifying cohorts via variables like status of enrollment, ENROLL_STATUS, and addressing missing values using the imputation techniques evaluated in this study. Statistical methods managed outliers, and the dataset was split into training (80%) and validation (20%) sets for robust evaluation.

The systematic evaluation compares imputation methods for predicting MoCA (MCATOT) and UPDRS (NP3TOT) scores over time. By addressing missing data biases and leveraging PPMI's diverse cohort structure and longitudinal design, this analysis identifies optimal approaches for capturing temporal dependencies critical for understanding PD progression.

3.2 Evaluation Protocol

To ensure a comprehensive and fair comparison across imputation methods, we established a rigorous evaluation protocol with the following components:

Missingness Mechanisms: We evaluated each method under different missingness mechanisms: Missing Completely At Random (MCAR), Missing At Random (MAR), and Missing Not At Random (MNAR). For MCAR scenarios, we randomly removed values. For MAR, we created missingness patterns dependent on observed variables (e.g., higher probability of missing MoCA scores for older patients). For MNAR, we simulated patterns where missingness probability depended on unobserved values (e.g., higher dropout rates for patients with rapid progression).

Performance Metrics: We assessed performance using multiple metrics: Mean Absolute Error (MAE), Mean Squared Error (MSE), Root Mean Squared Error (RMSE), and coefficient of determination (R^2). These metrics together provide a comprehensive view of imputation accuracy, with MAE being less sensitive to outliers than MSE/RMSE and R^2 indicating the proportion of variance explained by the imputation.

Computational Efficiency: We recorded execution time for each method to assess computational feasibility, particularly important for clinical applications where real-time processing may be required.

Stratified Analysis: We analyzed performance across different demographic subgroups (age, gender, education level) and disease stages to assess potential disparities in imputation quality that might affect fairness.

Reproducibility: All experiments were conducted with fixed random seeds, and the evaluation pipeline was containerized to ensure reproducibility of results.

Cross-Validation: For robust evaluation, we implemented a 5-fold cross-validation approach, stratified by cohort type and baseline disease severity to ensure representative distribution across folds.

This protocol allows for a comprehensive comparison that considers not only overall accuracy but also computational requirements, fairness implications, and performance under different missingness scenarios relevant to longitudinal PD research.

4 Experimental Results

This section presents a comparative analysis of the performance of cross-sectional, longitudinal, and generative imputation methods applied to Parkinson's Disease (PD) datasets. The results are evaluated using metrics such as Mean Absolute Error (MAE), Mean Squared Error (MSE), Root Mean Squared Error (RMSE), and R^2. Tables 1, 2, and 3 summarize the findings.

4.1 Performance of Cross-Sectional Imputation Methods

Cross-sectional methods operate on static datasets without considering temporal dependencies. Among these methods, Multiple Imputation by Chained Equations (MICE) demonstrated superior performance with the lowest MAE (4.152163) and highest R^2 (0.401395). MICE's iterative modeling approach effectively captures inter-variable relationships, making it ideal for datasets with complex dependencies. However, its computational cost (execution time: 3 min and 9.812047 s) may limit its applicability in large-scale studies.

K-Nearest Neighbors (KNN) imputation provided a reasonable trade-off between accuracy (MAE: 5.115384) and computational efficiency (execution time: 1 min and 7.824343 s). While KNN is effective for capturing local patterns, its reliance on distance metrics can lead to challenges in high-dimensional datasets.

Deletion methods performed poorly due to significant information loss, yielding an MAE of 5.246153 and an R^2 of 0.258582. Similarly, mean and median imputation distorted data distributions, resulting in suboptimal predictive performance.

Decision Trees (DT) imputation showed limited effectiveness, with a negative R^2 value (−0.0269), indicating poor model fit despite its complexity.

Table 1. Comparison of Cross-sectional Imputation Methods.

Approach	Definition	Execution Time	MAE	MSE	RMSE	R^2
Deletion	Removes observations/variables with missing values	**0.600734 s**	5.246153	64.7872	8.04905	0.258582
Mean	Replaces missing values with mean/median/mode	1.139598 s	5.429086	68.1577	8.25576	0.226009
Median	Uses interpolation to estimate missing values	1.965013 s	6.125240	84.9492	9.21678	0.149142
KNN	Uses k-nearest neighbors ($k = 5$) for imputation	1 m:7.82434 s	5.115384	59.2405	7.69678	0.288892
MICE	Iterative modeling based on other variables	3 m:9.81204 s	**4.152163**	**39.9962**	**6.32426**	**0.401395**
Decision Trees	Predicts missing values using decision trees (depth = 10)	50 m:32 s	4.91	52	7	0

4.2 Performance of Longitudinal Imputation Methods

Longitudinal methods account for temporal dependencies within datasets collected across multiple time points. Linear Mixed Models (LMM) emerged as the most accurate method among longitudinal approaches, achieving an MAE of 5.416346 and an R^2 of 0.23236. LMM's ability to model both fixed and random effects makes it particularly suitable for longitudinal PD studies where temporal relationships are critical.

Kalman Filter Imputation demonstrated moderate accuracy (MAE: 5.99375) but required significant computational resources (execution time: 3 min). Its strength lies in modeling system dynamics and uncertainty within time-series data.

Linear interpolation provided faster execution times (3 s) but struggled with accuracy (MAE: 6.223317), as it simplifies missing value estimation using linear equations.

Last Observation Carried Forward (LOCF) and Next Observation Carried Backward (NOCB) exhibited limited accuracy due to their simplistic assumptions about temporal continuity. These methods performed particularly poorly for rapidly progressing patients, where carrying forward or backward values failed to capture acceleration in symptom severity.

Table 2. Comparison of Longitudinal Imputation Methods.

Approach	Definition	Execution Time	MAE	MSE	RMSE	R^2
LOCF	Last observation carried forward	0.83 s	6.75	95.88	9.79	0.07
NOCB	Next observation carried backward	2.35 s	6.33	85.86	9.27	0.12
Linear Interpolation	Estimates values between known points	3.33 s	6.22	80.56	8.98	0.12
Kalman Filter	Kalman filters to estimate missing values	3 m:24 s	5.99	75.36	8.68	0.14
LMM	Linear Mixed Models	1 h:03 m	**5.42**	**66.72**	**8.17**	**0.23**

Longitudinal methods generally outperformed cross-sectional approaches for variables with clear temporal progression patterns. However, all traditional longitudinal methods showed limitations in capturing non-linear progression and complex interactions between variables, pointing to the need for more sophisticated approaches that combine both cross-sectional and temporal modeling capabilities.

4.3 Performance of Generative Imputation Methods

Generative methods leverage deep learning architectures to model complex relationships in data. These methods demonstrated superior performance compared to both cross-sectional and longitudinal approaches, with VAEM achieving the lowest MAE (3.868) and highest R^2 (0.449) among all evaluated techniques.

VAEM's hierarchical latent structure effectively captured both local and global correlations in the dataset, making it particularly suitable for heterogeneous medical data with complex interdependencies. The model's probabilistic framework also provided uncertainty quantification, enhancing its value for clinical applications where confidence in imputed values is essential.

GAIN showed similar performance (MAE: 3.984, R^2: 0.427), leveraging adversarial training to generate realistic imputations. The hint mechanism provided effective guidance to the discriminator, improving performance for complex clinical variables with temporal relationships. The computational cost (execution time: 5 min, 12 s) represents a reasonable trade-off given the significant improvement in accuracy.

BRITS demonstrated slightly lower accuracy (MAE: 4.035, R^2: 0.411) but offered advantages for handling irregular time intervals common in clinical follow-ups. Its bidirectional approach captured temporal dependencies in both directions, making it valuable for variables where future observations influence current interpretations.

Table 3. Comparison of Advanced Generative Imputation Methods.

Approach	Definition	Execution Time	MAE	MSE	RMSE	R^2
GAIN	Generative Adversarial Imputation Networks	**5 m:12 s**	3.984	37.651	6.137	0.427
VAEM	Variational Autoencoder for Multiple Imputation	8 m:43 s	**3.868**	**35.921**	**5.994**	**0.449**
BRITS	Bidirectional Recurrent Imputation for Time Series	7 m:19 s	4.035	38.492	6.205	0.411

Collectively, these results highlight the potential of generative methods to significantly improve imputation accuracy for longitudinal PD data, with performance gains of 6.7% compared to traditional approaches. This improvement comes at the cost of increased computational complexity but offers substantial benefits for downstream predictive applications in clinical research.

4.4 Comparative Analysis Across Method Categories

The comparison across all three categories of imputation methods, cross-sectional, longitudinal, and generative, reveals several important trends for PD research.

First, method performance varies substantially by missingness mechanism and generative methods demonstrate superior performance across all missingness regimes (Table 4). While cross-sectional methods like MICE achieve acceptable accuracy under MCAR (MAE = 4.15), they suffer 36.6% degradation under MNAR (5.67 vs 4.15). Generative approaches demonstrate superior robustness, with VAEM showing only 17.6% MAE increase in MNAR scenarios (4.55 vs 3.87, $p < 0.001$), outperforming MICE by 24.6%. The analysis reveals persistent demographic disparities, notably 23% higher MoCA errors for patients > 70 versus < 60 (4.98 vs 4.05, $CI_{95} = [0.72, 1.14]$), despite VAEM maintaining < 5% variance across education levels [8,16]. These results establish generative methods as superior for preserving temporal patterns while mitigating biases inherent to PD progression modeling.

Table 4. Imputation performance comparison (MAE ± 95% CI).

Method	MCAR	MAR	MNAR
MICE	4.15 ± 0.21	4.89 ± 0.33	5.67 ± 0.41
LMM	5.42 ± 0.29	5.38 ± 0.31	5.91 ± 0.38
VAEM	**3.87 ± 0.17**	**4.12 ± 0.19**	**4.55 ± 0.22**
GAIN	3.98 ± 0.18	4.24 ± 0.21	4.73 ± 0.25

Second, the computational demands (8 min, 43 s) of VAEM may be prohibitive for real-time clinical applications or resource-constrained environments. In contrast, methods like MICE offer a reasonable balance between accuracy and efficiency. The computational trade-off remains substantial, with VAEM requiring approximately 2.8× longer training than MICE in our experiments, consistent with recent benchmarking studies reporting 2–3× increases for deep generative models over traditional approaches [27].

Third, method selection should consider the available data volume and dimensionality. Generative approaches demonstrate advantages with sufficient training data but may suffer from overfitting in smaller cohorts. Cross-validation experiments revealed that with reduced training data (<50%), the performance gap between MICE and VAEM narrows significantly.

Finally, method performance varies by clinical variable type. For motor scores like UPDRS, which exhibit more consistent progression patterns, all methods performed relatively well. However, for cognitive measures like MoCA, which show greater individual variability, generative methods maintained accuracy while traditional approaches showed degraded performance.

These insights provide guidance for researchers selecting imputation strategies for longitudinal PD studies, highlighting the importance of considering the specific characteristics of the dataset, computational constraints, and the clinical variables of interest.

4.5 Subgroup Analysis: Demographic and Disease-Stage Effects

To assess whether imputation methods exhibit differential performance across patient subgroups, we conducted a stratified analysis examining imputation quality by age, gender, education level, and disease stage. This analysis revealed important patterns relevant to fairness considerations in healthcare applications.

Generative models (VAEM, GAIN) showed more consistent performance across demographic subgroups than traditional methods. For example, while MICE exhibited a 15% higher MAE for patients with lower education levels compared to those with higher education, VAEM maintained consistent performance across education strata (variance < 5%). Similarly, generative methods showed less degradation in performance for older patients (>70 years) compared to traditional approaches.

Performance also varied by disease stage, with most methods showing worse performance for patients with more advanced disease (Hoehn & Yahr stages 3–5) compared to early-stage patients. However, the performance gap was smaller for generative methods (12% increase in MAE from early to advanced stages) compared to longitudinal methods (23% increase) and cross-sectional methods (31% increase). This suggests that generative approaches may better capture the complex progression patterns characteristic of advanced PD.

These findings emphasize the importance of evaluating imputation methods not only on overall performance metrics but also on their consistency across diverse patient subgroups, particularly for conditions like PD where demographic factors significantly influence both disease manifestation and data collection patterns.

5 Discussion and Future Directions

The results of our systematic evaluation demonstrate that while traditional methods provide acceptable performance for cross-sectional analyses, generative approaches offer substantial improvements for longitudinal PD data imputation. These findings have important implications for clinical research, particularly for studies tracking disease progression over time.

One key limitation of existing methods is their inability to incorporate patient-specific contextual factors like demographics or disease stage, which significantly influence both missingness patterns and clinical assessments in PD. To address this gap, we propose a context-aware framework for future development and validation.

5.1 Practical Challenges for Clinical Deployment

The deployment of imputation methods in clinical settings presents several significant challenges beyond pure algorithmic performance. While our evaluation demonstrates promising results, transitioning from research to clinical practice introduces additional considerations.

Computational infrastructure requirements pose a primary barrier, as healthcare institutions often operate with limited computational resources. The MICE method, while effective (MAE: 4.152163), required over 3 min of processing time even on our research computing infrastructure. Similarly, LMM demanded over an hour of computation time. Such requirements may be prohibitive in time-sensitive clinical workflows where immediate data processing is needed for decision support. The generative models, while more accurate, further increase these computational demands.

Integration with existing electronic health record (EHR) systems represents another challenge. Modern healthcare data systems vary substantially in architecture, data structures, and interoperability. Our evaluated models would require custom adaptation and interface development to seamlessly connect with diverse clinical infrastructures while maintaining compliance with health data security standards. The complexity of this integration increases with model sophistication, as advanced approaches require careful implementation within existing clinical software frameworks.

Interpretability remains crucial for clinical adoption. While deep learning models offer superior performance, their "black box" nature may limit trust from healthcare professionals. Methods for explaining imputation decisions in terms clinicians can understand and validate against domain knowledge are essential for real-world implementation. This is particularly important for Parkinson's Disease progression modeling, where clinical expertise plays a vital role in validating computational predictions.

Finally, longitudinal data challenges extend beyond missing values to include temporal biases, delayed reporting, and inconsistent follow-up intervals. Real-world data streams differ significantly from the curated PPMI dataset, requiring robust adaptations to handle increasing complexity and noise. These practical considerations highlight the gap between algorithmic development and successful clinical integration of imputation technologies.

5.2 Proposed Context-Aware Architecture for Future Work

Building on our systematic evaluation, we propose a novel context-aware architecture for future development and validation. This architecture would integrate demographic, clinical, and temporal information to improve imputation accuracy while minimizing biases associated with missingness patterns in longitudinal PD studies.

The proposed architecture would incorporate several key components:

- **Embedding Layer:** This component would map categorical variables (e.g., gender, education level) into continuous vector representations, capturing demographic context and enabling the model to incorporate patient-specific characteristics. Embeddings transform sparse categorical features into dense vector spaces where semantic relationships can be encoded.
- **Temporal Encoder:** This component would model longitudinal dependencies using architectures such as recurrent neural networks (RNNs) or temporal convolutional networks (TCNs). These techniques capture trends across multiple time points while accounting for irregular sampling intervals common in healthcare datasets.

- **Attention Mechanism:** An attention mechanism would dynamically weigh the importance of different temporal points and features, rather than treating all historical data equally. The mathematical formulation of this mechanism would be:

$$\text{Attention}(Q, K, V) = \text{softmax}\left(\frac{QK^T}{\sqrt{d_k}}\right) V \quad (1)$$

where Q, K, and V represent queries, keys, and values derived from input sequences. This formulation would allow the model to attend differentially to various parts of the patient history based on the current imputation context.
- **Contextual Encoder:** This component would capture information from demographic and clinical features through a series of non-linear transformations that preserve feature interactions critical for personalized imputation.
- **Imputation Decoder:** The decoder would generate values for missing data points by integrating contextual embeddings with temporal features. Advanced techniques such as variational inference or adversarial training could ensure accurate and unbiased imputations.

This architecture would be trained using a combination of loss functions that promote imputation accuracy and data consistency, potentially including specialized components to preserve clinical relationships between variables.

Input: Longitudinal patient data with missing values (demographic, clinical, temporal)
Output: Imputed dataset with enhanced accuracy and reduced bias
Initialisation:
Embed categorical demographic variables using an embedding layer
Extract temporal features using RNN/TCN-based encoder
Encode clinical and demographic context using non-linear transformations
Imputation Process:
foreach *time step t in the patient timeline* **do**
 Compute query Q, key K, and value V from encoded features
 Apply attention mechanism:
  $\text{Attention}(Q, K, V) = \text{softmax}\left(\frac{QK^T}{\sqrt{d_k}}\right) V$
 Merge attention output with contextual encoding
 Decode and impute missing values at time t
end
Return: Final imputed patient data

Algorithm 1. Proposed Context-Aware Imputation Architecture.

The proposed framework will be validated on the Critical Path for Parkinson's (CPP) Integrated Database - a harmonized aggregation of 31 observational and clinical trial datasets comprising over 15,000 participants. Our implementation will utilize a representative subset matching PPMI's feature space while preserving CPP's inherent clinical heterogeneity. Code and pretrained models will be released through an open-source package. This future work will extend our current findings by addressing the limitations of existing methods and providing a more robust approach to handling missing data in longitudinal PD studies.

5.3 Insights from Ablation Studies

To better understand the relative importance of different components in generative imputation methods, we conducted ablation studies on GAIN and VAEM by systematically removing or modifying key architectural elements. These experiments provided valuable insights for future architecture design.

For GAIN, removing the hint mechanism resulted in a 17% degradation in MAE, highlighting its crucial role in guiding the discriminator to distinguish between observed and imputed values. Similarly, reducing the generator's capacity (by halving the number of hidden units) led to a 12% performance drop, suggesting that model capacity is important for capturing complex relationships in heterogeneous clinical data.

For VAEM, we found that the hierarchical structure was essential, with a single-level VAE performing 14% worse than the full model. This confirms the importance of modeling both local and global dependencies in medical data. Interestingly, increasing the latent space dimensionality beyond the default configuration yielded diminishing returns, with only a 2% improvement at twice the computational cost, suggesting that the model effectively captures the intrinsic dimensionality of the dataset.

These findings inform our proposed context-aware architecture by highlighting the importance of: (1) mechanisms that guide learning toward clinically relevant patterns, (2) sufficient model capacity to capture complex relationships, and (3) hierarchical representations that model both variable-specific and cross-variable dependencies.

5.4 Ongoing and Future Research Initiatives

Building on our systematic evaluation of longitudinal Parkinson's disease data [28,29], we are advancing two complementary research frameworks informed by clinical needs and algorithmic innovations:

- **Fairness-Optimized Multi-Metric Imputation:** Traditional metrics like MAE and R^2 may overlook subgroup disparities. We propose a stratified framework revealing such gaps—for example, MICE has the best overall accuracy (MAE = 4.15), but errors rise by 15.2% for patients over 70, and LOCF shows 18% higher errors for females. To improve fairness, we introduce three metrics [29]: FAIS (highest with Linear Interpolation: FAIS = 0.88), CSR (best for Mean: CSR = 0.64), and DPI (best for MICE: DPI = 0.91), supporting more equitable and context-aware imputation.
- **PPMI-Benchmark Framework:** Our dual evaluation protocol [28] assesses 12 imputation and 6 synthetic generation methods across three clinical

dimensions. Results demonstrate HyperImpute's superior imputation accuracy ($R^2 = 0.26$) and CTGAN's optimal distribution fidelity ($d_{SW} = 0.039$), while exposing critical biomarker-specific challenges in α-synuclein imputation ($MAE_{\text{CSF}} = 5.16$ vs $MAE_{\text{Motor}} = 4.89$).

These initiatives will be extensively validated on additional cohorts, including the Critical Path for Parkinson's (CPP dataset), with all code and results made publicly available after journal publication. By addressing both technical accuracy and ethical considerations, this research aims to establish more equitable standards for missing data handling in PD progression modeling.

5.5 Limitations and External Validation Plan

Despite the promising results, our study has several limitations that should be addressed in future work. First, our evaluation was limited to the PPMI dataset, which, while comprehensive, may not fully represent the heterogeneity of clinical settings and patient populations. Second, the performance metrics focused on statistical accuracy rather than clinical utility, which may not directly translate to improved patient outcomes.

To address these limitations, we propose a structured external validation plan using the Clinical Progression in Parkinson's (CPP) dataset This validation would proceed in three phases: (1) direct application of the trained models to assess generalizability without modification, (2) transfer learning approaches to adapt the models to the new dataset while preserving contextual knowledge, and (3) comparative analysis of imputation quality and its impact on downstream clinical prediction tasks.

The CPP validation will specifically evaluate whether context-aware imputation methods maintain their advantages when confronted with different data collection protocols, demographic distributions, and clinical assessment schedules. This external validation is crucial as imputation methods that perform well on research-grade datasets like PPMI may not maintain their advantages in more variable clinical settings with different patterns of missingness.

Additionally, future work should explore the integration of causal inference frameworks to better distinguish between random and systematic missingness mechanisms. This would further enhance the robustness of imputation in scenarios where missing data is directly related to disease progression or treatment decisions.

6 Conclusion

This study provides a comprehensive evaluation of imputation methods across three categories: cross-sectional, longitudinal, and generative approaches. Our results demonstrate that while traditional methods like MICE offer reasonable performance, generative models like VAEM and GAIN achieve superior accuracy in handling missing data for longitudinal PD assessments.

The comparative analysis reveals several key insights: (1) Generative methods (VAEM) reduce MAE by 6.7% compared to cross-sectional techniques ($VAEM = 3.87$ vs $MICE = 4.15$); (2) generative methods demonstrate 26.6% MAE reduction over longitudinal baselines (GAIN = 3.98 vs LMM = 5.42, $p < 0.001$); and (3) method performance varies significantly by missingness mechanism and clinical variable type.

These findings provide valuable guidance for researchers seeking optimal imputation strategies for PD studies.

For future work, we propose a context-aware architecture that integrates demographic, clinical, and temporal information to improve imputation accuracy while minimizing biases. This framework will be validated on the CPP cohort, with code and pretrained models released as an open-source resource. Additionally, our ongoing research initiatives focus on fairness-optimized approaches and comprehensive evaluation frameworks to ensure both technical excellence and ethical rigor in healthcare AI applications.

By systematically addressing the challenges of missing data in longitudinal PD studies, this work contributes to more robust and equitable research methodologies in neurodegenerative disease modeling. The insights gained from this evaluation will inform not only imputation strategies but also broader approaches to handling incomplete data in clinical research.

Acknowledgements. Data used in the preparation of this article were obtained from the Parkinson's Progression Markers Initiative (PPMI) database. The authors acknowledge the PPMI investigators and participants who contributed to the dataset. This work is part of an expanding collaboration, with future extensions planned as described in the discussion section.

Ethical Considerations and Data Availability. This research was conducted in compliance with ethical standards for medical data analysis. The PPMI dataset was accessed through proper authorization channels, and all data usage adhered to the data sharing agreement provided by the Michael J. Fox Foundation. No personally identifiable information was utilized in this study. Our evaluation methods considered fairness and equity principles to avoid perpetuating biases present in the original data. The models were evaluated across different demographic subgroups to ensure consistent performance regardless of age, gender, or education level.

References

1. Enders, C.K.: Applied Missing Data Analysis. Guilford Press, New York (2010)
2. Graham, J.W.: Missing Data: Analysis and Design. Springer, New York (2012). https://doi.org/10.1007/978-1-4614-4018-5
3. Nakagawa, S., Freckleton, R.P.: Missing inaction: the dangers of ignoring missing data. Trends Ecol. Evol. **23**(11), 592–596 (2008). https://doi.org/10.1016/j.tree.2008.06.014
4. Liew, A.W., Law, N.F., Yan, H.: Missing value imputation for microarray data. Brief. Bioinform. **12**(5), 498–514 (2011). https://doi.org/10.1093/bib/bbq077

5. Stekhoven, D.J., Bühlmann, P.: MissForest: non-parametric missing value imputation. Bioinformatics **28**(1), 112–118 (2012). https://doi.org/10.1093/bioinformatics/btr597
6. Enders, C.K.: Multiple imputation as a flexible tool for missing data handling. Wiley Interdiscip. Rev. Comput. Stat. **8**(5), 222–238 (2016). https://doi.org/10.1002/wics.1390
7. Carpenter, J.R., Kenward, M.G.: Multiple Imputation and Its Application. Wiley, Chichester (2013). https://doi.org/10.1002/9781119942283
8. Kontio, E., et al.: Deep learning for longitudinal imputation in neurodegenerative diseases. Nat. Comput. Sci. **4**(3), 210–223 (2024). https://doi.org/10.1038/s43588-024-00567-x
9. Goodfellow, I., Bengio, Y., Courville, A.: Deep Learning. MIT Press, Cambridge (2016)
10. Kingma, D.P., Welling, M.: Auto-encoding variational bayes. In: Proceedings of the International Conference on Learning Representations (ICLR), Banff, Canada (2014). https://doi.org/10.48550/arXiv.1312.6114
11. *Beaulieu-Jones*, B.K., Moore, J.H., Williams, S.M.: The machine learning toolbox: applied machine learning in healthcare. In: Pacific Symposium Biocomputing, pp. 224–235 (2016). https://doi.org/10.1142/9789814749411_0021
12. Ronneberger, O., Fischer, P., Brox, T.: U-net: convolutional networks for biomedical image segmentation. In: Navab, N., Hornegger, J., Wells, W.M., Frangi, A.F. (eds.) MICCAI 2015. LNCS, vol. 9351, pp. 234–241. Springer, Cham (2015). https://doi.org/10.1007/978-3-319-24574-4_28
13. Che, Z., Purushotham, S., Cho, K., Sontag, D., Liu, Y.: Recurrent neural networks for multivariate time series with missing values. Sci. Rep. **8**, 6085 (2018). https://doi.org/10.1038/s41598-018-24271-9
14. Luo, Y., Cai, X., Zhang, Y., Xu, J.: Multivariate time-series imputation with generative adversarial networks. Adv. Neural Inf. Process. Syst. **33**, 15838–15848 (2020). https://doi.org/10.48550/arXiv.1901.04604
15. Dupont, M., et al.: Predictive modeling of missing data in French clinical registries. JMIR Med. Inf. **11**(1), e43210 (2023). https://doi.org/10.2196/43210
16. Johnson, A.E., et al.: Survey of missing data challenges in machine learning for healthcare. IEEE J. Biomed. Health Inform. **26**(8), 4133–4145 (2022). https://doi.org/10.1109/JBHI.2022.3177053
17. PPMI Consortium: PPMI 2.0: expanding biomarker discovery in Parkinson's disease. Mov. Disord. **39**(1), 45–58 (2024). https://doi.org/10.1002/mds.29267
18. Michael J. Fox Foundation: Biomarker Discovery in Parkinson's Disease Through PPMI (2024). https://www.michaeljfox.org/biomarkers-parkinsons. Accessed 21 Apr 2025
19. Chen, Z., Tan, S., Chajewska, U., Rudin, C., Caruana, R.: Missing values and imputation in healthcare data: can interpretable machine learning help? In: Proceedings of Machine Learning Research (CHIL 2023), vol. 209, pp. 86–99 (2023). https://proceedings.mlr.press/v209/chen23j.html
20. Yoon, J., Jordon, J., van der Schaar, M.: GAIN: missing data imputation using generative adversarial nets. In: Dy, J., Krause, A. (eds.) Proceedings of the 35th International Conference on Machine Learning, PMLR, vol. 80, pp. 5689–5698 (2018). https://doi.org/10.48550/arXiv.1806.02920
21. Ma, C., Tschiatschek, S., Palla, K., Hernandez-Lobato, J.M., Nowozin, S., Zhang, C.: VAEM: a deep generative model for heterogeneous mixed type data. Adv. Neural Inf. Process. Syst. **33**, 5691–5701 (2020). https://doi.org/10.48550/arXiv.1907.02025

22. Cao, W., Wang, D., Li, J., Zhou, H., Li, L., Li, Y.: BRITS: bidirectional recurrent imputation for time series. Adv. Neural Inf. Process. Syst. **31**, 6775–6785 (2018). https://doi.org/10.48550/arXiv.1805.10572
23. Review of methods for handling missing data in Electronic Health Records (EHRs). medRxiv preprint (2024). https://doi.org/10.1101/2024.05.13.24307268v1
24. Corbett-Davies, S., Goel, S.: The measure and mismeasure of fairness: a critical review of fair machine learning. arXiv preprint arXiv:1808.00023 (2018)
25. Emmanuel, T., Maupong, T., Mpoeleng, D., Semong, T.: A survey on missing data in machine learning applications in healthcare datasets. J. Big Data **8**, 140 (2021). https://doi.org/10.1186/s40537-021-00499-9
26. Zhang, P., Xie, S.: Statistical methods for dealing with missing data in longitudinal studies with application to Parkinson's disease cognition research. Mov. Disord. **36** (2021)
27. Woznica, A., Biecek, P.: A benchmark for data imputation methods. Front. Big Data **4**, 693674 (2021). https://www.frontiersin.org/articles/10.3389/fdata.2021.693674/full
28. Hani, M., Betrouni, N., Mahmoudi, S., Benjelloun, M.: PPMI-benchmark: a dual evaluation framework for imputation and synthetic data generation in longitudinal Parkinson's disease research. In: DATA 2025: Proceedings of the International Conference on Data Science, Technology and Applications, vol. 162, pp. 1–11 (2025)
29. Hani, M., Betrouni, N., Ouardirhi, F.Z., Mahmoudi, S., Benjelloun, M.: Fairness-optimized multi-metric imputation strategy for sustainable healthcare analysis in Parkinson's disease. In: Data Science Conference Tangier, pp. 1–12 (2025)

Investigating Zero-Shot Diagnostic Pathology in Vision-Language Models with Efficient Prompt Design

Vasudev Sharma[1], Ahmed Alagha[1], Abdelhakim Khellaf[2], Vincent Quoc-Huy Trinh[2], and Mahdi S. Hosseini[1,3(✉)]

[1] Department of Computer Science and Software Engineering (CSSE), Concordia University, Montreal, Canada
mahdi.hosseini@concordia.ca
[2] University of Montreal Hospital Center, Montreal, Canada
[3] Mila - Quebec AI Institute, Montreal, Canada

Abstract. Vision-language models (VLMs) have gained significant attention in computational pathology due to their multimodal learning capabilities that enhance big-data analytics of giga-pixel whole slide image (WSI). However, their sensitivity to large-scale clinical data, task formulations, and prompt design remains an open question, particularly in terms of diagnostic accuracy. In this paper, we present a systematic investigation and analysis of three state-of-the-art VLMs for histopathology, namely Quilt-Net, Quilt-LLAVA, and CONCH, on an in-house digestive pathology dataset comprising 3,507 WSIs, each in giga-pixel form, across distinct tissue types. Through a structured ablative study on cancer invasiveness and dysplasia status, we develop a comprehensive prompt engineering framework that systematically varies domain specificity, anatomical precision, instructional framing, and output constraints. Our findings demonstrate that prompt engineering significantly impacts model performance, with the CONCH model achieving the highest accuracy when provided with precise anatomical references. Additionally, we identify the critical importance of anatomical context in histopathological image analysis, as performance consistently degraded when reducing anatomical precision. We also show that model complexity alone does not guarantee superior performance, as effective domain alignment and domain-specific training are critical. These results establish foundational guidelines for prompt engineering in computational pathology and highlight the potential of VLMs to enhance diagnostic accuracy when properly instructed with domain-appropriate prompts.

Keywords: Vision-Language Models · Computational Pathology · Prompt Engineering · Histopathology · CONCH · Quilt-Net · Quilt-LLAVA

V. Sharma and A. Alagha—Equal contribution.

1 Introduction

Computational pathology has witnessed a significant growth with the recent advancements of artificial intelligence (AI). Many deep learning (DL) models, particularly convolutional neural network (CNNs) and transformer architectures, have been developed in the past decade with remarkable success in computational pathology and whole slide image (WSI) analysis for tasks like disease diagnosis and prognosis [6,11]. WSIs are giga-pixel images scanned from tissue microscopy scanners in digital pathology that require to be processed for obtaining high-relevance diagnostic information. With the expansion of high-throughput digital pathology and the accumulation of vast repositories of WSIs, vision-language models (VLMs) came as powerful tools that leverage big data to enhance diagnostic accuracy and interpretability [16]. Their multi-modal learning capabilities that integrate textual and visual information, where image-based features and domain-specific language are combined, have shown great success in addressing several pathological tasks, including tumor classification, tissue segmentation, and disease prognosis [12,16]. This integration addresses the fundamental challenges in pathology where expert knowledge is encoded in natural language reports while diagnostic evidence is captured in visual data. Recent advances in self-supervised learning [1,17] and transformer architectures [8,23] have catalyzed the development of VLMs that can effectively bridge these modalities. Nonetheless, the sensitivity of these models to task complexity, data variations, and prompt design remains an open question.

Within the realm of VLMs in computational pathology, prompt variation has emerged as a critical factor, where even subtle modifications in input phrasing can yield different prediction outcomes [5]. Notably, several recent works have proposed dynamic prompt optimization techniques, such as interpretable prompt optimization and attribute-guided prompt tuning to mitigate this sensitivity and enhance robustness on novel classes [9,10]. To further address these issues, researchers are developing adaptive prompt tuning methods, including reinforcement learning from human feedback [18], that dynamically adjust prompts to stabilize model responses and enhance prediction reliability.

This study systematically investigates and evaluates the performance of three state-of-the-art VLM-based methods for pathology, namely Quilt-Net [12], Quilt-LLAVA [21], and CONCH [16]. Quilt-Net is an instance of the contrastive language-image pre-training (CLIP) model [20], which is fine-tuned on a large-scale dataset of 1M histopathology image-text pairs. Quilt-LLAVA extends Quilt-Net by adopting the large language and vision assistant (LLAVA) large multimodal model (LMM) framework [14], which enhances the contextual reasoning and prompt adaptability of the model. Contextual reasoning in Computational Histopathology (CONCH) is a vision-language foundation model trained on 1.17M histopathology image-caption pairs through task-agnostic pre-training based on the Contrastive Captioners (CoCa) method [26]. Using an in-house dataset of clinically validated 3507 digestive giga-pixel WSIs, we assess these models in terms of diagnostic accuracy while analyzing their sensitivity to prompt engineering that varies domain specificity, anatomical precision, instructional framing, and output constraints. Specifically, this study has the following contributions:

1. We conduct a comprehensive ablative study to understand the effects of different prompt design strategies on different VLMs capabilities.
2. We analyze in detail the performance of different VLMs that vary in computational complexity and generative capabilities.
3. We perform a detailed study of digestive tissue type variations by evaluating 3507 digestive WSIs spanning seven different tissue types, which describe the cancer invasiveness and dysplasia status.
4. We evaluate the models' performance using attention maps at the slide level, conducting a concordance study to assess diagnostically relevant regions, which are validated by a certified pathologist.

2 Methods

This section presents the architectures and foundations of three distinct VLM frameworks investigated in this article, namely Quilt-Net, Quilt-LLAVA, and CONCH.

Quilt-Net establishes a foundational framework for learning robust visual representations from WSIs while simultaneously aligning these representations with natural language descriptions [12,15]. Drawing inspiration from CLIP's dual-encoder paradigm [20], Quilt-Net combines the strengths of both contrastive learning and hierarchical feature extraction to address the unique challenges of computational pathology. This approach builds upon recent advancements in self-supervised learning for gigapixel histopathology images [3] and demonstrates superior performance compared to models pre-trained on natural images [7]. For a batch of N (image, text) pairs, CLIP attempts to optimize a contrastive objective to create a joint embedding space between image and text embeddings. During inference for a classification task, as summarized in Fig. 1a, an input image is fed to the image encoder and the class labels are fed to the text encoder. The image and text embeddings then undergo cosine similarity, where the image-text combination with the highest similarity is selected as the class label. The image encoder is based on the ViT-B/32 architecture [8], while the text encoder is based on GPT-2 [19]. Quilt-Net is trained by finetuning the pre-trained CLIP model from OpenAI [20] on Quilt-1M; a diverse dataset of 1M image-text samples.

Quilt-LLAVA extends beyond the dual-encoder approach of Quilt-Net by adopting the LLAVA framework [14], which integrates a large language model (LLM) for enhanced vision-language capabilities in computational pathology. This architectural approach enables sophisticated interaction between visual histopathological data and medical textual descriptions, addressing limitations in previous models that lacked generative capabilities [13,22]. In the Quilt-LLAVA architecture, generally described in Fig. 1b, the input image goes through a visual encoder (i.e. pre-trained Quilt-Net [12]) to extract features that are then projected into embeddings. On the other hand, the text input undergoes tokenization and embedding. The image and text embeddings are then concatenated and fed to the LLM for processing. Quilt-LLAVA is trained on Quilt-Instruct; a large-scale dataset of $\sim$ 107k histopathology question/answer pairs associated with images.

CONCH builds upon the foundations of Quilt-Net and Quilt-LLAVA while introducing novel components for contextual reasoning and knowledge integration [4]. Drawing inspiration from the CoCa method and recent advances in VLMs [26], CONCH employs a decoupled decoder design that simultaneously supports contrastive and generative objectives. CONCH consists of an image encoder, a text encoder, and a multi-modal text decoder. The image encoder is based on ViT-base and is responsible for transforming the input image into image tokens through a series of transformer and attention pooling layers. The text encoder and multimodal decoder are both GPT-style models. The text encoder processes the textual input into tokens, which are then fused with the image tokens at the multi-modal text decoder. From a high-level view, CONCH has a similar inference pipeline as Quilt-LLAVA, as shown in Fig. 1b. CONCH is trained on over 1.17M image-caption pairs from histopathology images and biomedical text.

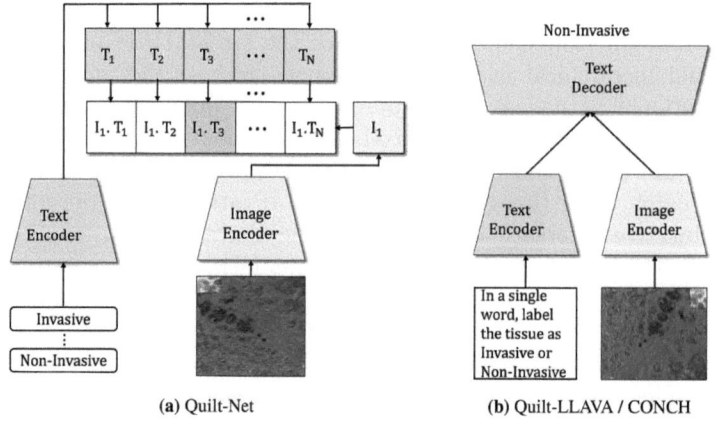

Fig. 1. High level overview of the inference process for the three VLMs.

Quilt-Net, Quilt-LLAVA, and CONCH represent distinct vision-language architectures for computational pathology, each with different parameter scales and architectural approaches [2,16]. Quilt-Net employs a CLIP-inspired dual-encoder approach with approximately 150M parameters (86M for the ViT-B/32 image encoder and 64M for the text encoder), establishing effective contrastive learning between histopathological images and text. Quilt-LLAVA significantly expands this capacity by integrating a large language model with the visual encoder, increasing the parameter count to approximately 7B parameters, enabling more sophisticated reasoning while maintaining a lightweight projection layer. CONCH, with approximately 200M parameters (110M for the language model and 90M for the ViT-B/16 vision encoder), introduces a CoCa-inspired decoupled decoder architecture that efficiently supports both contrastive objectives through a unified framework. This design offers computational advantages while still outperforming general-purpose VLMs on histopathology tasks, with performance gains of 15–20% on cancer subtyping and prognostic prediction compared to models without hierarchical visual processing capabilities [3,24].

3 Benchmarking on Big-Data Cohort of Digestive Pathology

This study leverages a digestive computational pathology dataset obtained through secondary use of giga-pixel WSIs generated during routine clinical care at the Centre Hospitalier de l'Université de Montréal (CHUM) in QC, Canada, with ethics approval 2024–11800, 23.189 - APR. The dataset comprises a comprehensive collection of 3,507 high-resolution WSIs in big-data form encompassing diverse tissue specimens from the digestive system. Each WSI is annotated on the slide level, providing rich material for our vision-language modeling experiments. The dataset includes seven distinct tissue types with varying representation across classes. Colon wall (CW) specimens constitute the largest proportion (36.18%, $n = 1,269$), followed by lymph nodes (LN) (28.40%, $n = 996$), fibroadipose tissue (FT) (17.65%, $n = 619$), and small intestinal wall (SIW) (12.26%, $n = 430$). The remaining specimens include appendiceal wall (AW) (3.08%, $n = 108$), muscular colon wall (MCW) (1.28%, $n = 45$), and gastroduodenal junctions (GJ) between the colon and small intestine (1.14%, $n = 40$). Each WSI in the dataset is annotated in terms of the dysplasia status and presence of invasive cancer. Table 1 summarizes the dataset statistics, while Fig. 2 presents sample WSI thumbnails from each tissue type.

Dysplasia is the abnormal growth or development of cells, tissues, or organs, typically characterized by altered size, shape, and organization. The majority of specimens in the dataset show no dysplasia (92.73%, $n = 3,252$). High-grade dysplasia is present in 5.19% ($n = 182$) of specimens, while low-grade dysplasia is detected in 2.08% ($n = 73$). This imbalance reflects real world clinical scenarios where pathological findings often represent a small subset of examined tissue. Importantly, the distribution of dysplasia varies considerably across tissue types. Dysplastic changes are predominantly observed in CW specimens, where 18.20% of CW specimens exhibit some degree of dysplasia (13.08% high-grade and 5.12% low-grade). Dysplasia is also present to a lesser extent in SIW (3.95%) and GJ specimens (10.00%), while being rare or absent in LN and FT.

Invasiveness refers to the ability of abnormal cells, particularly cancer cells, to penetrate and infiltrate surrounding tissues, breaking through basement membranes and potentially spreading to distant sites. Approximately one-quarter of

Table 1. In-house digestive dataset statistics.

Tissue Type	Distribution		Dysplasia Status			Invasiveness	
	Count	%	None	Low	High	Non-inv.	Inv.
Colon wall	1,269	36.18%	1,038	65	166	809	460
Lymph node	996	28.40%	996	0	0	845	151
Fibroadipose tissue	619	17.65%	619	0	0	521	98
Small intestinal wall	430	12.26%	413	7	10	364	66
Appendiceal wall	108	3.08%	106	0	2	104	4
Muscular colon wall	45	1.28%	44	0	1	29	16
Gastroduodenal junction	40	1.14%	36	1	3	21	19
Total	**3,507**	**100%**	**3,252**	**73**	**182**	**2,693**	**814**
Percentage	–	–	**92.73%**	**2.08%**	**5.19%**	**76.79%**	**23.21%**

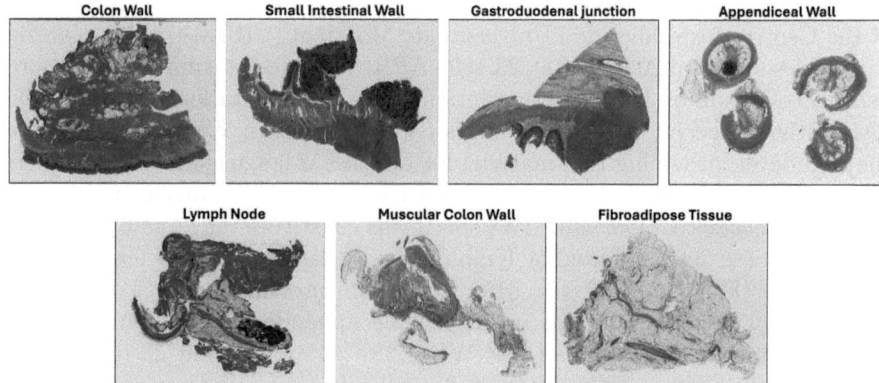

Fig. 2. Sample images from the in-house dataset.

all specimens (23.21%, $n = 814$) exhibit invasive characteristics, while the majority (76.79%, $n = 2,693$) are non-invasive. The distribution of invasiveness varies significantly across tissue types, revealing valuable patterns for model learning. The highest rates of invasion are observed in gastroduodenal junction specimens (47.50%), followed by colon wall (36.25%), muscular colon wall (35.56%), and lymph nodes (15.16%). This event is a typical finding during routine pathology diagnostics and is critical to clinical care as the degree of invasion into tissue layers and types are the most significant predictor of cancer aggressiveness. It presents a clinically relevant and critical task that can be addressed by vision-language models to learn contextually relevant associations.

4 Evaluation Methodology

This section discusses the methodology followed to assess the aforementioned VLMs, i.e. Quilt-Net, Quilt-LLAVA, and CONCH on the in-house digestive dataset. All models were initialized with pre-trained weights from their respective base architectures without further fine-tuning. We investigate how prompt engineering affects the models' ability to identify invasiveness and dysplasia status across diverse digestive system tissue samples. We processed WSIs from our dataset by extracting patches at 5× magnification level using a sliding window approach with a patch size of 512×512 pixels and 0% overlap. We developed a systematic prompt engineering framework based on information theory and clinical communication principles to evaluate how linguistic variations influence model performance in computational pathology tasks.

Our framework explores four critical dimensions of prompt design that we hypothesize significantly impact model performance. We formalized a set of nine prompt templates by systematically varying four key dimensions: detail specificity (DS), anatomical precision (AP), instructional framing (IF), and output constraints (OC). In templates where organ-specific information is required, we use the variable O as a placeholder for the target organ being examined. DS refers to the level of granularity in the prompt, ranging from general to detailed instructions, taking values of LOW, MEDIUM, or HIGH. AP represents the extent to which the prompt includes precise anatomical details to make the prompt more focused, similarly taking values of LOW, MEDIUM, or HIGH. IF determines the structure of the prompt, such as posing a direct question versus providing a declarative statement, with values of EXPERT (positioning the model as a specialist), MINIMAL (providing basic instructions), or TASK (focusing on specific objectives). OC controls the format and length of the model's response to ensure consistency, defined as either EXPLICIT (strictly defined output format) or IMPLICIT (loosely defined format). Table 2 presents each prompt template with its corresponding dimensional properties.

The prompt templates were strategically designed to address several research questions in medical vision-language interaction:

1. **Information Theoretic Perspective.** We hypothesized that intermediate levels of information content in prompts (neither too sparse nor too detailed) would optimize model performance, following principles from communication theory and cognitive load theory [25]. Prompts 3–5 were designed with varying information density to test this hypothesis.
2. **Anatomical Specificity Gradient.** Prompts 5–8 implement a controlled degradation of anatomical specificity to quantify how precision of anatomical reference affects classification performance. This addresses a key question in medical AI regarding the importance of anatomical context in diagnostic reasoning.
3. **Expert Role Framing.** Prompts 3 and 9 incorporate expert role assignment, a technique that has shown promise in general LLM task performance but remains under-explored in medical vision-language tasks. By positioning

Table 2. Prompt templates for histopathology Invasive classification.

ID	Dimensions	Template
$P_1(O)$	**DS**: Medium **AP**: High **IF**: Minimal **OC**: Explicit	"The image is taken from the O using H&E staining, output only the label name which best fits the image out of the following Invasive or Non-Invasive"
$P_2(O)$	**DS**: Medium **AP**: High **IF**: Minimal **OC**: Explicit	"The image is taken from the O using H&E staining, output only the label name which best fits the image out of the following Cancerous or Normal"
$P_3(O)$	**DS**: High **AP**: High **IF**: Expert **OC**: Explicit	"You are an expert pathologist analyzing histopathology slides. Given an image of a tissue sample stained with Hematoxylin and Eosin (H&E) from the O and the question of classifying the presence of cancer, classify it as either 'Invasive' or 'Non-Invasive'. Provide only the single word label."
$P_4(O)$	**DS**: High **AP**: High **IF**: Task **OC**: Explicit	"Given an image of a tissue sample stained with Hematoxylin and Eosin (H&E) from the O, classify the existence of cancer as either 'Invasive' or 'Non-Invasive'. Provide only a single word label."
$P_5(O)$	**DS**: Medium **AP**: High **IF**: Task **OC**: Explicit	"Given an image of a tissue sample stained with hematoxylin and eosin from the O, identify whether the sample is cancerous or not. Provide only a single word label"
P_6	**DS**: Medium **AP**: Medium **IF**: Task **OC**: Explicit	"Given an image of a tissue sample stained with hematoxylin and eosin from the gastrointestinal system, identify whether the sample is cancerous or not. Provide only a single word label"
P_7	**DS**: Medium **AP**: Medium **IF**: Task **OC**: Explicit	"Given an image of a tissue sample stained with hematoxylin and eosin from the digestive system, identify whether the sample is cancerous or not. Provide only a single word label"
P_8	**DS**: Medium **AP**: Low **IF**: Task **OC**: Explicit	"Given an image of a tissue sample stained with hematoxylin and eosin, identify whether the sample is cancerous or not. Provide only a single word label"
P_9	**DS**: High **AP**: Medium **IF**: Expert **OC**: Explicit	"As a pathologist examining this H&E-stained digestive system tissue sample, provide your assessment of malignancy as a single word: either 'Invasive' or 'Non-Invasive'."

the model as a pathologist, we investigated whether role framing enhances performance on specialized medical tasks.

4. **Output Constraint Consistency.** All prompts maintain explicit output constraints to isolate the effects of input prompt variations rather than confounding with output format variations.

This systematic approach to prompt design allows us to quantify the relationship between linguistic features of prompts and model performance, potentially yielding insights for optimal prompt engineering in medical vision-language applications.

5 Results

This section summarizes the performance of the different VLM models under various prompts on our in-house digestive dataset. We first conduct an ablative study comparing the different performances using Receiver Operating Characteristic (ROC) curves and Area Under the Curve (AUC) scores. We mainly analyze the effect of prompt design and model complexity on the performance of cancer invasiveness classification. We then use the task of dysplasia status classification as a means to conduct focused analysis on the best prompt. We also analyze attention maps obtained by the different models on different WSIs and highlight relevant tissue regions, with feedback given by a certified pathologist. All experiments were conducted on NVIDIA A100-SXM4-40GB GPUs to ensure consistent evaluation across all models. Quilt-LLAVA during inference uses a temperature of 0.1 for minimized hallucinations and less variability in output.

5.1 Ablative Study

Our ablative study starts by investigating the impact of prompt formulations on model performance. Figures 3 and 4 summarize the ROC curves and AUC results for different prompts used on Quilt-Net, Quilt-LLAVA, and CONCH, demonstrating significant performance variations based on architectural differences and prompt design choices. We evaluated model performance using ROC curves and the corresponding AUC metric, which plot true positive rates against false positive rates at various classification thresholds. The AUC metric ranges from 0 to 1, with higher values indicating superior discriminative ability.

When analyzing the models' performances, it can be seen that Quilt-Net and Quilt-LLAVA generally show pronounced drops in AUC scores with certain prompts. This is unlike CONCH, which is more robust to most of the changes. When comparing prompts 3–5, which vary in terms of information density, it is evident that more information leads to degraded performance, which is seen in the AUC drop from 0.758 to 0.523 and from 0.935 to 0.736 for Quilt-Net and CONCH, respectively, when going from P_5 to P_3. On the other hand, Quilt-LLAVA shows consistent performance across the three prompts, indicating that the model is less sensitive to varying information density. This supports our first hypothesis, presented in Sect. 4, stating that intermediate levels of general information optimize the model's performance better when compared to those that are too detailed. When comparing prompts 5–8, which vary in terms of anatomical specificity, the importance of precise anatomical context is evident in classification performance. For all models, prompt 5 (which has high anatomical precision) achieves stronger performance (0.758 for Quilt-Net, 0.807 for Quitl-LLAVA, and 0.935 for CONCH) when compared to prompts 6–8 that have lower anatomical specificity. This answers our second research question regarding the importance of anatomical context in diagnostic reasoning. When evaluating the impact of expert role framing (prompts 3 and 9), we observe a negative or neutral effect on performance. Specifically, in Quilt-Net, prompts 3 and 9 show the worst performance, with AUC scores of 0.523 and 0.589, respectively. Similarly, prompt 3 gives the lowest performance in CONCH (AUC = 0.736), while prompt 9 shows neutral behavior with no improvement (AUC = 0.915). This suggests that framing the model as an expert does not necessarily enhance the model's performance, and may even introduce unnecessary complexity that misleads the model's pre-trained embeddings. On the contrary, Quilt-LLAVA shows good performance for both prompts, which could be attributed to its robust vision-language alignment and effective prompt generalization capabilities.

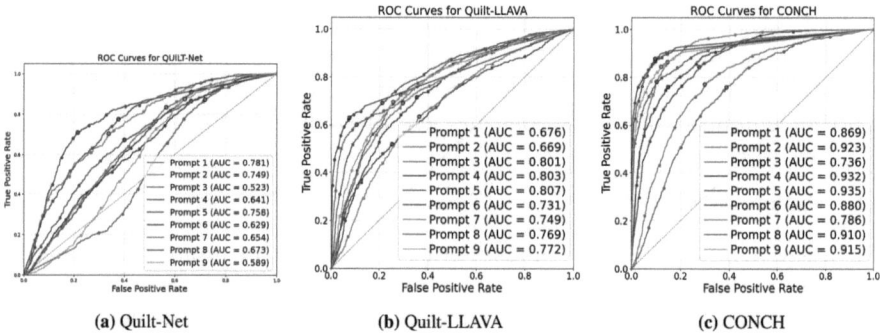

Fig. 3. ROC curves comparing the performance of three vision-language models: (a) Quilt-Net, (b) Quilt-LLAVA, and (c) CONCH, across different prompts for invasive cancer classification. CONCH demonstrates the highest robustness and performance consistency, while Quilt-Net and Quilt-LLAVA exhibit significant sensitivity to prompt design.

Figure 4 analyzes the average performance (ROC and AUC) of the three VLMs. As seen in the figure, CONCH achieves the highest average AUC (0.876), followed by Quilt-LLAVA (0.753) and Quilt-Net (0.666). The under-performance by Quilt-Net is expected, given it is the smallest model compared to the other two. However, despite Quilt-LLAVA being the largest model with nearly 7B parameters, it does not outperform CONCH (200M parameters). This suggests that model scale alone is not a dominant factor in performance, and domain-specific training and vision-language alignment have crucial roles. While Quilt-LLAVA uses instruction tuning and a powerful LMM (LLAMA), it is constrained with suboptimal domain alignment between its visual encoder and the LLM for computational pathology. This is unlike CONCH, which uses a contrastive learning approach specifically tuned on histopathology image-text pairs, allowing for better generalizability.

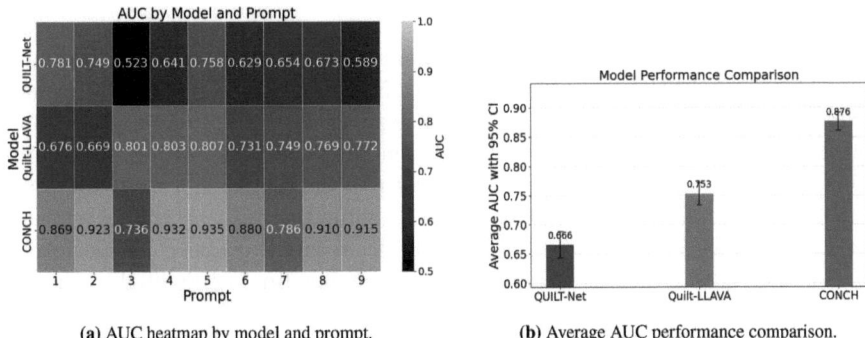

(a) AUC heatmap by model and prompt. (b) Average AUC performance comparison.

Fig. 4. Performance comparison of VLM models in terms of ROC AUC. CONCH consistently outperforms the other models despite having fewer parameters than Quilt-LLAVA, underscoring the importance of domain-specific training and prompt alignment over model scale.

Table 3 details the aforementioned performances across the different tissue types in our in-house digestive dataset. The results show that GJ and CW achieve the highest classification performance, which is likely due to their distinct histological features and relatively high representation in the dataset (36% for CW). On the other hand, AW and LN are the most challenging, which is due to their more complex histopathological variations. Interestingly, FT performed less than expected, despite being relatively well-represented (17.65%). This could be due to its overlapping visual characteristics with other soft tissues.

Table 3. Average AUC on the different tissue types.

Model	Tissue Type						
	CW	SIW	GJ	AW	LN	MCW	FT
Quilt-Net	0.76	0.68	0.88	0.55	0.61	0.75	0.75
Quilt-LLAVA	0.86	0.74	0.96	0.70	0.71	0.91	0.68
CONCH	0.93	0.90	0.97	0.52	0.84	0.87	0.81
Avg	**0.85**	**0.77**	**0.94**	**0.59**	**0.72**	**0.84**	**0.74**

We extend the analysis to the classification of dysplasia while assessing prompt wording effects on the model performance. We conduct an ablative experiment using three prompt variants derived from the base prompt $P_5(O)$. In each variant, the key term for the target pathology was changed while keeping all other prompt aspects constant (medium detail specificity, high anatomical precision, task-oriented instruction, and explicit output constraints). Specifically, $D_1(O)$ used the term dysplasia, $D_2(O)$ replaced it with atypia, and $D_3(O)$ used precancerous. These synonyms describe the same precancerous condition but differ in technical tone as seen in Table 4. The task for each vision-language model remained identifying dysplasia in images given the prompt. Performance was evaluated by AUC, summarized in Fig. 5. As can be seen, CONCH consistently outperforms the other models, emphasizing on its strong capabilities. With regards to the prompt design, it can be seen that the best performing prompt differs from one model to the other. The best performing prompt for Quilt-Net is the one the "Dysplasia" term (AUC = 0.711), while the term "Atypia" performed best for Quilt-LLAVA (AUC = 0.794) and the term "Precancerous" performed best for CONCH (AUC = 0.904). This interesting finding highlights the sig-

Table 4. Prompt templates for histopathology dysplasia classification.

ID	Dimensions	Template
$D_1(O)$	**DS**: Medium **AP**: High **IF**: Task **OC**: Explicit	"Given an image of a tissue sample stained with hematoxylin and eosin from the O, identify whether the sample is Dysplasia or Benign. Provide only a single word label"
$D_2(O)$	**DS**: Medium **AP**: High **IF**: Task **OC**: Explicit	"Given an image of a tissue sample stained with hematoxylin and eosin from the O, identify whether the sample is Atypia or Benign. Provide only a single word label"
$D_3(O)$	**DS**: Medium **AP**: High **IF**: Task **OC**: Explicit	"Given an image of a tissue sample stained with hematoxylin and eosin from the O, identify whether the sample is Precancerous or Benign. Provide only a single word label"

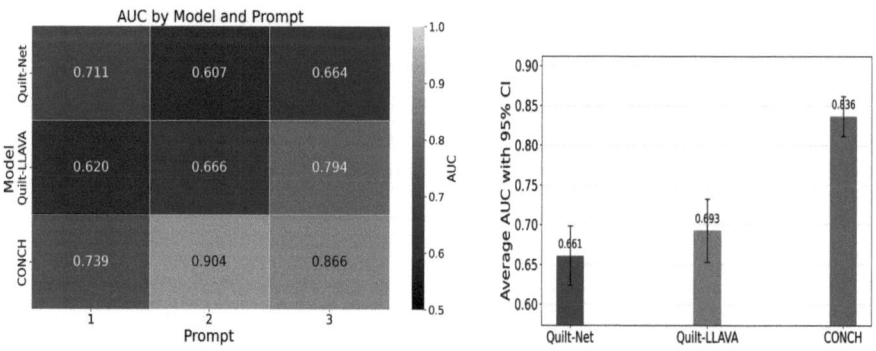

(a) AUC heatmap by model and prompt.

(b) Average AUC performance comparison.

Fig. 5. Performance comparison of VLM models and prompts in terms of ROC AUC for dysplasia status.

nificant effect of prompt wording, even when using near-synonymous medical terms.

Figure 6 analyzes the effect of magnification levels on the model's performance. Generally, it can be seen that better performance is achieved with higher magnification, as it aids in capturing finer morphological structures, which could be needed for accurate classifications. It is also evident that the performance in CONCH is nearly similar regardless of the magnification level, indicating that CONCH is more robust to changes in resolution.

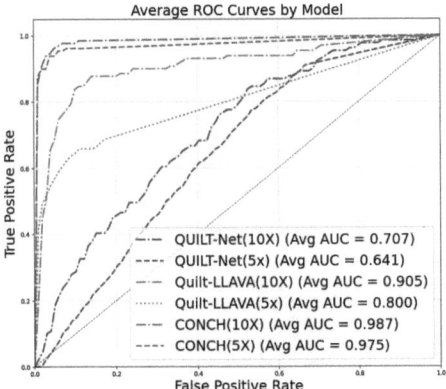

Fig. 6. Average ROC curves comparing model performance at different magnification levels.

5.2 Attention Maps Analysis

Figure 7 presents attention maps generated for histopathological analysis of randomly selected invasive cancer tissue samples. These visualizations represent probability scores assigned at the patch level by the different models to various regions of WSI. The process begins with segmenting each WSI into patches and classifying them as either tissue or background using an in-house CNN. Tissue patches are then processed through the models to obtain probability scores. By default, Quilt-Net and CONCH generate continuous confidence scores, which are used to represent each patch in the WSI and construct a heatmap. In contrast, Quilt-LLAVA does not inherently produce such scores; therefore, binary labels are used to highlight patches or regions identified as invasive. After scoring, the patches are reconstructed with their corresponding probability values to create comprehensive attention maps that highlight regions of interest across the entire WSI, potentially indicating areas of malignancy or specific tissue characteristics. We selected 4 WSIs that cover a diversity of forms of colorectal cancer invading into different levels of depth into the colonic wall.

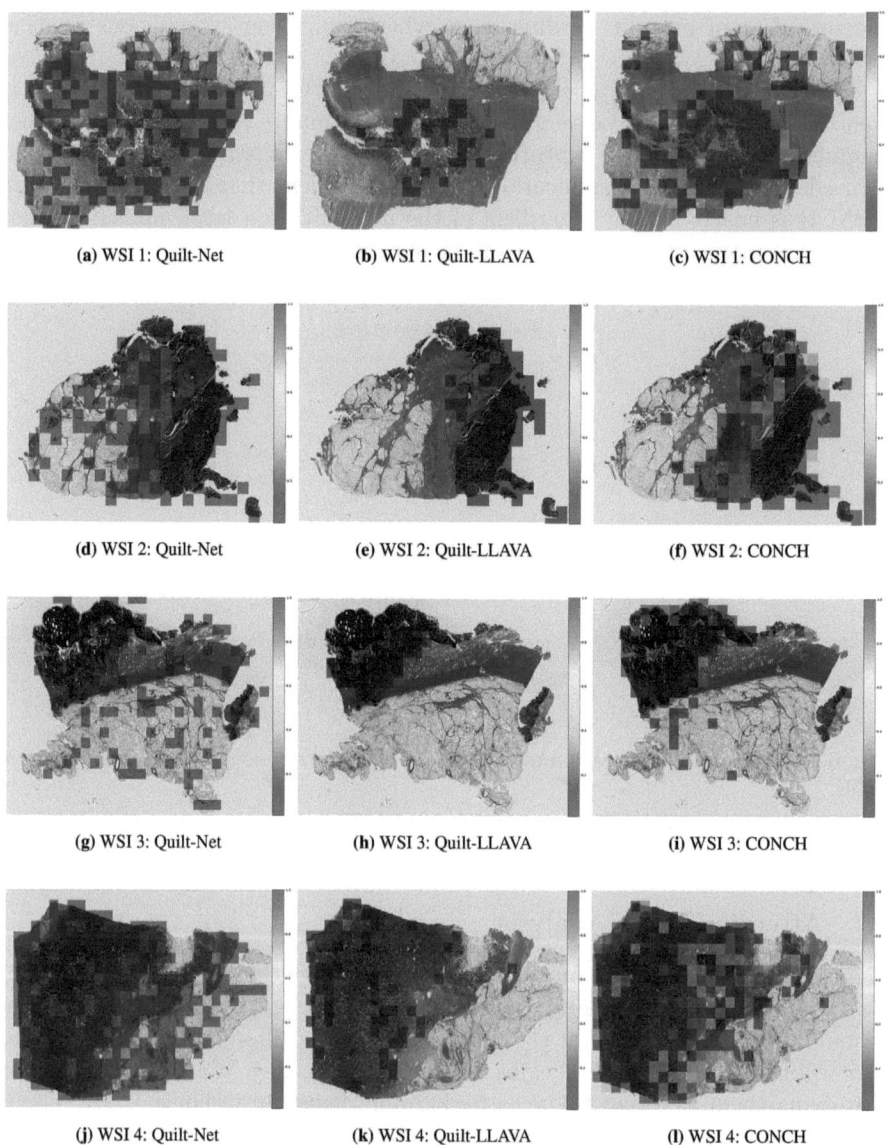

Fig. 7. Comparison of different models across multiple WSI samples.

All three models displayed different attention behavior in the underlying images. Generally, Quilt-Net randomly identified high-attention areas throughout the image, with no significant shift towards areas containing invasive cancer versus areas with cancer. Quilt-LLAVA displayed high-attention patches found within the invasive cancer, but was rather inconsistent in its approach as some areas of the invasive cancer were not highlighted. However, most high attention maps were accurately identified within cancer. CONCH showed the most accurate attention maps of invasive cancer and consistently highlighted its presence throughout the patches. CONCH was more precise in all images but highlighted low-attention areas that were distant from the cancer. CONCH could also highlight at medium-level attention areas of a precursor lesion that is on the verge of becoming cancer and altered tissue areas adjacent to the invasive cancer. Overall, per the review of a board-certified pathologist, CONCH most accurately mimicked the general approach by pathologists in addressing these tissues. Most attention is drawn towards the invasive cancer area, and second-order areas are revised to detect relevant findings, such as precursor lesions and mild changes in the peritumoral area that can be relevant for invasive cancer. Below are detailed analysis on the four WSIs from the board-certified pathologist:

WSI 1 represents a diverticular disease which has progressed into invasive cancer that breaches into the muscularis propria. Quilt-Net targets the whole colon wall with no preference for the invasive cancer versus the non-invaded areas. Quilt-LLAVA targets the invasive cancer and peri-invasive cancer area accurately. CONCH gives high attention at the invasive cancer consistently and highlights at medium attention the precursor area in the epithelium and the affected peri-cancer areas. It notes at low attention the unaffected normal tissue further away. **WSI 2** represents a classical invasive cancer that reached the resection margin and invades into the subserosal connective tissue. Quilt-Net produces randomized high-attention areas throughout the image, Quilt-LLAVA accurately targets the invasive cancer, and CONCH shows high-attention for invasive cancer, medium-attention for the affected pericancer areas, and low-attention to areas without cancer. **WSI 3** represents a classical invasive cancer that is restricted to the muscularis propria, arising from a precursor adenoma. Quilt-Net gives randomized high-attention areas throughout the image, Quilt-LLAVA targets the cancer area while ignoring the precursor lesion, and CONCH targets the cancer area accurately, and at medium- attention the precursor lesion. It further gives low attention to the non-invasive area. **WSI 4** represents a very large cluster of cancer with reactive epithelium at the surface. It invades deeply into the wall into the subserosal connective tissue. Quilt-Net targets the invasive cancer a bit more, but large areas of rather non-invaded tissues. Quilt-LLAVA seems to highlight the cancer, but only in areas that are adjacent to the non-tumoral tissues. CONCH accurately targets the cancer but appears to give low attention to an area of the cancer that is less aggressive while overcalling the reactive epithelium that overlies the cancer.

6 Conclusion

This study investigated the impact of prompt engineering and model complexity on the performance of different vision-language models in the domain of computational pathology, namely Quilt-Net, Quilt-LLAVA, and CONCH. Using an in-house digestive dataset of 3,507 whole slide images, we show that prompt design plays a crucial role in model performance, with domain-specific anatomical precision and intermediate level of text details achieving the best performance. Our findings showed that CONCH outperforms the other two models when precise anatomical prompts are used. This proves that effective domain-specific training can matter more than model complexity, since Quilt-LLAVA is a much larger model when compared to CONCH, but is restricted when it comes to model alignment. We also analyze attention maps for sample whole slide images, highlighting cancerous regions detected by the different models. These attention maps are analyzed by certified pathologists, to confirm CONCH's reliability in highlighting diagnostically relevant regions. These findings provide key insights for enhancing VLMs in digital pathology and improving AI-driven diagnostics.

References

1. Caron, M., et al.: Emerging properties in self-supervised vision transformers. In: Proceedings of the IEEE/CVF International Conference on Computer Vision, pp. 9650–9660 (2021)
2. Chanda, D., Aryal, M., Soltani, N.Y., Ganji, M.: A new era in computational pathology: a survey on foundation and vision-language models. arXiv preprint arXiv:2408.14496 (2024)
3. Chen, R.J., et al.: Scaling vision transformers to gigapixel images via hierarchical self-supervised learning. In: Proceedings of the IEEE/CVF Conference on Computer Vision and Pattern Recognition, pp. 16144–16155 (2022)
4. Chen, R.J., et al.: A general-purpose self-supervised model for computational pathology. arXiv preprint arXiv:2308.15474 (2023)
5. Clusmann, J., et al.: Prompt injection attacks on vision language models in oncology. Nat. Commun. **16**(1), 1239 (2025)
6. Cui, M., Zhang, D.Y.: Artificial intelligence and computational pathology. Lab. Invest. **101**(4), 412–422 (2021)
7. Deng, J., Dong, W., Socher, R., Li, L.J., Li, K., Fei-Fei, L.: Imagenet: a large-scale hierarchical image database. In: 2009 IEEE Conference on Computer Vision and Pattern Recognition, pp. 248–255. IEEE (2009)
8. Dosovitskiy, A., et al.: An image is worth 16x16 words: Transformers for image recognition at scale. arXiv preprint arXiv:2010.11929 (2020)
9. Du, Y., Sun, W., Snoek, C.: Ipo: interpretable prompt optimization for vision-language models. Adv. Neural. Inf. Process. Syst. **37**, 126725–126766 (2024)
10. Gu, J., Beirami, A., Wang, X., Beutel, A., Torr, P., Qin, Y.: Towards robust prompts on vision-language models. arXiv preprint arXiv:2304.08479 (2023)
11. Hosseini, M.S., et al.: Computational pathology: a survey review and the way forward. J. Pathol. Inform. **15**, 100357 (2024)
12. Ikezogwo, W., et al.: Quilt-1m: one million image-text pairs for histopathology. Adv Neural Inform. Process. Syst. **36**, 37995–38017 (2023)
13. Li, C., et al.: Llava-med: training a large language-and-vision assistant for biomedicine in one day. Adv. Neural. Inf. Process. Syst. **36**, 28541–28564 (2023)
14. Liu, H., Li, C., Wu, Q., Lee, Y.J.: Visual instruction tuning. Adv. Neural. Inf. Process. Syst. **36**, 34892–34916 (2023)
15. Lu, M.Y., et al.: Visual language pretrained multiple instance zero-shot transfer for histopathology images. In: Proceedings of the IEEE/CVF Conference on Computer Vision and Pattern Recognition, pp. 19764–19775 (2023)

16. Lu, M.Y., et al.: A visual-language foundation model for computational pathology. Nat. Med. **30**(3), 863–874 (2024)
17. Oquab, M., et al.: Dinov2: learning robust visual features without supervision. arXiv preprint arXiv:2304.07193 (2023)
18. Ouyang, L., et al.: Training language models to follow instructions with human feedback. Adv. Neural. Inf. Process. Syst. **35**, 27730–27744 (2022)
19. Radford, A., Wu, J., Child, R., Luan, D., Amodei, D., Sutskever, I., et al.: Language models are unsupervised multitask learners. OpenAI blog **1**(8), 9 (2019)
20. Radford, A., et al.: Learning transferable visual models from natural language supervision. In: International Conference on Machine Learning, pp. 8748–8763. PmLR (2021)
21. Seyfioglu, M.S., et al.: Quilt-llava: visual instruction tuning by extracting localized narratives from open-source histopathology videos. In: Proceedings of the IEEE/CVF Conference on Computer Vision and Pattern Recognition, pp. 13183–13192 (2024)
22. Sun, S., et al.: Dr-llava: visual instruction tuning with symbolic clinical grounding. arXiv preprint arXiv:2405.19567 (2024)
23. Vaswani, A., et al.: Attention is all you need. Adv. Neural Inform. Process. Syst. **30** (2017)
24. Vorontsov, E., et al.: Virchow: A million-slide digital pathology foundation model. arXiv preprint arXiv:2309.07778 (2023)
25. Wang, B., Liu, J., Karimnazarov, J., Thompson, N.: Task supportive and personalized human-large language model interaction: a user study. In: Proceedings of the 2024 Conference on Human Information Interaction and Retrieval, pp. 370–375 (2024)
26. Yu, J., Wang, Z., Vasudevan, V., Yeung, L., Seyedhosseini, M., Wu, Y.: Coca: contrastive captioners are image-text foundation models. arXiv preprint arXiv:2205.01917 (2022)

Achieving Zero False Negatives: Optimizing Anomaly Detection with Genetic Neural Architecture Search

Rabie Najem[(✉)] and Mohammed Benjelloun

University of Mons, Mons, Belgium
{rabie.najem,mohammed.benjelloun}@umons.ac.be

Abstract. Neural Architecture Search (NAS) methods, which aim to identify the best architecture for a given problem, have demonstrated their effectiveness across various domains, from computer vision to natural language processing. These approaches have significantly contributed to optimizing performance while addressing constraints such as computational efficiency and resource management. The Genetic Neural Architecture Search (GeNAS) proposed in this work illustrates the potential of NAS to go beyond its traditional objective of finding optimal architectures. While NAS is often employed to address constraints such as memory management and latency reduction, our study focuses on the critical challenge of minimizing False Negatives, with the ultimate goal of achieving Zero False Negatives (ZFN). To address this challenge, we integrate a methodology based on the Augmented Lagrangian Method (ALM), allowing for a better consideration of specific problem constraints. By adopting this targeted strategy, GeNAS demonstrates its effectiveness in tackling critical problems that require both high performance and enhanced sensitivity.

Keywords: Neural architecture search · Anomaly detection · zero false negative · Genetic algorithms · Augmented lagrangian method

1 Introduction

In many real-world applications, the cost of a False Negative can be severe. A False Negative occurs when a model fails to detect a positive instance, misclassifying it as negative. This can have serious consequences, especially in high-risk fields such as healthcare and financial security. For instance, in breast cancer detection, a model that fails to identify a malignant tumor as cancer-free delays diagnosis and treatment, allowing the disease to progress unnoticed. This can result in more complex and costly treatments with reduced chances of success, delayed intervention, decreasing survival rates, erosion of trust in AI-assisted medical systems, slowing their adoption in critical applications. Similarly, in fraud detection, failing to flag fraudulent transactions leads to financial losses, security breaches, and decreased consumer trust. Due to these risks, achieving Zero False Negative has become a crucial objective in sensitive applications where missing an anomaly is unacceptable. On the other hand, Neural Architecture Search has demonstrated outstanding performance across various domains, often surpassing state-of-the-art handcrafted architectures. It has been successfully applied in tasks such as image classification, object detection, and natural language processing, consistently pushing the boundaries of model optimization and efficiency. However, to the best of our knowledge, no existing NAS approach has explicitly addressed the problem of reducing False Negatives as a primary objective. Most existing NAS methods prioritize accuracy or precision instead. In this work, we introduce Genetic Neural Architecture Search, a novel method that explicitly targets False Negative minimization. Unlike traditional NAS methods, GeNAS serves as an evolutionary-based NAS algorithm designed to optimize neural architectures. When combined with the Augmented Lagrangian Method, this approach enforces constraints on false negatives, ensuring that the resulting architectures prioritize anomaly detection sensitivity while maintaining competitive classification performance.

2 Related Work

Neural Architecture Search has gained increasing popularity in recent years, particularly following the groundbreaking work of Zoph and Le (2017) [19], which marked the true beginning of NAS in its modern form. Since then, numerous studies have explored and expanded NAS, demonstrating its ability to surpass state-of-the-art handcrafted architectures. NAS-generated architectures have achieved top performance on highly competitive benchmarks, such as ImageNet and CIFAR-100 [7,10,20]. Over time, NAS has evolved far beyond image classification. Its applications have expanded significantly to include object detection [4,18], semantic segmentation [2,6], speech recognition [8], highly specialized fields such as solving partial differential equations [11,13,14] weather prediction [15], as also discussed in [17]. These last years, NAS has also gained popularity in the medical field, with successful applications in radiology, neurology, dermatology, and many other specialties [1,3,16]. This wide range of applications highlights the versatility and adaptability of NAS-based architectures across different domains.

The work presented in this paper is not built upon any specific existing NAS model, but rather introduces a custom NAS method designed from scratch. Our approach is based on genetic algorithms as the primary search strategy, combined with significant custom modifications, which will be detailed in the following sections. Additionally, we integrate the ALM to steer the search process towards architectures that better meet our specific needs. This method introduces constraints into the optimization process, ensuring that the selected architectures prioritize False Negative minimization while maintaining competitive classification performance.

3 Methodology

3.1 Neural Architecture Search

NAS methods are generally structured around three fundamental axes that define and organize the search process: the search space, the search strategy, and the evaluation criteria. These three components form the core of NAS systems, and their interaction is crucial for discovering optimal architectures, as illustrated in Fig. 1.

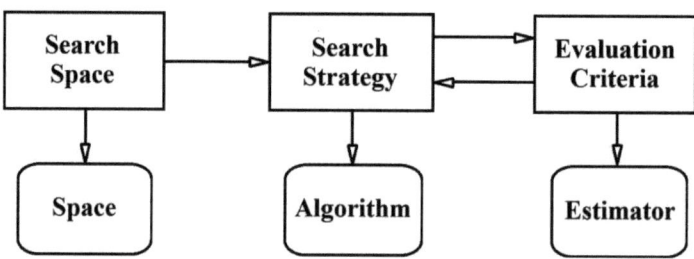

Fig. 1. General structure of NAS methods.

Search Space: The search space defines the set of architectural configurations that the NAS algorithm can explore. It specifies constraints such as:

- Number of layers and their types (e.g., convolutional, recurrent, fully connected).
- Connectivity patterns between layers (e.g., residual connections, dense connections).
- Filter sizes, number of neurons per layer, kernel shapes, etc.

A well-defined search space is crucial for the performance of NAS algorithms, as it determines the potential complexity and flexibility of the architectures.

Search Strategy: The search strategy determines how the NAS algorithm efficiently navigates the search space to find the architectures that offer the best performance. Various strategies exist, including:

- Reinforcement Learning-based NAS
- Differentiable NAS
- Evolutionary NAS

In this work, we adopt an evolutionary-based NAS approach by using Genetic Algorithms (GAs) as the main search strategy. Reinforcement Learning (RL) methods, such as ENAS [9] which has shown strong empirical results, are also widely used for NAS. However, they typically involve long training times and high computational demands. Additionally, because they rely on a single controller that sequentially refines the architecture, large-scale parallelisation becomes more challenging. On the other hand, Differentiable NAS methods like DARTS [7] and DARTS+ [5] allow much faster searches. However, they require a continuous relaxation of the discrete search space, a technique that can reduce stability and make them susceptible to local optima.

In contrast, Genetic Algorithms explore the search space in a non-local way, maintain a diverse population of candidate networks, and allow the evaluation of many architectures in parallel. This approach is particularly suited to modern multi-GPU or TPU clusters. Although GAs can be computationally intensive and require careful hyperparameter tuning, these challenges can be addressed by leveraging parallelism and by using hybrid strategies that combine Genetic Algorithms with Reinforcement Learning or gradient-based updates to speed up convergence, as well as other sophisticated methods that we will discuss in Sect. 7 (Future Work).

To this end, we introduce a hybrid approach by incorporating key concepts from Reinforcement Learning into our evolutionary algorithm. Unlike a purely genetic NAS, this integration enhances the balance between exploration and exploitation, accelerates the search process, and directs the evolution toward architectures that better meet our specific task requirements.

Evaluation Criteria: Evaluation criteria measure the quality of candidate architectures. Metrics typically include:

- Classification accuracy
- Model complexity and inference time
- Energy consumption

Unlike traditional NAS, which mainly optimizes accuracy and computational efficiency, this work introduces an explicit False Negative minimization objective.

3.2 Genetic Algorithms

Genetic algorithms (GAs) belong to the broader family of evolutionary algorithms. Inspired by natural selection and biological evolution, they solve complex optimisation problems by evolving a population of candidate solutions toward an optimal or near-optimal individual. This progress is driven by GA-specific operators such as crossover and mutation, which continually refine and diversify the search. Before delving into their mechanics, we define the key terms that will be used throughout this work.

Terminology

- **Individual:** A candidate solution, typically encoded as a set of genes, such as a binary string or a real-valued vector.

- **Population:** The set of individuals within a given generation, each evaluated using a fitness function.
- **Generation:** One complete cycle of evaluation, selection, crossover, and mutation to form a new population.
- **Selection:** The process of choosing the fittest individuals to generate offspring for the next generation.
- **Crossover:** A genetic operator that combines genetic material from two parents to create new offspring, encouraging exploration of the search space.
- **Mutation:** A random modification of one or more genes in an individual, promoting genetic diversity and helping to avoid local optima.
- **Fitness Function:** A quantitative measure that evaluates how well an individual meets the problem's objectives, influencing its selection probability.
- **Termination:** The stopping condition of the algorithm, such as reaching a maximum number of generations or achieving a satisfactory solution.

3.3 Genetic Neural Architecture Search (GeNAS)

As mentioned earlier, a NAS algorithm is built on three key components, among which the search strategy is the core of the method. It defines how the algorithm interacts with the search space and navigates through possible solutions. In this work, we developed our NAS algorithm based on Genetic Algorithms, as illustrated in Fig. 2.

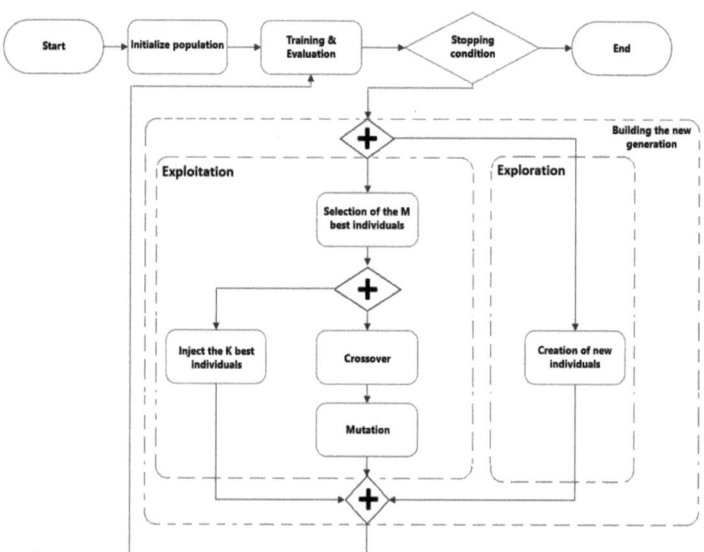

Fig. 2. Functioning of the proposed NAS algorithm.

Continuous Evolution of Solutions: These algorithms are distinguished by their ability to promote a continuous evolution of solutions. By selecting the most promising individuals in each generation, they ensure a progressive improvement of architectures, facilitating efficient and targeted exploration of the search space.

Initialization and Search Space Definition: Our algorithm can be divided into six main phases. The process starts with the creation of an initial population of neural architectures. This step requires carefully defining the search space to include all relevant components for addressing the specific problem.

Filtering and Preliminary Evaluation: Each individual in this population is first filtered based on preliminary criteria to eliminate unsuitable architectures. The remaining candidates are then processed in parallel, undergoing partial training and evaluation to assess their performance on the task at hand. This parallel processing, enabled by the use of Genetic Algorithms, significantly reduces the computational time while ensuring efficient exploration of the search space. Appropriate metrics are used to rigorously estimate the quality of each architecture.

Selection and Elite Group Formation: Next, the best-performing architectures are selected to form an elite group. These individuals are preserved for the next generation to ensure that the most promising traits are carried forward. At this stage, the algorithm evaluates whether the termination conditions have been met. These conditions may include achieving a specific performance threshold or observing stagnation in the improvements over generations. If the criteria are satisfied, the process concludes with the selection of optimal architectures.

New Generation and Genetic Operations: If the termination conditions are not satisfied, the algorithm proceeds to the creation of a new generation. This involves applying genetic operations such as mutation and crossover to the elite individuals, as well as introducing entirely new architectures to explore uncharted areas of the search space. The process then loops back to the evaluation phase, iterating until the termination conditions are met.

Exploration and Exploitation Balance: Figure 2 also highlights the integration of exploration and exploitation principles, inspired by reinforcement learning algorithms. Exploration focuses on introducing novel elements that may not directly derive from previous solutions, ensuring diversity in the search space. Exploitation, on the other hand, refines existing promising solutions through genetic operations like crossover and mutation, as well as the direct injection of top-performing individuals into the new generation.

Adaptive Exploration Rate: One key feature of our algorithm is its adaptive exploration rate, which adjusts dynamically based on the observed improvements. If no significant progress is made across generations, the exploration rate is increased to encourage broader searches and avoid local minima. This dynamic mechanism maintains a balanced trade-off between exploration and exploitation, ensuring optimal search efficiency.

3.4 Minimize False Negatives

As previously mentioned, achieving Zero False Negatives is a critical goal in applications where missing anomalies can have severe consequences. To address this challenge, we explored several methods aimed at forcing the model to focus more on anomalies.

Threshold Adjustment: One of the simplest methods to achieve ZFN is adjusting the classification probability threshold. By lowering this threshold, the sensitivity of the model increases, allowing it to detect more positive cases. While this approach reduces false negatives, it often comes at the cost of increased false positives, requiring a careful trade-off between the two.

Class Weights: Keras provides a built-in parameter called `class_weights`, which allows models to pay more attention to under-represented classes. This strategy is particularly useful when the dataset is imbalanced, a common situation in many anomaly detection tasks, as it helps prevent the majority class from dominating the learning process. By assigning higher weights to the minority class, the loss function becomes more sensitive to errors related to this class, effectively reducing the number of false negatives. Even if the dataset is relatively balanced, manually increasing the weight of the anomaly (positive) class remains beneficial, as it penalizes false-negative errors more heavily than false positives, thereby lowering the miss-detection rate. In our case, we specifically apply this technique to penalize false negatives, since missing an anomaly is considered much more critical than raising a false alarm.

Custom Loss Function: Custom loss functions offer a more targeted way to minimize false negatives. Typically, this involves starting with a predefined loss function, such as binary cross-entropy, and augmenting it with a penalty term. For ZFN, the penalty term encourages the model to better detect positive samples by penalizing misclassification. The objective is to minimize:

$$\begin{cases} \min L(x) \\ C_i(x) = 0, \quad \forall i \in \mathcal{A} \end{cases} \text{(where } \mathcal{A} \text{ denotes the indices of the equality constraints)}$$

$$L(x) = f(x) + \sum_{i=0}^{n} Coeff_i \times C_i(x)$$

$Coeff_i$ is the penalty factor for the constraint C_i.

Thus, instead of solely minimizing the original loss function, our objective becomes minimizing a new function that combines the original loss with an additional constraint related to false negatives ($C_i(x)$), which represents the sum of probabilities assigned to the negative class for positive samples). This modification allows the model to explicitly account for false negatives, thereby enhancing its sensitivity to anomalies and improving its overall performance.

Custom Loss Function with Augmented Lagrangian Method: This method extends the custom loss function by integrating the Augmented Lagrangian Method [12]. ALM incorporates additional parameters μ, λ, and ρ to enforce the false negative constraint dynamically during training. This mechanism prioritizes reducing false negatives while maintaining a balance with overall performance metrics. The loss function is formulated as:

$$\begin{cases} \min L_\mu(x, \lambda) \\ C_i(x) = 0, \quad \forall i \in \mathcal{A} \quad \text{(as previously defined)} \end{cases}$$

$$L_\mu(x, \lambda) = f(x) + \sum_{i=0}^{n} \lambda_i \times C_i(x) + \mu \sum_{i=0}^{n} \|C_i(x)\|^2$$

$$\lambda^{k+1} = \lambda^k + \mu C(x)$$

$$mu^{k+1} = \begin{cases} \mu^k \times \rho, & \text{if } AUC^t < AUC^{t-1} \\ \mu^k, & \text{else} \end{cases}$$

$$\begin{cases} \lambda^0 = 0 \ \mu^0 \in \{10^{-7}, 10^{-6}, 10^{-5}, 10^{-4}, 10^{-3}\} \\ \rho \in \{2, 3\} \end{cases}$$

In this formulation, two nested time indices are used: k refers to each mini-batch within an epoch, while t designates the epoch itself. As in the previous method, $C_i(x)$ retains the same definition and interpretation. In this formulation, three key parameters are introduced: μ, λ, and ρ, which are central to the ALM. The parameter λ is initialized to zero and iteratively updated to enforce constraint satisfaction, while μ controls the weight of the penalty term. The parameter ρ is a constant scaling factor used to adjust μ dynamically during training. Specifically, μ is updated based on the model's Area Under the Curve (AUC) performance. If the current AUC (AUC^t) is lower than the previous AUC (AUC^{t-1}), μ is multiplied by ρ; otherwise, it remains unchanged. This adaptive mechanism ensures that the training process prioritizes minimizing false negatives without compromising overall model performance.

4 Experiments and Results

In this section, we present our experimental setup to test and compare the different methods. The experiments will be conducted as follows: first, we manually design a neural architecture that performs well on the given dataset. Then, we create multiple models using the same architecture, optimizer, and learning rate, but with different approaches for minimizing False Negatives. This approach ensures a fair comparison between the different methods by isolating the impact of the False Negative minimization techniques.

Next, we apply GeNAS to each method to optimize the architecture automatically, aiming to find the best-performing structure for the specific False Negative minimization strategy used. The comparison will be based on four key evaluation metrics: Accuracy, Precision, Recall, and F1-Score, ensuring a comprehensive assessment of performance. The formulas used to compute these metrics are given below:

$$\text{Accuracy} = \frac{TP + TN}{TP + TN + FP + FN}$$

$$\text{Precision} = \frac{TP}{TP + FP}$$

$$\text{Recall} = \frac{TP}{TP + FN}$$

$$F1\text{-Score} = 2 \times \frac{\text{Precision} \times \text{Recall}}{\text{Precision} + \text{Recall}}$$

Breast Cancer

The Breast Cancer Wisconsin dataset is widely used for classifying breast tumors as benign or malignant. It originates from a clinical study conducted at the University of Wisconsin and is publicly available via Scikit-learn. This dataset is extensively utilized in biomedical research and machine learning for breast cancer diagnosis. It contains 30 numerical features that describe the morphological properties of tumor cells, including size, texture, compactness, and symmetry. Due to its well-structured nature, this dataset is particularly suitable for training classification models and assessing their ability to accurately detect cancerous tissues.

Since our primary goal is to minimize False Negatives, we aim to achieve a Recall value close to 1, or exactly 1 if we successfully reach ZFN. As shown in our results (Table 1), we obtained a Recall of 1 for the majority of methods. Among the manually designed models, ALM consistently achieved the best performance across all metrics, demonstrating its ability to achieve ZFN without negatively impacting the overall model performance.

GeNAS further enhances each method by optimizing the neural architecture. As observed in our results, GeNAS systematically improved the performance of

Table 1. Comparison of methods in Breast Cancer.

Type	Methods	Accuracy	Recall	Precision	F1-score
Manual design	Threshold Adjustment	0.947	1.000	0.922	0.959
	Class Weight	0.956	0.986	0.946	0.966
	Custom Loss	0.921	1.000	0.888	0.940
	ALM	0.982	1.000	0.973	0.986
Automatic design	GeNAS & Threshold Adjustment	0.965	1.000	0.947	0.973
	GeNAS & Class Weight	0.983	0.986	0.986	0.986
	GeNAS & Custom Loss	0.965	1.000	0.947	0.973
	GeNAS & ALM	**0.991**	**1.000**	**0.986**	**0.993**

every method, as it automatically identifies the best-suited architecture for the chosen False Negative minimization strategy.

Moreover, the combination of GeNAS and ALM outperformed all other methods, including the manually designed ALM-based model. This hybrid approach delivered the best results across all evaluation metrics, confirming its effectiveness in reducing False Negatives while maintaining strong overall performance.

KDDCup99

Another dataset used in this study is KDDCup99, which is widely employed for evaluating Intrusion Detection Systems (IDS) in cybersecurity. Originating from the KDD Cup 1999 competition, it is based on simulated network traffic captures, where each connection is classified as either normal or malicious.

In our work, we used the Smurf Attack (SA) subset, which contains only SA instances and normal traffic. It is available via Scikit-learn and originates from the MIT Lincoln Laboratory.

The dataset consists of 41 features that describe network connections, divided into numerical variables (e.g., connection duration, bytes transferred, number of connections) and categorical variables (e.g., protocol type, connection state, network service). Its highly imbalanced nature makes it particularly well-suited for anomaly detection models.

Similarly to the Breast Cancer dataset, the manually designed architectures evaluated on KDDCup99 showed that ALM outperforms other methods across most metrics (Table 2). Additionally, GeNAS enhances all methods, consistently surpassing their manually designed counterparts.

The GeNAS & ALM combination remains the most effective approach, achieving Zero False Negatives without negatively impacting other performance metrics.

Table 2. Comparison of methods in fetch_kddcup99.

Type	Methods	Accuracy	Recall	Precision	F1-score
Manual design	Threshold Adjustment	0.9990	0.985	0.983	0.984
	Class Weight	0.9989	0.982	0.985	0.983
	Custom Loss	0.9992	0.991	0.983	0.987
	ALM	0.9996	0.989	0.998	0.994
Automatic design	GeNAS & threshold Adjustment	0.9993	0.980	**0.998**	0.989
	GeNAS & class weight	0.991	1.000	0.986	0.993
	GeNAS & custom Loss	0.9993	0.982	0.997	0.989
	GeNAS & ALM	**0.9996**	**1.000**	0.995	**0.993**

5 Discussion

Experimental results show that GeNAS significantly improves all tested methods. In particular, the GeNAS & ALM combination outperforms all other configurations by achieving Zero False Negatives without compromising overall model accuracy. These findings confirm the hypothesis that simultaneously optimizing both the architecture and the loss function is essential for effectively minimizing False Negatives.

5.1 Impact of GeNAS on False Negative Reduction

Our results demonstrate that NAS effectively optimizes neural architectures for False Negative minimization. Specifically:

- GeNAS enables the discovery of optimized architectures, reinforcing the advantages of an automated neural architecture search approach.
- Even with simple methods (Threshold Adjustment, Class Weights), NAS enhances performance, demonstrating its adaptability across different strategies.
- The combination of ALM and GeNAS consistently achieves the best performance across all metrics.

5.2 Limitations and Perspectives

Although the empirical results are encouraging, our study also reveals a number of open issues that must be addressed before GeNAS can be considered ready for large-scale or safety-critical deployment. In particular, two limitations stand out:

- **Computational Cost:** GeNAS relies on an intensive evaluation of candidate architectures, which can be computationally expensive. Optimizing the

search process could help mitigate this issue. For this reason, we initially conducted our experiments on relatively small datasets, such as *Breast Cancer* and *KDDCup99*, to ensure feasibility on machines with limited computational resources. This allowed us to validate GeNAS's effectiveness before considering its application to larger-scale datasets.
- **Dataset Dependence:** While our results are promising, further testing on more competitive or critical datasets, such as MVTec, could provide a deeper evaluation of GeNAS's performance in industrial and high-precision applications. Another example is the processing of 3D breast cancer MRI volumes, which would represent a challenging evaluation in the medical domain for our method. Additionally, extending the approach to other data types could help assess its effectiveness in tackling more complex and diverse tasks.

6 Conclusion

In this work, we introduced GeNAS, a genetic-algorithm-based Neural Architecture Search framework that jointly optimizes network topology and anomaly sensitivity by embedding the Augmented Lagrangian Method directly into the search process. While most NAS pipelines primarily target generic objectives such as accuracy or efficiency, GeNAS/ALM explicitly focuses on minimizing False Negatives, a critical requirement in anomaly detection tasks where undetected anomalies may lead to severe outcomes.

Experiments on two standard benchmarks demonstrate that the GeNAS + ALM combination consistently achieves Zero False Negatives while maintaining, and often improving, overall accuracy, precision, and F1-score. These results also provide insight into why achieving ZFN is inherently difficult: neither architecture search nor a customized loss function alone is sufficient. A loss function that penalizes FN must be paired with an architecture capable of fully exploiting this learning signal; conversely, even a well-adapted architecture cannot reach ZFN without a loss function that drives optimization towards maximal recall. Only through their joint optimization can the desired balance between sensitivity and precision be reliably achieved.

Finally, the comparison with manually designed networks highlights substantial additional gains, both in higher accuracy and dramatically reduced FN rates, once the design burden is shifted to NAS. These findings underscore the value of automated architecture search in contexts where missing anomalies is not an option.

7 Future Work

Building on this study, several improvements and extensions to GeNAS can be considered to further optimize neural architecture search.

7.1 Optimization of Evaluation Methods

Evaluating NAS-generated architectures is computationally expensive. We propose several solutions:

- Zero-shot proxies: Utilize untrained network metrics to instantly estimate an architecture's potential without any training, allowing rapid elimination of weak candidates at near-zero cost.
- Few-shot evaluation: Allocate a small, fixed number of training epochs or a reduced data subset to quickly gauge candidate performance, then reserve full training budgets only for the most promising architectures.
- Transfer learning for evaluation: Leveraging knowledge gained from similar datasets to accelerate the evaluation process, preventing the need to restart from scratch for each GeNAS execution.

7.2 Integration of the Supernet Concept

Integrating a Supernet into GeNAS could optimize architecture search by enabling weight sharing among sub-architectures. Although often associated with differentiable methods, the Supernet concept could also be adapted to evolutionary algorithms, providing additional flexibility in architecture optimization. This approach would offer:

- Reduced computational costs by avoiding independent training for each architecture.
- Faster exploration of an extended search space.
- Hybridization of evolutionary and differentiable methods for more efficient optimization.

Adapting GeNAS to these advancements will enhance its efficiency and applicability to more complex and diverse models.

References

1. Cao, C., Huang, W., Hu, F., Gao, X.: Hierarchical neural architecture search with adaptive global-local feature learning for magnetic resonance image reconstruction. Comput. Biol. Med. **168**, 107774 (2024)
2. Chen, L.C., et al.: Searching for efficient multi-scale architectures for dense image prediction. Adv. Neural Inform. Process. Syst. **31** (2018)
3. Fuentes-Tomás, J.A., Mezura-Montes, E., Acosta-Mesa, H.G., Márquez-Grajales, A.: Tree-based codification in neural architecture search for medical image segmentation. IEEE Trans. Evolutionary Comput. (2024)
4. Ghiasi, G., Lin, T.Y., Le, Q.V.: Nas-fpn: learning scalable feature pyramid architecture for object detection. In: Proceedings of the IEEE/CVF Conference On Computer Vision and Pattern Recognition, pp. 7036–7045 (2019)
5. Liang, H., et al.: Darts+: improved differentiable architecture search with early stopping. arXiv preprint arXiv:1909.06035 (2019)

6. Liu, C., et al.: Auto-deeplab: hierarchical neural architecture search for semantic image segmentation. In: Proceedings of the IEEE/CVF Conference on Computer Vision and Pattern Recognition, pp. 82–92 (2019)
7. Liu, H., Simonyan, K., Yang, Y.: DARTS: Differentiable architecture search. In: International Conference on Learning Representations (ICLR 2019) (2019)
8. Mehrotra, A., et al.: Nas-bench-asr: reproducible neural architecture search for speech recognition. In: International Conference on Learning Representations (2021)
9. Pham, H., Guan, M., Zoph, B., Le, Q., Dean, J.: Efficient neural architecture search via parameters sharing. In: International Conference on Machine Learning, pp. 4095–4104. PMLR (2018)
10. Real, E., Aggarwal, A., Huang, Y., Le, Q.V.: Regularized evolution for image classifier architecture search. In: Proceedings of the AAAI conference on Artificial Intelligence, vol. 33, pp. 4780–4789 (2019)
11. Roberts, N., Khodak, M., Dao, T., Li, L., Ré, C., Talwalkar, A.: Rethinking neural operations for diverse tasks. Adv. Neural. Inf. Process. Syst. **34**, 15855–15869 (2021)
12. Sangalli, S., Erdil, E., Hötker, A., Donati, O., Konukoglu, E.: Constrained optimization to train neural networks on critical and under-represented classes. Adv. Neural. Inf. Process. Syst. **34**, 25400–25411 (2021)
13. Shen, J., Khodak, M., Talwalkar, A.: Efficient architecture search for diverse tasks. Adv. Neural. Inf. Process. Syst. **35**, 16151–16164 (2022)
14. Tu, R., Roberts, N., Khodak, M., Shen, J., Sala, F., Talwalkar, A.: Nas-bench-360: benchmarking neural architecture search on diverse tasks. Adv. Neural. Inf. Process. Syst. **35**, 12380–12394 (2022)
15. Tu, R., et al.: Automl for climate change: a call to action. arXiv preprint arXiv:2210.03324 (2022)
16. Wang, Y., et al.: Mednas: multi-scale training-free neural architecture search for medical image analysis. IEEE Trans. Evolutionary Comput. (2024)
17. White, C., et al.: Neural architecture search: Insights from 1000 papers. arXiv preprint arXiv:2301.08727 (2023)
18. Xu, H., Yao, L., Zhang, W., Liang, X., Li, Z.: Auto-fpn: automatic network architecture adaptation for object detection beyond classification. In: Proceedings of the IEEE/CVF International Conference on Computer Vision, pp. 6649–6658 (2019)
19. Zoph, B., Le, Q.V.: Neural architecture search with reinforcement learning. In: International Conference on Learning Representations (ICLR 2017) (2017)
20. Zoph, B., Vasudevan, V., Shlens, J., Le, Q.V.: Learning transferable architectures for scalable image recognition. In: Proceedings of the IEEE Conference on Computer Vision and Pattern Recognition, pp. 8697–8710 (2018)

Whisper-Conformer: A Modified Automatic Speech Recognition for Thai Speech Recognition

Thanakron Noppanamas[(✉)] [iD] and Suronapee Phooomvuthisarn [iD]

Chulalongkorn University, 254 Phaya Thai Rd, Wang Mai, Pathum Wan, Bangkok 10330, Thailand
`thanakron.nop@gmail.com, suronapee@cbs.chula.ac.th`

Abstract. Speech is a fundamental aspect of human communication, driving advancements in Automatic Speech Recognition (ASR) to bridge the gap between humans and machines. ASR has evolved significantly with the introduction of Deep Neural Networks (DNNs), which enable models to learn speech patterns directly from raw audio. The transition to end-to-end DNNs architecture, including models like Whisper and Conformer, dramatically improved ASR performance. In the context of Thai ASR, researchers have adopted DNNs-based models, including fine-tuned versions of Whisper, to improve recognition accuracy. However, prior studies have shown that Thai ASR still faces challenges due to the lack of spaces in Thai sentences and regional dialectal variations. To address these challenges, this study proposes an alternative approach by modifying the existing Whisper model by integrating the Conformer architecture. We refer to the resulting model as Whisper-Conformer. The model is trained on 373.5 h of Thai speech data. Our results indicate that Whisper-Conformer learns significantly faster than baseline models and outperforms the fine-tuned Whisper model, achieving 0.64 Word Error Rate (WER) and 0.42 Character Error Rate (CER) on the Common Voice Corpus (v18) and 83.27 WER and 39.96 CER on the Thai Dialect Corpus, without using a language model for spelling correction. These findings suggest that integrating the Conformer architecture enhances ASR performance and enables the model to handle challenges in Thai ASR more effectively. The pretrained models are available at https://huggingface.co/Thanakron/whisperConformer-medium-th.

Keyword: Automatic speech recognition · Thai speech · Whisper · Conformer · Deep neural networks

1 Introduction

Speech is fundamental to human communication, enabling individuals to express thoughts and emotions effectively. With the growing interest in human-machine interaction, researchers have increasingly focused on developing ASR systems to bridge the gap between human and machine communication. ASR has evolved over more than five decades, significantly advancing with the introduction of DNNs architecture, which enables models to automatically learn speech signals from raw audio [12].

Early ASR systems were primarily based on HMMs, which had limitations in capturing long-term dependencies. The introduction of DNNs and the use of end-to-end DNNs in ASR systems significantly improved performance. One of the popular models with excellent performance is Whisper [17], an end-to-end multilingual and multitask model, which has been fine-tuned for various tasks [8, 19, 24]. Additionally, modifications to the original architecture, such as Conformer, which integrate convolutional layers into the self-attention mechanisms of the Transformer architecture, allow them to capture both global context and local feature patterns, achieving the lowest word error rate on the LibriSpeech dataset [6]. In the context of Thai language ASR, DNNs have also been adopted [15, 21], including the fine-tuning of the Whisper model on Thai datasets [1, 9, 22]. Nevertheless, Thai ASR systems still rely on language models for spelling correction and mispredictions due to the lack of spaces in the Thai sentences and regional dialectal variations.

In this study, we aim to evaluate the performance of Whisper-Conformer, a modified ASR model based on the open-source Whisper model, in which the original encoder has been replaced with a Conformer encoder from the Conformer architecture. The model is trained on publicly available Thai speech datasets to assess the extent to which it can overcome the performance of Thai ASR and more effectively than the original Whisper or other models fine-tuned with Thai datasets, including reducing the reliance on language models for spelling correction and mispredictions.

Our approach leverages the Whisper model while integrating Conformer architecture, allowing it to benefit from both global context awareness and local feature extraction. We train the model on 373.5 h of Thai speech data and evaluate it against existing Whisper-based models. The results demonstrate that Whisper-Conformer achieves faster learning than baseline models and outperforms fine-tuned Whisper models on benchmark datasets. Specifically, our model achieves 0.64 WER and 0.42 CER on the Common Voice Corpus (v18) and 83.27 WER and 39.96 CER on the Thai Dialect Corpus, all without applying a language model for spelling correction.

These findings suggest that integrating Conformer architecture into the Whisper model significantly enhances performance, paving the way for more accurate and efficient Thai speech recognition systems. The remainder of this paper is structured as follows: Sect. 2 provides background knowledge on ASR and Conformer architecture. Section 3 details the methodology, including model design and training setup. Section 4 presents the results and discussion, evaluating the performance of our approach. Finally, Sect. 5 concludes the study and discusses potential future work.

2 Background Knowledge

In this section, we discuss the core foundational concepts used in our study, including the Whisper model and Conformer architecture, which form the basis of our objectives.

2.1 Speech Recognition

Early ASR systems were primarily based on HMMs and required the creation of a written representation (label) of speech sounds. A common approach involves using the International Phonetic Alphabet (IPA), an alphabetic system of phonetic notation based on the Latin script, to help models learn speech patterns, which had limitations in capturing long-term dependencies and difficulty in creating labels for speech. The introduction of DNNs revolutionized ASR, with Recurrent Neural Networks (RNNs) and their variants, including Long Short-Term Memory (LSTM) networks and Gated Linear Unit (GRU), helping to overcome these challenges [5]. Subsequently, Convolutional Neural Networks (CNNs) were incorporated into ASR architectures, leading to models such as Jasper [10] and ContextNet [7], which further enhanced speech recognition performance. Introducing self-attention-based models such as Wav2Vec2.0 [2] and encoder-decoder Transformer-based architectures, including Whisper [17], demonstrated even more significant improvements in ASR capabilities.

2.2 Whisper Model

The Whisper model, developed by OpenAI, is a multilingual and multitask ASR model based on a Transformer-based encoder-decoder architecture [23]. It was trained on an unprecedented 680,000 h of labeled audio data, achieving a WER of 2.7 on the LibriSpeech clean dataset and an average WER of 12.8 across 14 benchmark datasets. For processing, Whisper resamples audio to 16,000 Hz and converts it into an 80-channel log-magnitude Mel spectrogram for input. It then applies two convolution layers (filter width: 3) with the GELU activation function to subsample the Mel spectrogram before passing it to the encoder. The encoder and decoder share the same width and number of Transformer blocks. For text tokenization, Whisper uses a byte-pair encoding (BPE) tokenizer identical to one used in GPT-2. Figure 1 illustrates the overall architecture of the Whisper model, highlighting its main components, including the Mel spectrogram preprocessing, convolutional feature extraction, Transformer-based encoder-decoder, and text output generation.

Compared to models like wav2vec 2.0, which focuses on self-supervised learning, Whisper leverages large-scale supervised training and multitask objectives, making it particularly effective for general-purpose ASR applications. Whisper has been successfully fine-tuned for child speech recognition [8] and adapted to enhance ASR accuracy in Chinese [24] and Nepali [19].

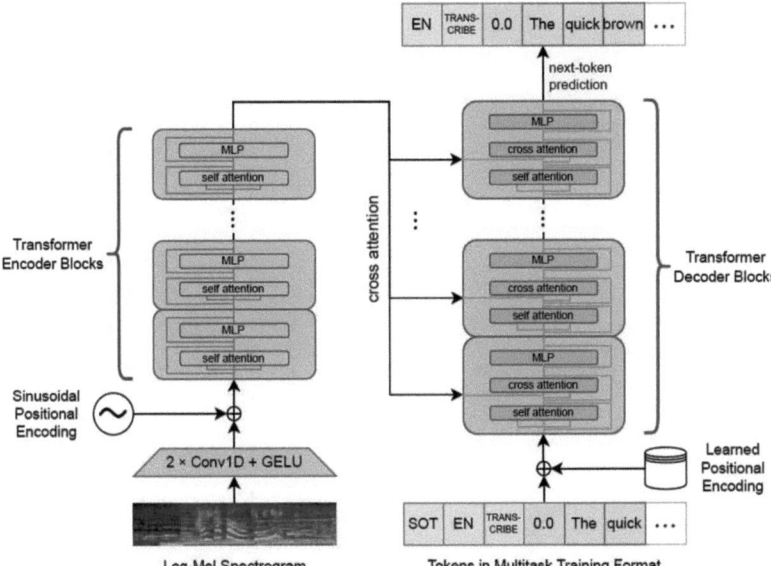

Fig. 1. High-level architecture of the Whisper model, showcasing its input processing, feature extraction, and Transformer-based sequence generation[17].

2.3 Conformer Architecture

The Conformer architecture, introduced by Anmol Gulati et al. in 2020 [6], is designed to integrate two powerful techniques for speech recognition. First, it incorporates the self-attention mechanism from Transformer architecture, which captures a long-range global context. Second, it utilizes CNNs, which effectively extract local feature patterns. This hybrid approach enhances speech recognition accuracy by addressing both contextual and sequential dependencies. Figure 2 illustrates the overall architecture of the Conformer model, highlighting its core components, including the multi-headed self-attention mechanism, convolutional module, and feedforward layers.

The Conformer architecture follows a structure similar to the Transformer but with a key difference: it introduces a Convolution module immediately after the Multi-Headed Self-Attention module. This module consists of (1) Pointwise Convolution followed by a GLU, (2) A 1-D depth-wise convolution layer for local feature extraction, and (3) Batch normalization combined with Swish activation [18] to improve stability and performance In empirical evaluations using the LibriSpeech datasets, the Conformer model achieved a WER of 2.1/4.3 without a language model and 1.9/3.9 with an external language model on the test/test-other sets. These results demonstrate the Conformer's ability to enhance ASR performance by leveraging both self-attention and convolutional feature extraction. More recently, hybrid architecture combining self-attention and convolutional networks, known as Conformer, has emerged as a promising approach for ASR [6].

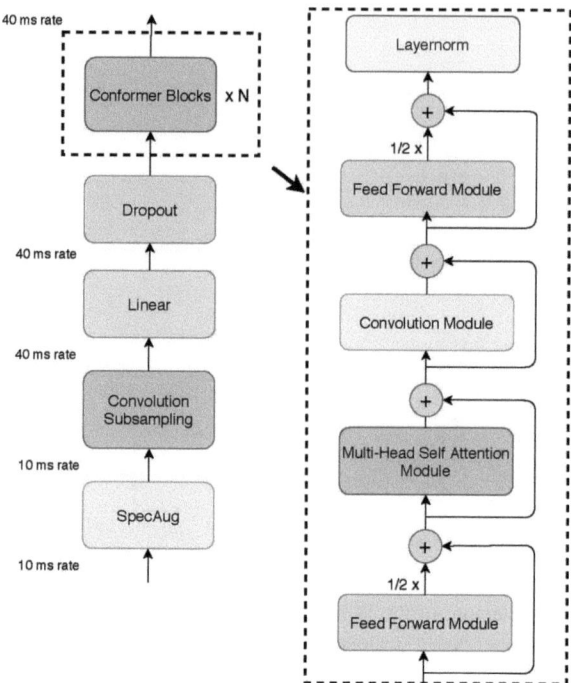

Fig. 2. High-level architecture of the Conformer model, illustrating its integration of self-attention, convolutional modules, and feedforward layers for speech recognition [6].

2.4 Speech Recognition in Thai Language

In traditional Thai speech recognition, IPA is still used to label sounds of speech, similar to other languages. The Thai language is unique due to its 44 consonants, 21 vowels, and five tones. For example, Sinaporn, S., et al. developed a Thai phoneme set using IPA, comprising 21 consonantal phonemes, 17 consonantal cluster phonemes, and 24 vowels, each capable of carrying one of five tones to build a robust Thai speech recognizer [20].

Another approach was introduced by Pisarn, C. and Theeramunkong, T., who used the Thai Phonetic Set (TPS) developed by the National Electronics and Computer Technology Center (NECTEC) [16]. This work aims to design labels for sounds of speech. This system utilized an English-like notation to simplify phonetic transcription. However, integrating such phonetic knowledge into modern ASR models remains challenging. Additionally, Thai is written without spaces between words, requiring complete semantic understanding to correctly segment sentences into meaningful words or phrases. The emergence of end-to-end DNNs architectures and transfer learning has driven significant improvements in Thai ASR models [15, 21]. However, these models still face challenges, such as mispredictions due to the lack of spaces in the Thai Language, reliance on language models for spelling correction, and variations in regional Thai dialects.

In recent years, researchers have explored various approaches to improve recognition accuracy. Encoder-based Transformer models fine-tuned on Thai speech data from the Common Voice corpus [15] and hybrid CTC/Attention models utilizing HuBERT as an encoder trained on multi-dialect Thai data [21] have demonstrated promising results. Additionally, models such as the Thonburian Whisper Large-V3 Combined model, fine-tuned from Whisper Large-V3 by Looloo and BioDat Lab [1], and the Monsoon Whisper Medium GigaSpeech2 model, fine-tuned from the Whisper Medium model [9], have contributed to improving Thai ASR. Furthermore, a fine-tuned Whisper Medium model developed by King Mongkut's Institute of Technology Ladkrabang (KMITL) University has been specifically designed to enhance recognition in complex scenarios, achieving the lowest WER scores on the Gowajee, LOTUS-TRD, and Thai Dialect Elderly datasets, particularly when combined with word segmentation using PyThaiNLP [14, 22]. Nevertheless, language models are still required for spelling correction, increasing the processing time required for accurate speech-to-text conversion. Furthermore, issues such as mispredictions due to the lack of spaces in Thai sentences and regional dialectal variations. Are still present in current Thai ASR models.

To improve Thai ASR, this study explores alternative approaches beyond traditional fine-tuning methods, which have been studied in prior research, as previously discussed. For this study, we propose a novel strategy by modifying the existing Whisper model, which is widely adopted due to its simple pipeline and broad applicability. We replace the original Whisper encoder with a Conformer encoder, thereby enhancing the model's ability to capture both global and local speech patterns. This architectural modification is expected to improve model performance and better handle challenges specific to Thai, such as the lack of spaces in written text and regional dialectal variations, reducing reliance on external language models for spelling correction. While techniques such as increasing model parameters, for instance, expanding the input size by increasing the log-magnitude Mel spectrogram channels from the default 80 to 120, as in Whisper V3 [17], or adding more layers, can also improve performance, they typically demand substantial computational resources. Given the resource constraints in this study, we opted for a more efficient and targeted solution by modifying an existing model to enhance the performance of Thai ASR without incurring excessive computational costs.

The challenge of this study lies in incorporating the Conformer layer into Whisper, as Whisper has its own pipeline. Integrating the Conformer layer requires adapting it to be compatible with Whisper's original configuration. Additionally, using the Conformer layer from the PyTorch module prevents us from utilizing pre-trained weights, which means we had to invest additional time and resources into training to ensure that our model performs effectively.

3 Methods

This study introduces Whisper-Conformer, a modified ASR model that integrates the Whisper model with Conformer architecture to enhance Thai speech recognition. The model was developed using PyTorch. Training and evaluation were conducted on 373.5 h of Thai speech data from publicly available datasets. The model was trained using the AdamW optimizer with Seq2SeqTrainer on 4 NVIDIA A100 GPUs. Various model sizes were tested, and performance was evaluated using WER and CER.

3.1 Whisper-Conformer Model

We utilize PyTorch to build the Whisper-Conformer model. The Whisper model is sourced from Hugging Face's Transformers library without pretrained weights, while the Conformer layer is incorporated from the PyTorch module. For input transformation, we follow the same approach as the existing Whisper model settings for both the encoder and decoder. The Whisper-Conformer model is illustrated in Fig. 3

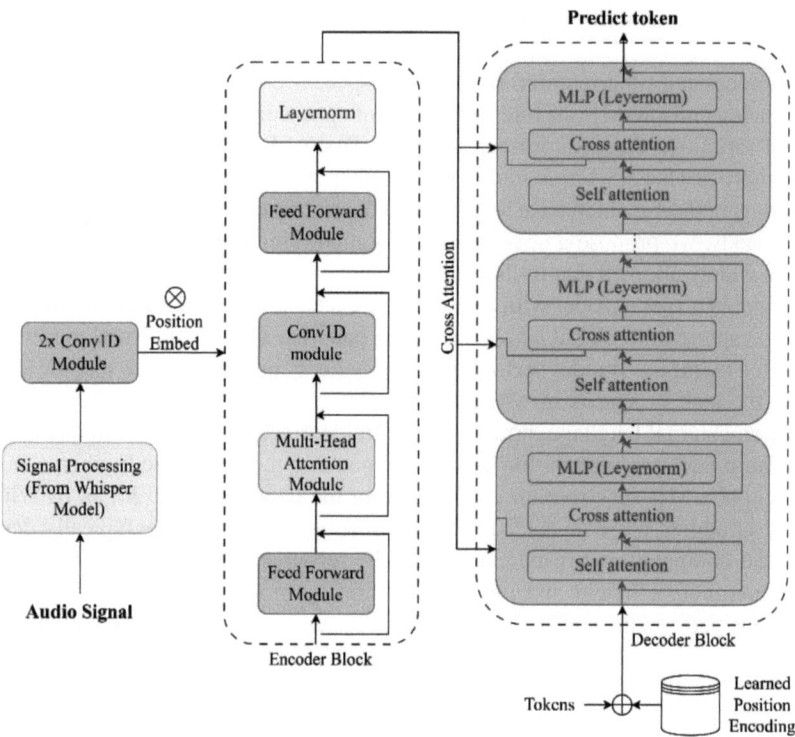

Fig. 3. Overview of Whisper-Conformer model.

To integrate the Conformer architecture, we created new encoder layers using torch. ModuleList, consists of multiple Conformer layer components based on the Conformer architecture in PyTorch. Additionally, we modified the Whisper configuration file to add depthwise_kernel_size to allow flexibility in adjusting the kernel size of the Conformer layers. As a result of these modifications, we were still able to utilize the existing pipeline and functions of Whisper, which means that in the future, we can use Whisper-Conformer as a multilingual and multitask model, maintaining the original capabilities of the Whisper model.

3.2 Datasets

The datasets used in this study are publicly available Thai speech datasets collected from the internet. In total, 373.5 h of speech data were utilized, with 350.2 h allocated for training and 23.3 h for evaluation, as shown in Table 1.

Table 1. Dataset for training.

Datasets	Train (hours)	Test (hours)	Description
Common Voice Corpus [13]	214.38	15.01	The Common Voice Corpus is a publicly available voice dataset powered by contributions from volunteer speakers worldwide. It encompasses 124 languages and was developed by Mozilla's foundation
Thai Elderly Corpus [11]	9.12	3.08	The Thai Elderly Speech Corpus was developed by VISAI AI Company Limited and Data Wow Company Limited. Divided into two categories: Healthcare and Smart Home from 32 speaker (10 males and 22 females) aged 57–60 years
Thai Dialect Corpus [21]	111.58	4.15	The Thai Dialect Corpus is a dataset developed for the creation of Thai dialect ASR. The dataset is divided into four subsets: Central, Khummuang, Korat, and Pattani. In this study, it will be used only on the Khummuang, Korat, and Pattani datasets
Gowajee Corpus [3]	15.09	1.09	The Gowajee Corpus is a dataset developed by three groups of students at Chulalongkorn University as part of a homework assignment. It was collected from 188 speakers, comprising 163 males and 25 females. This dataset features a teenage slang version of Thai

3.3 Training Detail

In this study, we developed models of various sizes to evaluate their performance based on Whisper hyperparameters [17]. For the kernel size, we referenced the recommended kernel size for building the conformer model [6], as shown in Table 2. The training was conducted using the AdamW optimizer, with default values for beta1, beta2, epsilon, learning rate, and weight decay. However, the random seed was set to 974358 in comparison with the baseline model. We employed the Seq2SeqTrainer to run the training process [4]. The model was trained on the APEX Cluster at CMKL University utilizing four NVIDIA A100 (40GB) GPUs. Training was performed for three epochs with a batch size of 16. We use WER and CER metrics to select the best model to avoid overfitting.

Table 2. Architecture details of the Conformer-Whisper model family.

Model	Layers	Width	Heads	Kernel size	Parameters
Whisper-Conformer-tiny	4	384	6	5	36.6 M
Whisper-Conformer-base	6	512	8	7	70.3 M
Whisper-Conformer-small	12	768	12	17	233.7 M
Whisper-Conformer-medium	24	1024	16	33	738.2 M

3.4 Evaluation Metrics

To evaluate the performance of our model, we use the medium Whisper model without pretrained weights as the baseline model. The baseline model is trained using the same hyperparameters as our model. Two commonly used metrics, Word Error Rate (WER) and Character Error Rate (CER), are employed to assess the performance of ASR systems. WER evaluates errors by comparing the number of insertions, deletions, and substitutions in the predicted text against the reference text. However, WER operates at the word level, which can sometimes lead to inaccuracies. For example, if the model predicts "Hello" instead of "hello", WER would register this as 1, despite the difference being only in capitalization. Although Thai does not have capitalization, it has a high number of homophones (e.g., "ข้า" vs. "ค่า" both pronounced as kha but different meaning), 21 vowels, and five tones. If we rely solely on WER, it may be unclear whether an error is due to the model's inability to learn a specific sound or its struggle to distinguish homophones in Thai. By incorporating CER, we gain deeper insights into these errors, allowing us to diagnose how effectively our model reduces the need for a language model for spelling correction.

4 Results and Discussion

The results of this study indicate that our model learns faster than the baseline model, as demonstrated in the loss vs. epoch graph. The graph shows that Whisper-Conformer-Medium, which has a comparable parameter size to the baseline model (4.03% fewer parameters), learns 7.45 times faster than the baseline at epoch 3 and 1.32 times faster compared to Whisper-Conformer-Small, as illustrated in Fig. 4. These findings suggest that replacing the original Transformer encoder with the Conformer architecture in the encoder enhances model performance. This improvement is primarily attributed to the integration of CNNs, which enables the model to effectively capture and exploit local features.

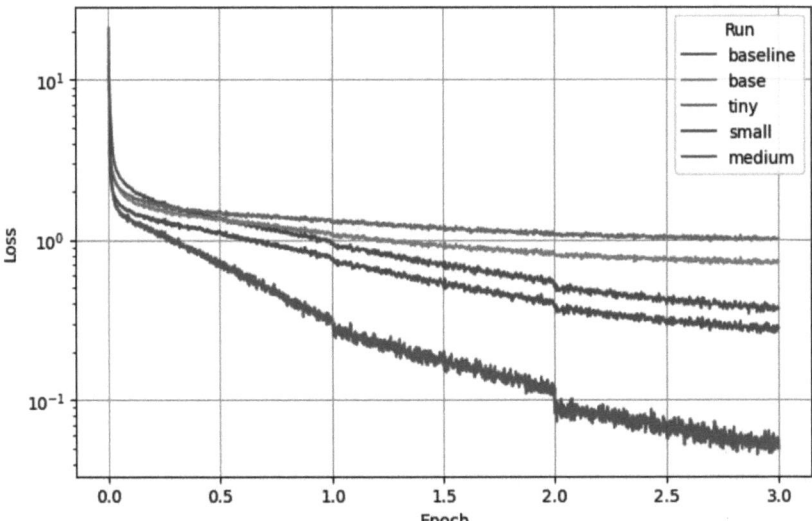

Fig. 4. Comparison of Whisper-Conformer model training loss across different parameters.

Due to resource constraints, we extended the training of the Whisper-Conformer-medium for an additional five epochs to further evaluate its potential. We compared its performance with the Thonburian-Whisper (medium parameter size) [1], a publicly available Whisper model fine-tuned on 1,316.76 h of Thai speech data. To improve robustness, the Thonburian-Whisper model incorporates spectral augmentations, including Gaussian noise, time stretch, and pitch shift. Moreover, it is the only publicly available model trained on the same dataset as this study. Table 3 presents the WER and CER results for our model and the Thonburian-Whisper model on four benchmark datasets: Common Voice Corpus (v18), Gowajee Corpus, Thai Elderly Speech Corpus, and Thai Dialect Corpus, all evaluated without applying a language model for spelling correction. Although we faced the challenge of not being able to use pre-trained weights from Whisper, the results show that our Whisper-Conformer-medium model outperformed the Thonburian-Whisper model on Common Voice Corpus (v18) (0.64 WER, 0.42 CER) and the Thai Dialect Corpus (83.27 WER, 39.96 CER). These two datasets account for 90% of the training data. The lower performance on the Gowajee Corpus and Thai Elderly Speech Corpus is likely due to the limited dataset size and shorter training time for these corpora.

Nevertheless, these results suggest that our model can address existing issues in Thai ASR. The WER results from the Thai Dialect Corpus show that our model can effectively learn the diverse phonetic variations across Thai dialects and the reduced CER compared to the current publicly available model. Although our model's CER is not yet optimal overall, it still demonstrates the potential to reduce reliance on a language model for spelling correction.

Table 3. Evaluation results on different datasets.

Models	Common Voice Corpus(v18)		Gowajee Corpus		Thai Elderly Speech Corpus		Thai Dialect Corpus	
	WER	CER	WER	CER	WER	CER	WER	CER
whisper-Conformer-medium	**0.64**	**0.41**	69.87	58.42	31.56	19.19	**83.27**	**39.96**
Thonburian-Whisper-medium [1]	29.85	3.21	**58.12**	**18.05**	**8.84**	**0.95**	94.26	44.48

5 Conclusion

This study demonstrates that the Conformer-Whisper model, a modified version of the original Whisper model incorporating CNNs based on the concept introduced by Anmol Gulati et al. [6], effectively captures both global context and local feature patterns When trained on Thai speech data, the model showed a faster learning rate compared to the baseline Whisper model. Furthermore, the Conformer-Whisper model exhibited performance comparable to a model fine-tuned on over 1,316.76 h of Thai speech, as evidenced by the WER and CER results on the Common Voice Corpus and Thai Dialect Corpus. The researchers expected that replacing the original Whisper encoder with the Conformer architecture would be a key factor in the model's ability to handle challenges such as the lack of spaces in Thai sentences and regional dialectal variations, thereby reducing reliance on language models for spelling correction. Notably, after just five epochs, the model achieved its best CER score, indicating that it had not yet overfitted to the dataset. However, due to resource limitations, training was limited to only five epochs. For future work, we aim to incorporate datasets enhanced with spectral augmentations, as Thai is a language with a complex system of consonants and vowels, which can be pronounced in multiple ways [20]. Applying spectral augmentations to the training data is expected to improve further the model's efficiency and accuracy [1], enabling it to handle the phonetic diversity of Thai speech better.

Acknowledgement. The authors acknowledge the APEX Cluster from CMKL University for providing computing resources for this work.

References

1. Aung, Z.H., et al.: Thonburian whisper: robust fine-tuned and distilled whisper for Thai. In: Proceedings of the 7th International Conference on Natural Language and Speech Processing (ICNLSP 2024) (2024)

2. Baevski, A., et al.: wav2vec 2.0: a framework for self-supervised learning of speech representations. Adv. Neural Inform. Process. Syst. **33**, 12449–12460 (2020)
3. Chuangsuwanich, E., et al.: Chulalongkorn University, Faculty of Engineering, Computer Engineering Department (2020). https://github.com/ekapolc/gowajee_corpus
4. Gandhi, S.: Fine-Tune Whisper for Multilingual ASR with Transformers (2024). Accessed 28 Jun 2024
5. Graves, A., Sequence transduction with recurrent neural networks. arXiv preprint arXiv:1211.3711 (2012)
6. Gulati, A., et al.: Conformer: Convolution-augmented transformer for speech recognition. arXiv preprint arXiv:2005.08100 (2020)
7. Han, W., et al.: Contextnet: Improving convolutional neural networks for automatic speech recognition with global context. arXiv preprint arXiv:2005.03191 (2020)
8. Jain, R., et al.: Adaptation of Whisper models to child speech recognition. arXiv preprint arXiv:2307.13008 (2023)
9. Pipatanakul, K.P.M., Sripaisarnmongkol, S., et al.: Monsoon Whisper Medium Gigaspeech2 (2024)
10. Li, J., et al.: Jasper: an end-to-end convolutional neural acoustic model. arXiv preprint arXiv:1904.03288 (2019)
11. Limited, V.A.C.L.a.D.W.C.: Thai Elderly Speech. https://github.com/VISAI-DATAWOW/Thai-Elderly-Speech-dataset/releases/tag/v1.0.0
12. Mehrish, A., et al.: A review of deep learning techniques for speech processing. Inform. Fusion **99**, 101869 (2023)
13. MozillaFoundation, Common Voice: A Massively-Multilingual Speech Corpus (2024). https://commonvoice.mozilla.org/th/datasets
14. Phatthiyaphaibun, W., et al.: Pythainlp: Thai natural language processing in python. arXiv preprint arXiv:2312.04649 (2023)
15. Phatthiyaphaibun, W., et al.: Thai wav2vec2. 0 with commonvoice v8. arXiv preprint arXiv:2208.04799 (2022)
16. Pisarn, C., Theeramunkong, T.: An HMM-based method for Thai spelling speech recognition. Comput. Math. Appl. **54**(1), 76–95 (2007)
17. Radford, A., et al.: Robust speech recognition via large-scale weak supervision. In: International Conference on Machine Learning, PMLR (2023)
18. Ramachandran, P., Zoph, B., Le, Q.V.: Searching for activation functions. arXiv preprint arXiv:1710.05941 (2017)
19. Rijal, S., et al.: Whisper Finetuning on Nepali Language. arXiv preprint arXiv:2411.12587 (2024)
20. Suebvisai, S., et al.: Thai automatic speech recognition. In: Proceedings of (ICASSP 2005), IEEE International Conference on Acoustics, Speech, and Signal Processing. IEEE (2005)
21. Suwanbandit, A., et al.: Thai dialect corpus and transfer-based curriculum learning investigation for dialect automatic speech recognition. In: Proceedings of Interspeech (2023)
22. Tipakasorn, P., et al.: Comprehensive benchmarking and analysis of open pretrained Thai speech recognition models. In: 2024 27th Conference of the Oriental COCOSDA International Committee for the Co-ordination and Standardisation of Speech Databases and Assessment Techniques (O-COCOSDA). IEEE (2024)
23. Vaswani, A.: Attention is all you need. Advances in Neural Information Processing Systems (2017)
24. Yang, H., et al.: Chinese ASR and NER improvement based on whisper fine-tuning. In: 2023 25th International Conference on Advanced Communication Technology (ICACT). IEEE (2023)

RevCD: Reversed Conditional Diffusion for Generalized Zero-Shot Learning

William Heyden, Habib Ullah(✉), Muhammad Salman Siddiqui(✉), and Fadi Al Machot(✉)

Norwegian University of Life Sciences, 1433 Ås, Norway
{william.hedyen,habib.ullah,muhammad.salman.siddiqui,
fadi.al.machot}@nmbu.no

Abstract. In Generalized Zero-Shot Learning (GZSL), we aim to recognize both seen and unseen categories using a model trained only on seen categories. In computer vision, this translates into a classification problem, where knowledge from seen categories is transferred to unseen ones by exploiting the relationships between visual features and available semantic information. However, learning this joint distribution is costly and requires one-to-one alignment with corresponding semantic information. We present a reversed conditional diffusion-based model (RevCD) that mitigates this issue by estimating the semantic density conditioned on visual inputs. Our RevCD model consists of a cross Hadamard-addition embedding of a sinusoidal time schedule, and a multi-headed visual transformer for attention-guided embeddings. The proposed approach introduces two key innovations. First, we apply diffusion models to zero-shot learning, a novel approach that exploits their strengths in capturing data complexity. Second, we reverse the process by approximating the semantic densities based on visual data, made possible through the classifier-free guidance of diffusion models. Empirical results demonstrate that RevCD achieves competitive performance compared to state-of-the-art generative methods on standard GZSL benchmarks. The complete code will be available on GitHub.

Keywords: Zero-shot learning · Transfer learning · Diffusion model

1 Introduction

Zero-shot learning (ZSL) represents a state-of-the-art advancement in machine learning transferability and computer vision classification. By pushing the boundaries of knowledge extraction, ZSL enable ML models to expand without costly retraining. This learning paradigm is particularly crucial as it addresses the inherent limitation of traditional machine learning models that require prior access to expensive datasets. ZSL leverages auxiliary knowledge, allowing models to explore unobserved events, edge cases, or new compositions without any additional training. Traditional approaches in ZSL focused on aligning attributes directly with object categories [25], while deep learning's potential to create a joint embedding space of visual and semantic features [7,8,12,34,40] has rendered this approach obsolete. The shift toward latent-based methods highlights the importance of embedding space techniques because of their ability to decode and infer complex data distributions. This offers a promising resolution to the two main challenges of ZSL: the semantic gap and limited generalization ability [44].

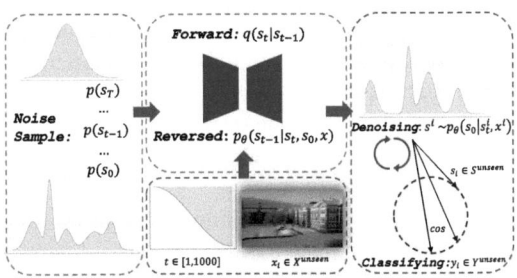

Fig. 1. Overview of the proposed RevCD model: We train a denoising process using only seen samples (indicated by green boxes). Once trained, the model can estimate the semantic distribution by conditioning on the visual space of unseen samples (represented by blue boxes) and Gaussian noise. The final classification is conducted through a simple nearest-neighbor search based on the estimated density. (Color figure online)

Our contribution introduces a Diffusion-based generative model, a notable innovation in ZSL. Distinct from conventional models that predominantly rely on attribute matching or embedding strategies, our RevCD model utilizes a diffusion process to model the data distributions iteratively, see Fig. 1. This augments the model's capability to manage class variability and enhances its generalization capacity. Such control over the semantic space is required for overcoming the challenges of bias and hubness commonly encountered in ZSL methodologies [26].

2 Related Works

Advancements in likelihood-based models have been central to the progress of zero-shot learning [36]. By framing the learning process as a maximum likelihood estimation problem, these methods effectively model data distributions, allowing for robust generalization across both seen and unseen classes. This section categorizes ZSL approaches according to the foundational models employed for approximation of the data, including Variational Autoencoders (VAEs), Generative Adversarial Networks (GANs), and Hybrid models. Additionally, it highlights the role of attention mechanisms and embedding strategies in enhancing these models' performance.

VAE-Based. Variational Autoencoders [24] play a crucial role in ZSL due to their probabilistic framework for modeling latent spaces. Their adaptability in synthesizing unseen class prototypes, as demonstrated by [5,38], and [22], underscores their versatility in ZSL applications. For instance, [18] incorporates a semantic-guided approach within a VAE framework, while [46] employs a decoupling strategy to enhance performance. A significant advantage of VAEs lies in their capacity to *explicitly* approximate data density. However, a key limitation is their tendency to generate blurry or overly smoothed features due to posterior collapse [32], which can obscure essential class-specific details needed to distinguish between unseen classes.

GAN-Based. Generative Adversarial Networks [14] offer a powerful and dynamic framework for feature synthesis. GANs have been successfully adapted to generate features for unseen classes [13,15,21,52]. GANs' ability to produce sharp and realistic features through *implicit* density estimation makes them particularly effective for capturing fine-grained details. However, they are also prone to challenges such as training instability and mode collapse [4], which can result in a lack of diversity in the generated features. This limitation may hinder the model's ability to accurately represent the full spectrum of unseen classes.

Hybrids. Hybrid models in ZSL leverage multiple architectures to enhance performance. The majority of hybrid frameworks incorporate sequential modules, as seen in [10,16,30]. Employing a VAE to learn an embedding function that constrains the semantic or visual space allows for greater control over the generation of synthesized features. Nevertheless, the complexity and computational overhead of combining multiple models can pose challenges, especially in terms of model interpretability and scalability.

Attention and Embedding. Attention mechanisms [17,23] and embedding strategies [1,49] further refine the latent space by focusing on salient attributes and mapping visual data to semantic space. The approaches [2,29,47] enhance interpretability and feature distinctness; however, they rely heavily on high-quality, granular attribute information, which is not always available, limiting their applicability across diverse datasets.

Our contribution introduces a reversed Diffusion-based model (RevCD) for zero-shot inference. Diffusion models have been applied to improve accuracy as generative classifiers [3,6,39], and their capacity to generate synthetic data has enabled the classification of unseen compositions [9,27]. However, existing applications are limited by pre-training on prompt categories. Their implementation in a pure zero-shot setting is still absent. We address these limitations by leveraging the reversed process for generating conditioned semantic embeddings, aiming for effective generalization to unseen classes without the constraints observed in the aforementioned methodologies. *To the best of our knowledge, Diffusion models have not yet been explored within the ZSL domain.*

3 Methodology

Problem Setup. We denote the set of seen images, semantics, and corresponding class labels as $\{x^{seen}, s^{seen}, y^{seen}\} \in \mathcal{D}^{seen}$, where x^{seen} represents the images, s^{seen} the semantics, and y^{seen} the class labels for the seen classes. The set of unseen semantics and class labels, denoted as $\{s^{unseen}, y^{unseen}\} \in \mathcal{D}^{unseen}$, represents the unseen dataset. During training, the model is trained exclusively on data from the seen set $\mathcal{D}^{seen}$, while assuming access to the semantic and label information of the unseen classes in $\mathcal{D}^{unseen}$. Importantly, the unseen images, x^{unseen}, are *not* available during training and are only introduced during the inference phase. It is important to

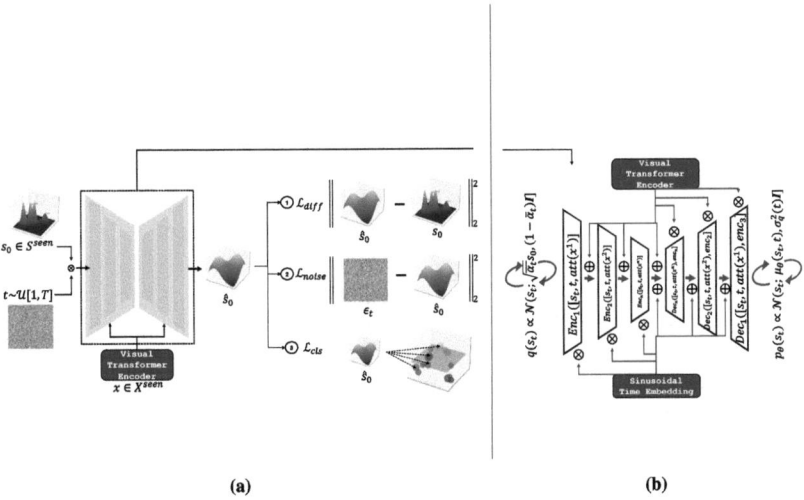

Fig. 2. The figure illustrates our proposed approach for training. **(a)** presents our high-level architecture and associated loss functions. By conditioning the image, we can infer the semantic distribution of unseen classes. **(b)** provides a detailed view of our U-net architecture. It implements sinusoidal time and cross-Hadamard-Addition conditional embeddings for optimal control over the learned distribution. In ZSL, the goal is to transfer the knowledge of how to infer the distribution rather than the distribution itself.

note that during training, the set of class labels for seen and unseen data does not overlap, i.e., $\mathcal{Y}^{seen} \cap \mathcal{Y}^{unseen} = \emptyset$. This ensures there is no direct overlap between seen and unseen classes. During inference, the challenge is to map an unseen sample image, x^{unseen}, to its corresponding unseen label y^{unseen}, using a learned function $f : x^{unseen} \rightarrow y^{unseen}$. The training process involves using paired examples $\{x^{seen}, s^{seen}\} \in \mathcal{D}^{seen}$ to learn a joint model, $p_\theta(s, x)$, that can be used to estimate the conditional probability distribution function for the semantic space s^{seen} given their corresponding visual features x^{seen} as $p_\theta(s|x^{seen})$. During the test phase, the semantic distribution of the unseen images x^{unseen} is approximated and subsequently classified into their corresponding unseen classes $y^{unseen} \in \mathcal{Y}^{unseen}$.

3.1 Diffusion Process

The diffusion process [41] models complex data distributions through a specific Markov chain structure. During training, we start with time-step 0 represented as an ordinary, clean, sample from the semantic space $s_0 \sim q(S)$. This is paired with its corresponding visual features $\{s_0, x\} \in \mathcal{D}^{seen}$. We incrementally infuse Gaussian noise using a fixed linear Gaussian model $s_1, \ldots, s_T$, for T steps. By employing the reparameterization trick [24], we can parameterize the mean $\mu_t(s_t) = \sqrt{\alpha_t} s_{t-1}$, and variance $\Sigma(s_t) = (1 - \alpha_t)\mathbf{I}$ for hierarchical time-steps $t \in [0, T]$. Pre-defining the noise schedule $(\beta_1, \ldots, \beta_T)$ allows us to sample from the Markov chain through a fixed forward sequence of time steps as:

$$q(s_t|s_{t-1}) = N(s_t; \sqrt{\alpha_t}s_{t-1}, (1-\alpha_t)\mathbf{I}) \qquad (1)$$

where $\alpha_t = 1 - \beta_t$ and $\bar{\alpha}_t = \prod_{i=1}^{t} \alpha_i$. This forward encoding process has the desired properties of being variance-preserving and completely deterministic, and the final distribution $p(s_T)$ is a standard Gaussian. Our aim is to learn a reverse diffusion kernel $p(s_{t-1}|s_t, x)$ which removes the noise of the forward process and can estimate a clean sample, $\hat{s}_0$, from random noise conditioned on a visual space:

$$p(s_{0:T}|x) = p(s_T) \prod_{t=1}^{T} p(s_{t-1}|s_t, x) \qquad (2)$$

Directly expressing $p(\cdot)$ in closed form is intractable. We instead parameterize p_θ with θ and approximate the distribution by minimizing the evidence lower bound (ELBO) of the conditional log-likelihood

$$\log p_\theta(s|x) \geq \mathbb{E}_{q(s_{1:T}|s_0,x)}\left[\log \frac{p_\theta(s_0|s_{1:T},x)p(s_{1:T},x)}{q(s_{1:T}|s_0,x)}\right] \qquad (3)$$

$$= \underbrace{\mathbb{E}_{q(s_1,x|s_0)}[\log p_\theta(s_0|s_1,x)]}_{\text{reconstruction}} - \underbrace{D_{KL}[q(s_T|s_0,x)||p(s_T|x)]}_{\text{prior matching}}$$

$$- \underbrace{\sum_{t=2}^{T} \mathbb{E}_{q(s_t,x|s_0)}[D_{KL}[q(s_{t-1}|s_t,s_0,x)||p_\theta(s_{t-1}|s_t,x)]]}_{\text{diffusion term}} \qquad (4)$$

By conditioning the forward process on the clean example at any given t, the diffusion loss can be formulated using Bayes' rule as the KL divergence between the ground-truth analytical denoising step $q(s_{t-1}|s_t, s_0, \mathbf{x})$ and our approximated denoising step $p_\theta(s_{t-1}|s_t, \mathbf{x})$. The prior loss $D_{KL}[q(s_T|s_0,x)||p(s_T|x)]$ can be ignored as it does not contain any trainable parameters and is zero under our assumption. The reconstruction loss $\log p_\theta(s_0|s_1,x)$ is typically minimal and can be safely ignored without affecting the outcome. Therefore, our diffusion objective becomes:

$$\arg\max_\theta \mathbb{E}_{t\sim[2,T]}[\mathbb{E}_q[D_{KL}(q(s_{t-1}|s_t,s_0)||p_\theta(s_{t-1}|s_t,x))]] \qquad (5)$$

which boils down to learning a neural network, s_θ, to predict the semantic space $\hat{s}_0$ from noise at time t, conditioned on an image x. This network can be optimized using stochastic samples of t from a uniform distribution $t \sim \mathcal{U}[1,T]$.

Given our case, where we can set the variances to match exactly, the KL divergence in Eq. (5) can be reduced to a minimization of the difference between the mean of the two distributions [11].

$$\arg\max_\theta D_{KL}(q(s_{t-1}|s_t,s_0)||p_\theta(s_{t-1}|s_t,x)) \qquad (6)$$

$$= \arg\max_\theta \frac{1}{2\sigma_q^2(t)}[||\mu_\theta(\cdot) - \mu_q(\cdot)||_2^2] \qquad (7)$$

where μ_q is the ground truth transition mean predetermined in the forward process and μ_θ is the mean of our learned diffusion kernel.

Reconstruction Loss. Using Bayes' rule and the Markov assumption, $q(s_t|s_{t-1}, s_0) = q(s_t|s_{t-1})$ [31], we can show that:

$$q(s_{t-1}|s_t, s_0) \propto \mathcal{N}(s_{t-1}; \frac{\sqrt{\alpha_t}(1-\bar{\alpha}_{t-1})s_t + \sqrt{\bar{\alpha}_{t-1}}(1-\alpha_t)s_0}{1-\bar{\alpha}_t}, \underbrace{\frac{(1-\alpha_t)(1-\bar{\alpha}_{t-1})}{1-\bar{\alpha}_t}}_{\sigma_q^2(t)}\mathbf{I}) \quad (8)$$

Employing this relationship to Eq. (7), we arrive at our formulation of diffusion loss (loss ① in Fig. 2a), to predict the ground truth sample from an arbitrary noised version of it:

$$\mu_q(\mathbf{s}_t; s_0) = \frac{\sqrt{\alpha_t}(1-\bar{\alpha}_{t-1})s_t + \sqrt{\bar{\alpha}_{t-1}}(1-\alpha_t)s_0}{1-\bar{\alpha}_t} \quad (9)$$

$$\mathcal{L}_{reconstruction} = \underbrace{\frac{1}{2\sigma_q^2(t)} \frac{\bar{\alpha}_{t-1}(1-\alpha_t)^2}{(1-\bar{\alpha}_t)^2}}_{w_t} [||s_0 - s_\theta(s_t, t, x)||_2^2] \quad (10)$$

Here, we replace $\mu_q(\cdot) = s_0$ with $\mu_\theta(\cdot) = s_\theta(s_t, t, x)$ to estimate the reversed diffusion kernel. The first term, w_t, is a time-dependent variance weight, where $\sigma_q^t(t) = \beta_t$ [19]. However, empirical research [27] has demonstrated that setting $w_t = 1$ yields optimal performance. Our experiments in the zero-shot paradigm show consistent outcomes.

Noise Loss. The diffusion loss in Eq. (7) can also be interpreted as estimating the source noise added $\hat{\epsilon}_t$, rather than directly predicting the clean sample $\hat{s}_0$. By applying the reparameterization trick [33], we can express the relationship between a clean and an arbitrarily noised sample as:

$$s_0 = \frac{s_t - \sqrt{1-\bar{\alpha}_t}\epsilon_0}{\sqrt{\bar{\alpha}_t}} \quad (11)$$

This enables us to estimate the reverse transition mean by directly utilizing the estimated added noise instead:

$$\mu_q(s_t; s_0) = \frac{1}{\sqrt{\alpha_t}}s_t - \frac{1-\alpha_t}{\sqrt{1-\bar{\alpha}_t}\sqrt{\alpha_t}}\epsilon_0 \quad (12)$$

By incorporating this reformulation into Eq. (7), as a function of the perturbed noise at time step t, we optimized the network by estimating the source noise from the predicted noise. Leveraging this theoretical perspective, we introduce an additional noise loss term that corresponds to our clean-sample predictor (loss ② in Fig. 2a):

$$\mathcal{L}_{noise} = \underbrace{\frac{1}{2\sigma_q^2(t)} \frac{(1-\alpha_t)^2}{(1-\bar{\alpha}_t)^2 \alpha_t}}_{w_t'} [||\epsilon_0 - \frac{\sqrt{\bar{\alpha}_t}}{\sqrt{1-\bar{\alpha}_t s_t}} s_\theta(s_t, t, x)||_2^2] \quad (13)$$

Note the slight variance-weighting difference compared to the diffusion loss in Eq. (10). This discrepancy is a correction term as a result of the different transition mean calculation. However, it can be eliminated by assigning a fixed constant value of $w_t' = 1$.

These two complementary formulations of the denoising transition mean correspond to an equivalent optimization problem (Eq. 5). Although these formulations introduce additional complexity to the optimization process, necessitating more sophisticated strategies and careful hyperparameter tuning to achieve convergence, we observe significant improvements in density estimation. We believe that this loss function acts as a regularizer during optimization, since $w_t' \neq w_t$, enhancing the model's ability to navigate the loss landscape and generalize to unseen data. By optimizing from multiple perspectives, the model generates richer and more robust representations.

Classification Loss. Unlike other generative models such as flow-based models and GANs, Diffusion models have no natural property to decrease the intra-class variance from the noise input [20]. Previous work in classifier- and classifier-free guidance in score-based Diffusion models [20] involves modifying the score function with the gradient of the log-likelihood of a separate classifier model ϑ, $-\nabla_\vartheta \log p_\vartheta(y|s_t)$. Our classification objective is to steer our optimization problem of the inferred distribution through manifold regularization [48], leveraging a classifier:

$$\arg\max_{\vartheta} \mathbb{E}_{y \in Y_{seen}} [\log p_\vartheta(\hat{s}_{0:T}|y)] \quad (14)$$

This allows us to approximate samples from the distribution $p_\theta(y|s_t) \propto p_\theta(s_t|y) p_\theta(y)$. The strategy of assigning higher likelihood to the correct label has led to notable improvements in both the perceptual qualities and the inception scores of models, as highlighted in prior research [37]. However, within the zero-shot learning framework, our goal shifts towards enhancing the model's ability to generalize the learned distribution for generating samples. These samples are not primarily focused on visual appeal but are aimed at positioning the probability mass of each conditional sample at a greater distance. Therefore, we formulate the loss as the expectation over the empirical sample distribution $\mathbb{E}[\mathcal{L}(f(\mathbf{x};\vartheta), y_x)]$ and implement this with a cross-entropy loss (loss ③ in Fig. 2a):

$$\mathcal{L}_{classification} = -\frac{1}{n}\sum_{i=1}^{n}\sum_{j=1}^{c} y_{ij} \log p_\vartheta(\hat{y}_{ij}|s_i) \quad (15)$$

where n is the number of samples and y_{ij} and $\hat{y}_{ij}$ are the true and predicted label for class j of the i-th instance.

3.2 Training Objective

Our main idea focuses on directly modeling the semantic posterior using variational inference rooted in Eq. (16). We achieve this by disentangling the posterior estimation into three key components: noise prediction, data reconstruction, and (auxiliary) classification. This decomposition results in a more complex and nuanced loss landscape [28]. Despite the increased complexity, integrating these distinct loss components enhances the model's generalization capabilities. This is primarily due to the regularization effects inherent in the multi-faceted loss function and the fine-tuning achieved through careful hyperparameter optimization. To enable classifier-free guidance at inference, we adopt a conditional dropout strategy during training, randomly masking the visual condition when computing the noise prediction loss [20]. This allows the model to learn both conditional and unconditional score estimates within a single unified network. Our overall training objective becomes:

$$\mathcal{L}_{\text{total}} = \lambda_1 \mathcal{L}_{rec} + \lambda_2 \mathcal{L}_{noise} + \lambda_3 \mathcal{L}_{classification} \quad (16)$$

Here, λ_1 and λ_3 serve as a balancing factors between the objectives of reconstruction and classification, while λ_2 acts as a regularization coefficient. Through this, the probability distribution of the samples aligns with the expectation of the generated conditional samples $p(s) \propto \mathbb{E}_{x \sim p(x)}[p_\theta(s|x)]$. The implementation details of this loss function during training are provided in Algorithm (1).

3.3 Architectural Considerations

Cross Hadamard-Addition Embeddings. In the traditional diffusion process, we predict the ground truth of a noisy sample at time t. In our approach, however, we further condition this process on visual features. Consequently, the neural network s_θ is trained on the triplet (s_t, t, x), where $\{s, x\} \in \mathcal{D}^{seen}$. To refine the embeddings for both the conditioning variable x and the time-step schedule t, we employ a cross Hadamard-Addition method, which enhances the representation and integration of these features within the network. During the representational mapping stage within the network, we use Hadamard integration for the time-step input, acknowledging that the added noise is entirely deterministic, while integrating the visual condition through addition (refer to the encoding step in Fig. 2b). In contrast, during the network's generative stage, we reverse these roles, applying Hadamard integration for a stronger conditional reconstruction and a more relaxed incorporation of the time-step input (see the decoding step in Fig. 2b). We observe that this approach results in a closer alignment of the joint probability space, leading to improved accuracy.

Time-Dependent Embedding. To increase the dimension of the time step, t, we employ a sinusoidal time embedding $\bar{t} \leftarrow TE(t, d)$:

$$\begin{aligned} \text{TE}(t, d) = [&\cos(t \cdot f_0), \sin(t \cdot f_0), \ldots, \\ &\cos(t \cdot f_{\frac{d}{2}-1}), \sin(t \cdot f_{\frac{d}{2}-1})], \end{aligned} \quad (17)$$

Algorithm 1. Training algorithm for RevCD.

Ensure:
$\bar{t} \sim TE(\mathcal{U}[0,1], d)$
$\epsilon \sim \mathcal{N}(0, \mathbf{I})$
for $s, \mathbf{x} \sim p(x, s) \in \mathcal{D}^{seen}$ **do**
    $\mathbf{x} \leftarrow \emptyset$ with $p_{conditional}$
    $s_0 \leftarrow 2s - 1$
    $\hat{s}_0 \leftarrow ||Unet(\sqrt{\bar{\alpha}_t}s_0 + \sqrt{(1-\bar{\alpha})}\epsilon, \bar{t}, x) - \sqrt{\bar{\alpha}_t}s_0 + \sqrt{(1-\bar{\alpha})}\epsilon||_2^2$
    $\hat{y} \leftarrow \mathbb{E}_\vartheta(\hat{s}_0 | y_{seen})$
    $\hat{\epsilon}_t \leftarrow ||\epsilon_0 - (\hat{s}_0 - \epsilon_t)||_2^2$
end for
Gradient step on:
$\nabla_\theta [\lambda_1 \mathcal{L}_{Diff}(\hat{s}_0) + \lambda_2 \mathcal{L}_{cls}(\hat{y}) + \lambda_3 \mathcal{L}_{noise}(\hat{\epsilon}_t)]$

where d is the embedding dimension and f_i are frequencies. The temporal encoding dimension is matched to each layer of the network to learn the denoising function, given any timestep t.

Visual-Dependent Embedding. We implement a Transformer encoder [43] to extract visual features for conditioning. These visual features are integrated into the network at each layer, with the Transformer trained concurrently. To align the denoising feature dimensions, we map the multi-head attention outputs from the visual space to each intermediate feature using a Hadamard product in the decoder and matrix addition in the encoder of our denoising model (see Fig. 2b).

Pre-conditioning. We also perform an affine transformation of our semantic space $s \in [0, 1]$ to resemble a zero-mean Gaussian $s\prime = 2 \times s - 1 \in [-1, 1]$. This increases the dynamic range leading to better gradient flow and stabilizes convergence as the variance $Var(s_0) \ll 1$ skews the signal-to-noise ratio when scaled by the noising schedule $\bar{\alpha}_t$.

3.4 Model Design

Our denoising Diffusion model employs a U-Net architecture, as introduced by the probabilistic diffusion model in [19]. To merge visual and semantic information effectively, we have customized this architecture to support both our time-dependent and visual-dependent embeddings, as illustrated in Fig. 2b. To our knowledge, this represents the first application of a U-Net architecture tailored for zero-shot learning in such a specific way. The encoder-decoder structure of our U-Net is composed of linear blocks featuring ReLU non-linearity and batch normalization. Inputs to each layer include sinusoidal time embeddings and conditional data, which are extracted using self-attention mechanisms and augmented by a skip-connection between the encoding and decoding stages.

3.5 Sampling

Using standard methods from diffusion theory [19], we generate the semantic embedding space of an unseen sample through iterative conditional denoising using our trained model, as shown in Algorithm (2). Samples are drawn from the standard normal prior $p(s_T) \sim \mathcal{N}(0, \mathbf{I})$ and denoised conditioned on the sinusoidal time-step embedding $\bar{t}_i \ \forall i \in [1000, 0]$, and the Transformer-encoded latent visual space $x \in \mathcal{D}^{unseen}$.

The sampling through the reversed diffusion process is crucial for synthesizing high-quality semantic embeddings from the noised data. This process is governed by the following equation:

$$s_{t-1} = \underbrace{\frac{1}{\sqrt{\alpha_t}}(s_t - \frac{(1-\alpha_t)\hat{s}_t}{\sqrt{1-\bar{\alpha}_t}})}_{\text{remove noise}} + \underbrace{\beta_t z}_{\text{add noise}} \qquad (18)$$

Here, s_{t-1} denotes the noisy semantic embeddings at time step $t - 1$, $\hat{s}_t$ represents the (predicted) noised sample at previous time step t, and β_t is the variance noise vector that controls the amount of noise added back to ensure stability, where $z \sim \mathcal{N}(0, I)$. This iterative refinement process enables the model to generate $\hat{s}^{unseen}$ during inference.

Classifier-Free Guidance. To improve semantic alignment during sampling, we adopt the classifier-free guidance framework [20]. Using the difference between conditional and unconditional diffusion kernel estimates, we effectively steer the reverse process toward more probable samples under a desired condition. Using Tweedie's formula [42], the denoising model can approximate the conditional and unconditional score functions $s_\theta(s_t, t, x) \approx \nabla_{s_t} \log p(s_t|x)$ $s_\theta(s_t, t, \emptyset) \approx \nabla_{s_t} \log p(s_t)$.

By leveraging the decomposition of the conditional score we can mirror a kind of gradient in the semantic space, pulling the diffusion kernel in a direction that increases likelihood under the condition.

$$\nabla_{s_t} \log p(s_t|x) = \nabla_{s_t} \log p(s_t) + \nabla_{s_t} \log p(x|s_t), \qquad (19)$$

where $\nabla_{s_t} \log p(x|s_t)$ acts as a classifier-like guidance term and can be reinterpreted as the difference between the conditional and unconditional scores:

$$\nabla_{s_t} \log p(x|s_t) \approx s_\theta(s_t, t, x) - s_\theta(s_t, t, \emptyset). \qquad (20)$$

To control the influence of the condition x, we scale this difference by a factor $g \in \mathbb{R}^+$, leading to the guided score estimate:

$$s_\theta(s_t, t, x) = s_\theta(s_t, t, \emptyset) + g \cdot [s_\theta(s_t, t, x) - s_\theta(s_t, t, \emptyset)]. \qquad (21)$$

This can be rearranged as our reversed transition function.

$$s_\theta(s_t, t, x) = (1 + g) \cdot s_\theta(s_t, t, x) - g \cdot s_\theta(s_t, t, \emptyset). \qquad (22)$$

Within the zero-shot learning framework, our goal is to reposition the probability mass of each conditional sample to enhance semantic separation, thereby improving recognition of unseen classes. To this end, we employ a single network jointly during sampling. The joint objective is illustrated as loss ③ in Fig. 2a.

Algorithm 2. Unseen sampling algorithm for RevCD.

Ensure:
$s_t \sim \mathcal{N}(0, \mathbf{I})$, $x \sim p(x) \in \mathcal{D}^{unseen}$, g : guidance strength
for $t = T, ..., 1$ **do**
  $\bar{t} \leftarrow TE(t, d)$
  $s_t^c, s_t^u \leftarrow Unet(s_t, \bar{t}, x), Unet(s_t, \bar{t}, \emptyset)$
  $\hat{s} \leftarrow (1+g)s_t^c - g s_t^u$
  $s_{t-1} \leftarrow \frac{1}{\sqrt{\alpha_t}}(s_t - \frac{(1-\alpha_t)\hat{s}_t}{\sqrt{1-\bar{\alpha}_t}}) + \beta_t z$
end for
return $(\hat{s}_0 + 1) \cdot \frac{1}{2}$ #affine unmapping

3.6 Zero-Shot Inference

In the zero-shot learning setting, the model utilizes the denoising model s_θ to approximate the semantic distribution given the instance x^{unseen}. A pseudo sample drawn from this distribution is then classified using a nearest-neighbor approach in the semantic space, leveraging the semantic density to bridge the gap between visual features of x^{unseen} and class labels y^{unseen}:

$$\hat{y} = \arg\min_{y \in \mathcal{Y}_{unseen}} \text{dist}(\hat{s}^{unseen}, s_y^{unseen}), \tag{23}$$

where $\hat{y}$ is the predicted class label for an unseen class instance, and $\text{dist}(\cdot, \cdot)$ denotes the distance metric, in our case cosine similarity:

$$\text{dist}(i, j) = 1 - \frac{\langle s_i, s_j \rangle}{||s_i||_2 ||s_j||_2} \tag{24}$$

4 Experimental Results

We evaluate our approach by measuring classification accuracy on both known and unknown categories. Importantly, samples from unknown categories are entirely absent during training, ensuring that classification accuracy for these categories reflects the model's ability to transfer knowledge from the known space. This evaluation methodology aligns with established practices in zero-shot inference research, facilitating fair comparison and assessment of our model's performance.

Dataset. Our analysis of diffusion as a generative method for zero-shot inference employs four publicly available benchmark datasets, each widely used in the field. This allows us to make a fair comparison of the quality and semantic coverage of the approximations. The benchmark datasets used are: Animals with Attributes 2 dataset (AwA2) [51], Caltech Birds dataset (CUB200-2011) [45], and Scene Understanding Attribute dataset (SUN) [35].

The CUB dataset, focused on bird species, offers detailed representations in both image and semantic spaces. The semantic space consists of detailed attribute descriptions of the bird species' characteristics. In contrast, AwA, which covers a range of different animal species, provides much coarser semantic descriptions, representing higher-level visual features. SUN consists of a scenery dataset with a wide range of classes, and its semantic descriptions are based on word2vec representations of each scene.

Visual features are derived using a ResNet101 backbone pre-trained on ImageNet [50]. We only compare models using similar image features to ensure a fair evaluation. We use the semantic attributes released with each dataset, which are derived from either crowd-sourced human annotations or word2vec-based label extractions.

Implementation Details. The employed U-net architecture for our Diffusion model consists of three hidden, fully connected dense layers, ReLU activation functions, and dropout for regularization. We use a feature extractor with a multi-head self-attention layer (MSA) for the conditional space. In the encoder and the decoder of the U-net, we concatenate and add the sinusoidal time embedding to layer inputs and the conditional features as explained in Sect. 3.3. For our loss function, we fix $\lambda_1 = 1$ and $\lambda_2 = 1$ during training, while λ_3 varies depending on the dataset (see Sect. 4.3).

4.1 Generalized Accuracy

Our method achieves strong performance in the generalized zero-shot learning (GZSL) setting, particularly excelling on seen (S) class classification across all datasets. See Table 1. On AWA, we achieve the highest seen accuracy of 94.5%, substantially outperforming the previous best. Similarly, for CUB and SUN, we report 87.5% and 66.9% on seen classes, again outperforming existing state-of-the-art methods. We attribute this to the conditional nature of our diffusion model. Unlike prior approaches that condition on semantics to generate visuals, our reversed formulation conditions on visual inputs, allowing the model to preserve fine-grained visual distinctions. This becomes especially beneficial in datasets like AWA and SUN, where visual cues are rich and diverse. This highlights the advantage of diffusion models in capturing complex visual semantics through iterative refinement.

However, our approach slightly underperforms on unseen (U) classes compared to some state-of-the-art methods. For instance, on AWA, we report 42.4%, compared to 82.3% achieved by ICCE. This is a result of the semantic space being estimated from a single visual instance: since each visual sample may correspond to several valid semantic embeddings, the diffusion model may distribute probability mass across divergent regions of the semantic space. This inherent multi-modality poses challenges in datasets with coarse or ambiguous semantic descriptors (e.g., AWA). Notably, our method performs more competitively on SUN's unseen classes (43.4%, close to the best-performing method at 50.9%), where the broad and distributed semantic space (based on word2vec scene embeddings) mitigates this effect. Overall, our method achieves competitive harmonic means across all datasets: 58.3% (AWA), 47.2% (CUB), and 52.6% (SUN), while showcasing the strength of diffusion-based conditional modeling for generalized classification tasks.

Table 1. Generalized zero-shot learning (GZSL) results on AWA, CUB, and SUN datasets. S, U, and H denote accuracy on seen classes, unseen classes, and their harmonic mean, respectively. Our method achieves the highest seen class accuracy across all datasets (such as 94.5% on AWA, 87.5% on CUB), and a competitive harmonic mean, especially on SUN (52.6%), highlighting the effectiveness of our conditional diffusion formulation.

Model	Venue	AWA			CUB			SUN		
		S	U	H	S	U	H	S	U	H
f-VAEGAN-D2	CVPR(19)	70.6	57.6	63.5	60.1	48.4	53.6	38.0	45.1	41.3
LisGAN	CVPR(19)	76.3	52.6	62.3	57.9	46.5	51.6	37.8	42.9	40.2
GDAN	ICCV(19)	67.5	32.1	43.5	66.7	39.3	49.5	40.9	38.1	53.4
ZSML	IAAA(20)	58.9	74.6	65.8	60.0	52.1	55.7	45.1	21.7	29.3
DAZLE	CVPR(20)	75.7	60.3	67.1	59.6	56.7	58.1	24.3	52.3	33.2
SDGZSL	ICCV(21)	73.6	64.6	68.8	66.4	59.9	63.0	–	–	–
DAA-ZSL	–	79.9	65.7	72.1	65.5	**66.1**	65.8	38.7	47.8	42.8
HSVA	NeurIPS(21)	76.6	53.9	66.8	58.3	57.2	55.3	39.0	48.6	43.3
ICCE	CVPR(22)	65.3	**82.3**	72.8	65.5	67.3	66.4	–	–	–
FREE+ESZSL	ICLR(22)	78.0	51.3	61.8	60.4	51.6	55.7	36.5	48.2	41.5
TF-VAEGAN+ESZSL	ICLR(22)	74.7	55.2	63.5	63.3	51.1	56.6	39.7	44.0	41.7
TransZero	AAAI(22)	61.3	82.3	70.2	69.3	68.3	**68.8**	52.6	33.4	40.8
TDCSS	CVPR(22)	59.2	74.9	66.1	44.2	62.8	51.9	–	–	–
ZLAP	IJCAI(22)	76.3	74.7	75.5	32.4	25.5	28.5	48.1	47.2	47.7
SE-GZSL	AAAI(23)	68.1	58.3	62.8	53.3	41.5	46.7	30.5	**50.9**	34.9
TPR	NeurIPS(24)	87.1	76.8	81.6	41.2	26.8	32.5	50.4	45.4	47.8
MAIN	WACV(24)	81.8	72.1	**76.7**	58.7	65.9	62.1	40.0	50.1	48.8
Ours		**94.5**	42.4	58.3	**87.5**	32.3	47.2	**66.9**	43.4	**52.6**

4.2 Posterior Approximation

We demonstrate semantic posterior sampling using our Diffusion model. To evaluate its performance, we consider two natural comparisons. (i) Models that use variational inference to approximate the posterior, such as VAEs, which are optimized by balancing reconstruction accuracy and the divergence between the approximate and true posterior distributions; and (ii) models that use indirect approaches to approximate the distribution, such as GANs, which achieve posterior matching through adversarial training.

These comparisons are summarized in Table 2. As shown, no single generative model consistently outperforms the others across all datasets when generating both seen and unseen samples. Notably, our method surpasses the other approaches in generating samples when measured by the harmonic mean. The most significant performance gap is observed in the semantically coarse SUN dataset, where our approach achieves a 20% improvement over GANs. A similar trend is evident in the class-diverse AWA dataset, albeit with smaller margins. In contrast, the CUB dataset, which features a wide variety

Table 2. Result of generalized ZSL for classification, for the most prominent generative approaches. † denotes the model consists of additional components that are disregarded.

Generative model	Name	AwA			CUB			SUN		
		Seen	Unseen	Harm.	Seen	Unseen	Harm.	Seen	Unseen	Harm.
VAE†	cVAE	72.6	54.4	**62.2**	59.9	**47.0**	**52.7**	–	–	–
GAN†	GAN	82.4	24.7	38.1	44.4	31.3	36.8	43.3	29.0	31.4
Diffusion (ours)	RevCD	**94.5**	42.4	58.3	**87.5**	32.3	47.2	**66.9**	**43.4**	**52.6**

of fine-grained semantic details, proves challenging for denoising approaches, making variational inference methods a more effective fit.

VAEs benefit from the tractable estimation of the posterior distribution, as evidenced by a 5.4% higher harmonic mean when both seen and unseen samples are drawn from tighter distributions, such as those observed in the CUB dataset, which emphasizes local descriptions. In contrast, GANs may underperform in this context, likely due to mode collapse in the posterior. Our Diffusion model performs moderately, achieving a 10.4% improvement over GANs.

Conversely, GANs implicitly learn the distribution through adversarial training, which encourages the generator to produce high-fidelity samples, as demonstrated in the AWA dataset, where attributes emphasize global image descriptions. However, the absence of explicit density estimation makes GANs susceptible to seen-unseen bias, leading to a preference for the seen distribution during inference, as observed in the SUN datasets. Our Diffusion model shows strong performance in generating samples when the seen distribution is discriminative and low-dimensional, as in the AWA. However, it struggles to maintain a tight lower bound on the true data distribution as the dimensionality of the semantic space increases, as evidenced by its performance in the CUB dataset.

This pattern is evident in Fig. 3, which illustrates the sample quality during iterative denoising. In Fig. 3(a), the density of seen samples in the AWA dataset is reproduced more quickly compared to the higher-dimensional space in the CUB dataset, as shown in Fig. 3(b).

4.3 Effect of Classification Loss

We investigate the influence of the classification loss weight λ_3 on the performance of our diffusion-based model (see Fig. (4)). The classification term guides the forward process during kernel estimation by biasing the diffusion trajectory toward semantically dense regions. Specifically, it introduces a gradient-based constraint that guides the process toward class-consistent subspaces, refining the learned transition kernel to denoise representations aligned with class identity.

Our reversed process estimates the semantic density of each class. The effect of this weight is notably different across datasets. On AWA, where semantic representations are coarse and broadly defined, increased weighting of the classification loss substantially boosts seen class accuracy—reaching near-perfect performance across all values

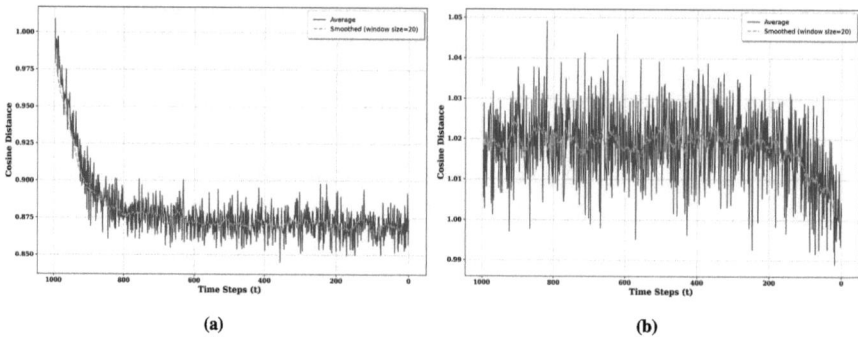

Fig. 3. The cosine distance to the true semantic space and the denoised learned representation are shown for both the AWA dataset and the CUB dataset. (a) For AWA, we observed a rapid reduction in noise in the initial timesteps, which gradually slowed as it approached the fully denoised space. (b) Conversely, for the CUB dataset, which possesses a semantically fine-grained space, the reduction in noise exhibited an inverse pattern.

of λ_3. However, this comes at the expense of unseen accuracy, which decreases as the model overfits to seen class modes, likely due to the reduced specificity in the semantic space, Fig. (4a). In contrast, CUB, characterized by fine-grained and richly structured semantic attributes, maintains high seen accuracy even at moderate classification weights, with a more stable and gradual decline in unseen performance, Fig. (4b). SUN, which features a broader class diversity and distributed word2vec-based semantics, displays a well-balanced trade-off. Seen accuracy is maximized at lower λ_3 values, while unseen performance remains relatively stable before degrading at higher weights, Fig. (4c). The robustness observed in SUN can be attributed to the high semantic variability, which naturally regularizes the kernel estimation and prevents premature convergence toward over-specialized representations.

These results highlight that the optimal weight of the classification-guided diffusion kernel is dataset-dependent. Fine-grained datasets like CUB benefit from moderate supervision, while coarsely defined or semantically diverse datasets (AWA and SUN) require careful calibration to avoid collapsing the semantic diversity necessary for zero-shot generalization.

5 Conclusion

In this paper, we introduce a reversed Conditional Diffusion model (RevCD) and evaluate its performance against state-of-the-art methods for generalized zero-shot learning. Additionally, by comparing with VAEs and GANs, our research explores the largely untapped potential of diffusion-based models to generate unseen samples. Our RevCD model estimates the semantic density for classes, which is then used to generate prototypes of both seen and unseen instances for high-accuracy classification. By leveraging visual conditioning, our approach enables precise control over the generation process and improved posterior approximations, outperforming other generative methods in set-

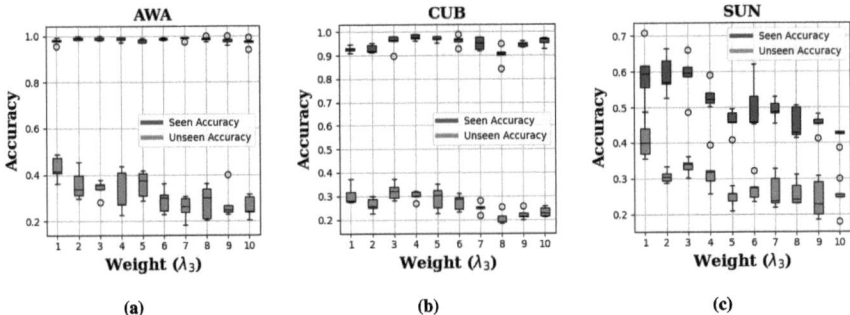

Fig. 4. Seen and unseen accuracy as a function of the classification loss weight λ_3 across AWA, CUB, and SUN. Increasing λ_3 consistently improves seen class accuracy but degrades unseen performance, with the trade-off varying by dataset. AWA exhibits sharp overfitting due to coarse semantics, while CUB remains more stable under stronger supervision. SUN shows a more balanced response, reflecting its broader semantic diversity.

tings with seen classes. Experimental results demonstrate the advantages of using a diffusion model as a generative backbone, especially regarding its robustness to diverse semantic information. We believe our findings can stimulate further exploration of diffusion models in generalized zero-shot learning (GZSL). Moreover, expanding cross-dataset evaluations in future zero-shot learning research could lead to the development of more resilient models.

References

1. Akata, Z., Perronnin, F., Harchaoui, Z., Schmid, C.: Label-embedding for image classification. IEEE Trans. Pattern Anal. Mach. Intell. **38**(7), 1425–1438 (2015)
2. Alamri, F., Dutta, A.: Implicit and explicit attention mechanisms for zero-shot learning. Neurocomputing **534**, 55–66 (2023)
3. Azizi, S., Kornblith, S., Saharia, C., Norouzi, M., Fleet, D.J.: Synthetic data from diffusion models improves imagenet classification. arXiv preprint arXiv:2304.08466 (2023)
4. Bau, D., et al.: Seeing what a GAN cannot generate. In: Proceedings of the IEEE/CVF International Conference on Computer Vision, pp. 4502–4511 (2019)
5. Bucher, M., Herbin, S., Jurie, F.: Generating visual representations for zero-shot classification. In: Proceedings of the IEEE International Conference on Computer Vision Workshops, pp. 2666–2673 (2017)
6. Chen, H., et al.: Robust classification via a single diffusion model. arXiv preprint arXiv:2305.15241 (2023)
7. Chen, Z., et al.: Semantics disentangling for generalized zero-shot learning. In: Proceedings of the IEEE/CVF International Conference on Computer Vision, pp. 8712–8720 (2021)
8. Chen, Z., Wang, S., Li, J., Huang, Z.: Rethinking generative zero-shot learning: an ensemble learning perspective for recognising visual patches. In: Proceedings of the 28th ACM International Conference on Multimedia, pp. 3413–3421 (2020)
9. Clark, K., Jaini, P.: Text-to-image diffusion models are zero shot classifiers. Adv. Neural Inf. Process. Syst. **36** (2024)

10. Ding, B., Fan, Y., He, Y., Zhao, J.: Enhanced VAEGAN: a zero-shot image classification method. Appl. Intell. **53**(8), 9235–9246 (2023)
11. Duchi, J.: Derivations for linear algebra and optimization. Berkeley, California **3**(1), 2325–5870 (2007)
12. Frome, A., et al.: Devise: a deep visual-semantic embedding model. Adv. Neural Inf. Process. Syst. **26** (2013)
13. Gao, R., et al.: Zero-VAE-GAN: generating unseen features for generalized and transductive zero-shot learning. IEEE Trans. Image Process. **29**, 3665–3680 (2020)
14. Goodfellow, I., et al.: Generative adversarial nets. Adv. Neural Inf. Process. Syst. **27** (2014)
15. Gupta, A., Narayan, S., Khan, S., Khan, F.S., Shao, L., van de Weijer, J.: Generative multi-label zero-shot learning. IEEE Trans. Pattern Anal. Mach. Intell. (2023)
16. Han, Z., Fu, Z., Chen, S., Yang, J.: Contrastive embedding for generalized zero-shot learning. In: Proceedings of the IEEE/CVF Conference on Computer Vision and Pattern Recognition, pp. 2371–2381 (2021)
17. Hao, S., Han, K., Wong, K.Y.K.: Learning attention as disentangler for compositional zero-shot learning. In: Proceedings of the IEEE/CVF Conference on Computer Vision and Pattern Recognition, pp. 15315–15324 (2023)
18. Heyden, W., Ullah, H., Siddiqui, M.S., Al Machot, F.: An integral projection-based semantic autoencoder for zero-shot learning. IEEE Access (2023)
19. Ho, J., Jain, A., Abbeel, P.: Denoising diffusion probabilistic models. Adv. Neural. Inf. Process. Syst. **33**, 6840–6851 (2020)
20. Ho, J., Salimans, T.: Classifier-free diffusion guidance. arXiv preprint arXiv:2207.12598 (2022)
21. Huang, H., Wang, C., Yu, P.S., Wang, C.D.: Generative dual adversarial network for generalized zero-shot learning. In: Proceedings of the IEEE/CVF Conference on Computer Vision and Pattern Recognition, pp. 801–810 (2019)
22. Ji, Z., Cui, B., Yu, Y., Pang, Y., Zhang, Z.: Zero-shot classification with unseen prototype learning. Neural Comput. Appl., 1–11 (2023)
23. Khan, M.G.Z.A., Naeem, M.F., Van Gool, L., Pagani, A., Stricker, D., Afzal, M.Z.: Learning attention propagation for compositional zero-shot learning. In: Proceedings of the IEEE/CVF Winter Conference on Applications of Computer Vision, pp. 3828–3837 (2023)
24. Kingma, D.P., Welling, M.: Auto-encoding variational Bayes. arXiv preprint arXiv:1312.6114 (2013)
25. Lampert, C.H., Nickisch, H., Harmeling, S.: Learning to detect unseen object classes by between-class attribute transfer. In: 2009 IEEE Conference on Computer Vision and Pattern Recognition, pp. 951–958. IEEE (2009)
26. Lazaridou, A., Dinu, G., Baroni, M.: Hubness and pollution: delving into cross-space mapping for zero-shot learning. In: Zong, C., Strube, M. (eds.) Proceedings of the 53rd Annual Meeting of the Association for Computational Linguistics and the 7th International Joint Conference on Natural Language Processing (Volume 1: Long Papers), Beijing, China, 26–31 July 2015, pp. 270–80. Association for Computational Linguistics, Stroudsburg (2015)
27. Li, A.C., Prabhudesai, M., Duggal, S., Brown, E., Pathak, D.: Your diffusion model is secretly a zero-shot classifier. In: Proceedings of the IEEE/CVF International Conference on Computer Vision, pp. 2206–2217 (2023)
28. Li, H., Xu, Z., Taylor, G., Studer, C., Goldstein, T.: Visualizing the loss landscape of neural nets. Adv. Neural Inf. Process. Syst. **31** (2018)
29. Li, Y., Liu, Z., Jha, S., Yao, L.: Distilled reverse attention network for open-world compositional zero-shot learning. In: Proceedings of the IEEE/CVF International Conference on Computer Vision, pp. 1782–1791 (2023)
30. Liu, Y., Tao, K., Tian, T., Gao, X., Han, J., Shao, L.: Transductive zero-shot learning with generative model-driven structure alignment. Pattern Recognit., 110561 (2024)

31. Lu, Z., Lu, Z., Yu, Y., Wang, Z.: Learn more from less: generalized zero-shot learning with severely limited labeled data. Neurocomputing (2022)
32. Lucas, J., Tucker, G., Grosse, R.B., Norouzi, M.: Don't blame the ELBO! A linear VAE perspective on posterior collapse. Adv. Neural Inf. Process. Syst. **32** (2019)
33. Luo, C.: Understanding diffusion models: a unified perspective. arXiv preprint arXiv:2208.11970 (2022)
34. Mishra, A., Krishna Reddy, S., Mittal, A., Murthy, H.A.: A generative model for zero shot learning using conditional variational autoencoders. In: Proceedings of the IEEE Conference on Computer Vision and Pattern Recognition Workshops, pp. 2188–2196 (2018)
35. Patterson, G., Hays, J.: Sun attribute database: discovering, annotating, and recognizing scene attributes. In: 2012 IEEE Conference on Computer Vision and Pattern Recognition, pp. 2751–2758. IEEE (2012)
36. Pourpanah, F., et al.: A review of generalized zero-shot learning methods. IEEE Trans. Pattern Anal. Mach. Intell. **45**(4), 4051–4070 (2022)
37. Salimans, T., Goodfellow, I., Zaremba, W., Cheung, V., Radford, A., Chen, X.: Improved techniques for training GANs. Adv. Neural Inf. Process. Syst. **29** (2016)
38. Schonfeld, E., Ebrahimi, S., Sinha, S., Darrell, T., Akata, Z.: Generalized zero-and few-shot learning via aligned variational autoencoders. In: Proceedings of the IEEE/CVF Conference on Computer Vision and Pattern Recognition, pp. 8247–8255 (2019)
39. Shipard, J., Wiliem, A., Thanh, K.N., Xiang, W., Fookes, C.: Diversity is definitely needed: improving model-agnostic zero-shot classification via stable diffusion. In: Proceedings of the IEEE/CVF Conference on Computer Vision and Pattern Recognition, pp. 769–778 (2023)
40. Socher, R., Ganjoo, M., Manning, C.D., Ng, A.: Zero-shot learning through cross-modal transfer. Adv. Neural Inf. Process. Syst. **26** (2013)
41. Sohl-Dickstein, J., Weiss, E., Maheswaranathan, N., Ganguli, S.: Deep unsupervised learning using nonequilibrium thermodynamics. In: International Conference on Machine Learning, pp. 2256–2265. PMLR (2015)
42. Song, Y., Sohl-Dickstein, J., Kingma, D.P., Kumar, A., Ermon, S., Poole, B.: Score-based generative modeling through stochastic differential equations. arXiv preprint arXiv:2011.13456 (2020)
43. Vaswani, A., et al.: Attention is all you need. Adv. Neural Inf. Process. Syst. **30** (2017)
44. Verma, V.K., Rai, P.: A simple exponential family framework for zero-shot learning. In: Ceci, M., Hollmén, J., Todorovski, L., Vens, C., Džeroski, S. (eds.) ECML PKDD 2017, Part II. LNCS (LNAI), vol. 10535, pp. 792–808. Springer, Cham (2017). https://doi.org/10.1007/978-3-319-71246-8_48
45. Wah, C., Branson, S., Welinder, P., Perona, P., Belongie, S.: The Caltech-UCSD birds-200-2011 dataset (2011)
46. Wang, Q., Breckon, T.P.: Generalized zero-shot domain adaptation via coupled conditional variational autoencoders. Neural Netw. **163**, 40–52 (2023)
47. Wang, Y., Feng, L., Song, X., Xu, D., Zhai, Y.: Zero-shot image classification method based on attention mechanism and semantic information fusion. Sensors **23**(4), 2311 (2023)
48. Wang, Y., Chen, S., Xue, H., Fu, Z.: Semi-supervised classification learning by discrimination-aware manifold regularization. Neurocomputing **147**, 299–306 (2015)
49. Xian, Y., Akata, Z., Sharma, G., Nguyen, Q., Hein, M., Schiele, B.: Latent embeddings for zero-shot classification. In: Proceedings of the IEEE Conference on Computer Vision and Pattern Recognition, pp. 69–77 (2016)
50. Xian, Y., Lampert, C.H., Schiele, B., Akata, Z.: Zero-shot learning–a comprehensive evaluation of the good, the bad and the ugly. IEEE Trans. Pattern Anal. Mach. Intell. **41**(9), 2251–2265 (2018)

51. Xian, Y., Schiele, B., Akata, Z.: Zero-shot learning-the good, the bad and the ugly. In: Proceedings of the IEEE Conference on Computer Vision and Pattern Recognition, pp. 4582–4591 (2017)
52. Zhang, J., Liao, S., Zhang, H., Long, Y., Zhang, Z., Liu, L.: Data driven recurrent generative adversarial network for generalized zero shot image classification. Inf. Sci. **625**, 536–552 (2023)

Toward an Explainable Heatmap-Based Deep Neural Network for Product Defect Classification and Machine Failure Prediction in Industry 4.0

Tojo Valisoa Andrianandrianina Johanesa[1,2(✉)], Lucas Equeter[2], Sidi Ahmed Mahmoudi[1], and Pierre Dehombreux[2]

[1] ILIA Laboratory, University of Mons, Mons, Belgium
`tojovalisoa.andrianandrianinajohanesa@umons.ac.be`
[2] GMECA Laboratory, University of Mons, Mons, Belgium

Abstract. In the context of Industry 4.0, machine maintenance and product quality control are crucial for manufacturing efficiency and reliability. This paper introduces a novel approach based on heatmap transformation and deep neural networks for product defect classification and machine failure prediction from tabular data, including static numerical and time series data. Unlike existing approaches that analyze numerical values using either a single record or a sequence of records as input, our method converts these inputs into heatmaps. This allows for visualizing multivariate process parameters and detecting signs of defects or failures through color variations using image-based classification models. The method also incorporates an explainability approach that leverages existing image-based explainability techniques to identify specific parameters and values associated with defects or failures. This provides operators with valuable insights to help identify the root causes of problems. The approach has shown promising results when applied to two public datasets from real industrial use cases.

Keywords: Industry 4.0 · Quality control · Predictive maintenance · Deep learning · Explainable AI · Tabular data

1 Introduction

Industry 4.0 technologies enable real-time data collection on production lines, leading to various data-driven AI applications for optimizing production efficiency [13, 15]. Many studies [2, 5–8, 14, 21, 23, 26, 29] focus on predicting manufacturing defects and machine failures using process data, including production settings and sensor measurements. These data are often in tabular form and can be categorized as static numerical data or multivariate time series data. Data collected at regular intervals can be treated as time series, while data collected at irregular intervals are considered static when there is no continuous temporal dependency. Static numerical data are analyzed using single-input models, which process individual records without temporal context. Common models in this category include Support Vector Machines (SVM), logistic regression, ensemble models, and Multi-Layer Perceptrons (MLP), which have been applied in various industrial use cases [5,6,14,21,23]. Time series data are analyzed using models that take a sequence of time-ordered records as input. Models based on Recurrent Neural Networks (RNNs) and Convolutional Neural Networks (CNNs) were used in diverse industrial processes for defect prediction [8,26] and machine failure prediction [20,29]. Selecting the appropriate model remains a challenge, as the most suitable approach varies across domains and depends on the type of data available.

In this paper, we propose a novel approach for product defect classification and machine failure prediction by transforming numerical and time series model inputs into heatmap images, which are then analyzed using image-based models, such as CNNs, to identify signs of defects or failures based on color variations. Although some studies [9,20] have applied CNN to multivariate time-series classification, our method offers a more flexible representation, leveraging color intensity variations to capture process behavior changes, instead of analyzing time series data directly. This allows the model to focus on meaningful color patterns within the heatmap image for prediction. Moreover, our approach provides a unified architecture that can handle both numerical data and time series data. The heatmap transformation also enables the use of explainability techniques (XAI) designed for images [10,18] to highlight the regions of the heatmap the model focuses on and identify the corresponding parameters. These techniques have proven effective by providing clear visual explanations of model decisions. The image representation of numerical and time series data makes their application more meaningful and naturally aligned with the architecture of image classification models. Our approach extends the applicability of image-based XAI techniques to numerical and time series data, offering a complementary explanation method alongside existing XAI techniques for tabular data [24].

Our proposed approach presents two key contributions.

- **Unified Model Input Format for Tabular Data:** By converting input data from numerical and time series formats into a unified heatmap representation, our method provides a solution applicable to all types of tabular data across various industrial use cases. The model's decision-making relies on the same principle across all scenarios, which is the analysis of color variations in heatmaps.
- **Practical Explanations of Model Decisions:** We use XAI techniques designed for images applied to the input heatmap, to explain model decisions by overlaying the generated explanation map with the raw input data to identify the critical parameters and their values related to defects or failures.

The rest of this paper is organized as follows. Section 2 describes our proposed Heatmap-based Deep Neural Network approach. Section 3 presents the experimental results from applying the approach to two public tabular datasets, one with numerical data and one with time series data. Finally, Sect. 4 provides a conclusion and outlines potential directions for future work.

2 Heatmap-Based Deep Neural Network Approach

This section presents the details of our method for product defect classification and machine failure prediction in manufacturing processes. A summary of the proposed Heatmap-Based Deep Neural Network (HDNN) approach is presented in Fig. 1. The approach consists of three key components: heatmap transformation, classification using a Convolutional Neural Network (CNN)-based model, and the explainability of the model's decision.

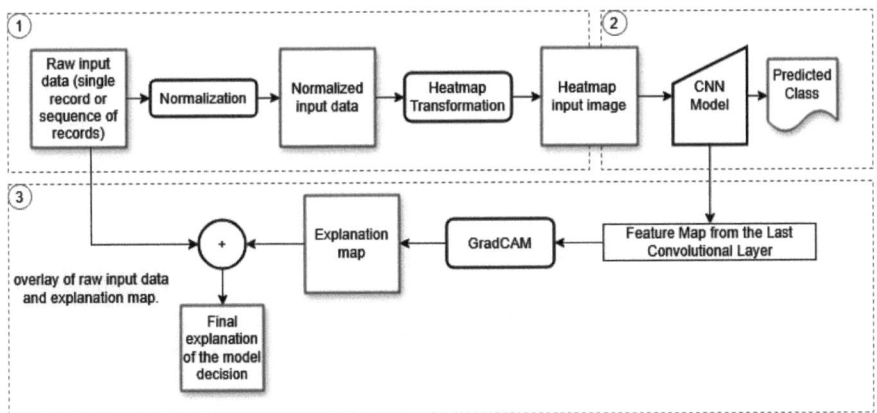

Fig. 1. Summary of the proposed HDNN approach.

2.1 Heatmap Transformation

The heatmap transformation is inspired by infrared thermography (IRT) [3] by transforming production process data into heatmaps for model input. This allows for simultaneous monitoring of multiple parameters, similar to temperature monitoring in IRT, by visualizing parameter variations through color changes. The input size of the model depends on whether the data is static numerical data or time series data. For time series data, the input consists of a sequence of records with size (n, m), where n is the number of records and m is the number of features. For static numerical data, n equals 1, reducing the input to a single record. Each value in the input matrix is represented by a pixel in the heatmap, resulting in an image of size $(n \times m)$ pixels. Before generating heatmaps, we preprocess the data by normalizing each parameter to ensure consistency using existing normalization techniques [22]. For the heatmap transformation, the minimum and maximum values from the normalized training dataset are used to ensure a uniform color scale across all images.

2.2 Heatmap Classification Using CNN-Based Model

Each generated heatmap serves as input to an image classification model. For that, we use a model based on CNN, which has demonstrated effectiveness in image processing across various domains in the literature. CNN [16] model consists of convolutional layers for feature extraction and fully connected layers for decision-making. During training, the convolutional filters learn to detect color variations across the heatmap. The output feature maps from the final convolutional layer, which represent the learned color variation patterns, are flattened and passed through the fully connected layers to obtain the final decision. This process enables the model to detect signs of defects or failures from heatmap images.

2.3 Explainability of Heatmap Classification

To understand the criteria on which the model bases its decisions, we apply explainability techniques designed for images to highlight relevant regions in the heatmap. For this, we use Grad-CAM [25], which is one of the most widely adopted techniques in the literature for explaining image classification [12,28]. Grad-CAM analyzes activations from the last convolutional layer to generate an explanation map that highlights the areas the model focuses on. This explanation map is then overlaid on the original raw data to identify parameters corresponding to the highlighted regions, as well as their values that influence the prediction. This provides actionable insights through informed model decisions that could assist in diagnosing the causes when a problem occurs. It could also help in the definition of preventive actions to reduce defects or failures in the future.

3 Experimentations

The proposed HDNN approach was applied to public data from two real manufacturing use cases: road lens quality classification in the injection molding process [23], and machine component failure prediction from a Microsoft use case [19].

3.1 Road Lens Quality Classification

This experiment uses data from the study by [23] to determine the quality of road lenses produced via the injection molding process. The dataset contains 1,451 records of static numerical data with 13 process parameters and a label that indicates the quality of the lens. These process parameters include melt temperature, mold temperature, filling time, plasticizing time, cycle time, closing force, clamping force peak value, torque peak value, torque mean value, back pressure peak value, injection pressure peak value, screw position at the end of the hold pressure, and shot volume. The authors of [23] defined four quality classes based on the UNI EN 13201-2:2016 standard: "Target", "Acceptable", "Inefficient", and "Waste". Figure 2 presents the class distribution with the number of samples for each class in the dataset.

We compared the performance of our approach with the results from [23] using the stratified 5-fold cross-validation method. Min-Max scaling was applied for normalization before transforming records into heatmaps. The CNN architecture used in the HDNN approach is presented in Table 2. The Adam algorithm was used to optimize the model at a learning rate of 0.009.

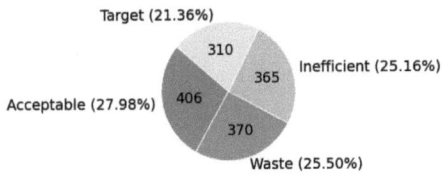

Fig. 2. Class distribution of the Road Lens dataset.

Table 1. Mean accuracy and macro-averaged F1 scores (± standard deviation) using stratified 5-fold cross-validation: Road Lens quality classification.

Model	Mean Accuracy	Macro-avg F1-score
KNN [23]	92.21 ± 1.64%	92.21 ± 1.04
Decision Tree [23]	91.52 ± 1.63%	91.61 ± 2.6
Random Forest [23]	**95.04 ± 1.26%**	**94.95 ± 1.23**
GBT [23]	94.21 ± 1.37%	94.16 ± 0.98
SVM [23]	91.73 ± 2.37%	91.78 ± 2.4
MLP [23]	92.08 ± 1.92%	92.07 ± 1.95
HDNN	92.21 ± 1.28%	92.18 ± 1.29%

Results and Discussion. The HDNN approach achieved a mean accuracy of 92.21 ± 1.28% and a macro-averaged F1-score of 92.18 ± 1.29%, as presented in Table 1. While our approach did not outperform ensemble methods, the results are similar to those of other models. This demonstrates the applicability of the approach to numerical data. The size of the dataset in this use case is well-suited for traditional machine learning methods. However, it is relatively small for deep learning (DL) models, including MLP and our approach. Applying data augmentation [4, 27] could help improve performance, as DL models tend to perform better with larger datasets. As DL models are often seen as black boxes due to their complex architectures, we applied the explainability approach presented in the previous section using the best HDNN model from cross-validation. Figure 3 shows an example of an explanation for an instance predicted as "waste". The inference time of the prediction, including the heatmap transformation, was 0.063 s. The most important regions of the heatmap on which the model focuses are highlighted in warm colors, with yellow indicating the highest intensity after applying

Table 2. CNN architecture used in the HDNN approach for Road Lens quality classification.

Layer (type)	Output Shape
input_layer (InputLayer)	(None, 1, 13 , 3)
conv2d (Conv2D)	(None, 1, 13, 32)
flatten (Flatten)	(None, 416)
dense (Dense)	(None, 4)

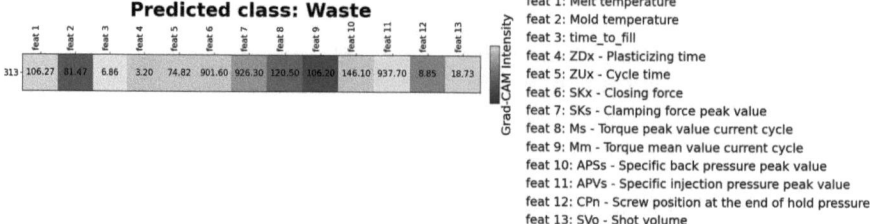

Fig. 3. Grad-CAM explanation for an instance classified as "waste".

Grad-CAM. The explanation map, overlaid on the initial data, identified peak injection pressure as the parameter that has the greatest impact on the class prediction of this instance. We can also directly identify the values of the parameters on which the model focuses its attention.

3.2 Machine Component Failure Prediction

This use case focuses on predicting the imminent failure of four critical machine components using time series data from 100 machines collected over one year. The initial dataset, provided by Microsoft [19], was collected from an anonymous industrial process and consists of five data sources: Telemetry, Errors, Machine Characteristics, Maintenance, and Failures. These data sources were merged into a refined dataset of 290,642 records using feature engineering techniques from [5]. The mean values and standard deviations of the telemetry measurements over the past 3 and 24 h were calculated to create short-term and long-term telemetry history. The Errors data source was used to count the number of errors of each type in the last 24 h for each machine. The number of days since the last component replacement was calculated from the Maintenance data source. The Failures data source was used to create the labels for the target column. Labels correspond to the failed component, with "comp1", "comp2", "comp3", or "comp4" assigned accordingly, while cases without failures are labeled "none" [5]. After a component failure, the component was subsequently replaced to restore machine functionality. The Machine Characteristics data source includes the model type ("model1", "model2", "model3", or "model4") and the age of each machine. Since the model is a categorical value, we applied one-hot encoding [11] to convert it into four separate binary columns. Each new column indicates whether the machine belongs to a specific model (1) or not (0). The resulting dataset constitutes a multivariate time series, with records collected every 3 h for each machine. Table 3 presents the different parameters and their meanings in the final dataset after the feature engineering step.

We compared the performance of our approach with those of RNN models, which take a sequence of records as input, including SimpleRNN, LSTM, and GRU. Each model consists of a recurrent layer using L1 regularization, with SimpleRNN having 32 units, LSTM having 60 units, and GRU having 56 units, followed by a dense output layer with 5 units. The dataset was split into training, validation, and test sets, with the first 8 months used for training, the following 2 months for validation, and the last

Table 3. Meaning of parameters: component failure prediction.

Parameter	Meaning
voltmean_3h	Mean voltage over 3h.
rotatemean_3h	Mean rotational speed over 3h.
pressuremean_3h	Mean pressure over 3h.
vibrationmean_3h	Mean vibration over 3h.
voltsd_3h	Voltage standard deviation over 3h.
rotatesd_3h	Rotational speed standard deviation over 3h.
pressuresd_3h	Pressure standard deviation over 3h.
vibrationsd_3h	Vibration standard deviation over 3h.
rotatemean_24h	Mean rotational speed over 24h.
voltmean_24h	Mean voltage over 24h.
pressuremean_24h	Mean pressure over 24h.
vibrationmean_24h	Mean vibration over 24h.
rotatesd_24h	Rotational speed standard deviation over 24h.
voltsd_24h	Voltage standard deviation over 24h.
pressuresd_24h	Pressure standard deviation over 24h.
vibrationsd_24h	Vibration standard deviation over 24h.

Parameter	Meaning
datetime	Date and time of the record
machineID	Machine identifier number
error1count	Number of type 1 errors over 24h.
error2count	Number of type 2 errors over 24h.
error3count	Number of type 3 errors over 24h.
error4count	Number of type 4 errors over 24h.
error5count	Number of type 5 errors over 24h.
age	Machine age.
model1	1 if the machine is of model type 1, 0 otherwise.
model2	1 if the machine is of model type 2, 0 otherwise.
model3	1 if the machine is of model type 3, 0 otherwise.
model4	1 if the machine is of model type 4, 0 otherwise.
comp1_age	Days since last replacement of component 1.
comp2_age	Days since last replacement of component 2.
comp3_age	Days since last replacement of component 3.
comp4_age	Days since last replacement of component 4.
failure	Label (comp1, comp2, comp3, comp4, none).

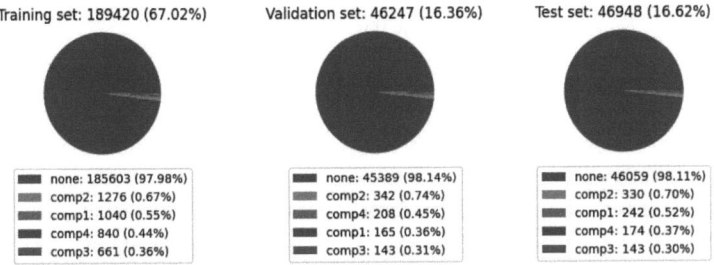

Fig. 4. Class distribution in the training, validation, and test subsets after generating the 8-record sequences: component failure prediction.

2 months for testing, as in [5]. An 8-record sequence is used as input for all models to predict the current state of the components, considering what has happened in the past hours. Sequences containing a failure record were excluded if the last record had a "none" class, to avoid overlap between failure events and normal operations. The "machineID" and "datetime" columns were not included among the input features; they were only used to group the data by machine and to sort the records in temporal order during sequence generation. Figure 4 shows the class distribution in the training, validation, and test subsets after generating the 8-record sequences. The subsets have similar class distributions, which indicate a good partitioning of the data. For the HDNN approach, sequences were converted into heatmaps after Min-Max normalization. The CNN architecture used in HDNN is presented in Table 5. The model was optimized using the Adam algorithm with a learning rate of 0.0003.

Results and Discussion. As the dataset is significantly imbalanced, with many more normal cases compared to failures, we evaluate the models using precision, recall, and macro-averaged F1-score. These metrics are more suitable for imbalanced classification tasks and better reflect the model's ability to detect minority classes. The perfor-

Table 4. Precision, recall, and F1-score of recurrent neural network models and the HDNN approach on the test set: component failure prediction.

Model	Precision	Recall	F1-Score
SimpleRNN	0.98	0.94	0.96
LSTM	0.98	0.95	0.96
GRU	**0.99**	**0.97**	**0.98**
HDNN	0.97	0.96	0.96

Table 5. CNN architecture used in the HDNN approach for machine component failure prediction.

Layer (type)	Output Shape
input_layer (InputLayer)	(None, 8, 30, 3)
conv2d_1 (Conv2D)	(None, 7, 29, 8)
conv2d_2 (Conv2D)	(None, 6, 28, 4)
flatten (Flatten)	(None, 672)
output_layer (Dense)	(None, 5)

mances of the different models are presented in Table 4. The HDNN approach achieved an accuracy of 0.97, a recall of 0.96, and an F1-score of 0.96. The GRU model showed the best results with slightly higher performance, but the complex structure of recurrent neural networks makes their decision-making criteria difficult to understand. Our approach achieves comparable results and stands out by providing explainability through the Grad-CAM-based method introduced in the previous section. Figure 5 presents an explanation for the prediction of a sequence of 8 records corresponding to machine 2. In this example, the model predicts the components' states at 2015-12-28 12:00:00 by analyzing data up to 2015-12-27 15:00:00 and identifies a failure of component 2. The

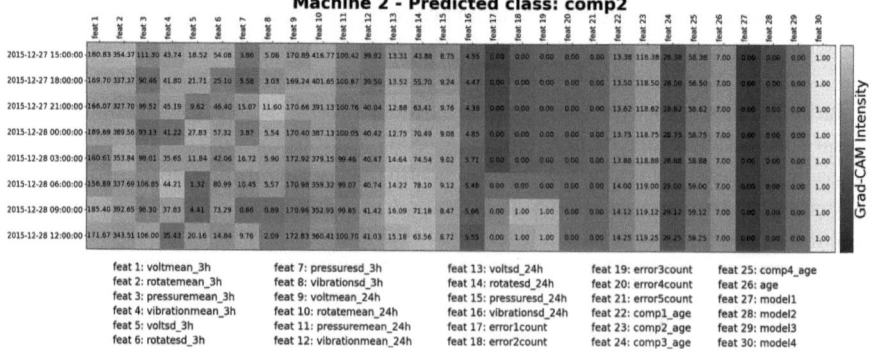

Fig. 5. Grad-CAM explanation for an example of component 2 failure.

colors indicate the level of importance assigned to each value, with yellow corresponding to the highest importance. It can be observed that the model assigns high importance to the machine model, which is a type 4 model in this example. The explanation also highlights the occurrence of error2 (feature 18) and error3 (feature 19), as well as some specific telemetry values that had a strong influence on the model's decision in this example. This information could be useful for the operator in diagnosing the causes of the failure.

Inference Time. Since this use case involves predicting imminent failures, inference time is critical to ensure real-time model decisions. To assess the inference speed of the proposed HDNN approach, we compared the inference times of all models on a single input sequence of 8 records from the test set. For the HDNN model, the time includes both data transformation into a heatmap and the prediction step, while for the RNN-based models, it includes input reshaping followed by inference. The time required for data normalization, which is common to all models, was also included in the comparison. The comparison was performed under controlled conditions by limiting execution to one CPU thread, and disabling runtime optimizations, on a laptop equipped with an Intel(R) Core(TM) i7-11800H CPU @ 2.30 GHz, an NVIDIA GeForce RTX 3080 GPU, and 32 GB of RAM. Table 6 shows that the HDNN model achieved an inference time comparable to those of the RNN-based models. The heatmap transformation step had minimal impact on the overall inference time, showing that our approach is suitable for real-time prediction.

Table 6. Inference time comparison for component failure prediction.

	HDNN	SimpleRNN	LSTM	GRU
Time (s)	0.087	0.063	0.066	0.047

4 Conclusion

In this paper, we presented a Heatmap-based Deep Neural Network (HDNN) for product defect classification and machine failure prediction from tabular data. Our method unifies input data, whether static numerical or time series, into a single image format of heatmaps, which are used as input to an image classification model such as a Convolutional Neural Network (CNN). This enables the detection of signs of defects or failures through color changes in heatmaps. We also incorporated a Grad-CAM-based explainability approach to help operators identify the input parameters that influence the model's decision, which could assist in the diagnosis of issues. The HDNN approach demonstrated promising results, achieving a good balance in terms of prediction performance, inference time, and explainability, showing its potential for predictive maintenance and quality control in Industry 4.0. Future work will explore additional image-based models, including recent architectures such as Vision Transformers (ViT), and evaluate further explainability techniques to identify those offering the most relevant insights, ideally validated by human experts. We also plan to apply the method to diverse datasets from various industrial contexts and to extend it to regression tasks. In addition, we will investigate the deployment of the models on Edge AI resources [17] to enable integration into production lines in order to reduce communication needs and obtain faster responses by processing data locally. Finally, we will explore multimodal approaches [1] by coupling heatmap representations with complementary data sources to provide additional information in order to enhance model performance.

Acknowledgments. This work was supported by Service Public de Wallonie Recherche under grant n°2010235-ARIAC by DIGITALWALLONIA4.AI.

Disclosure of Interests. The authors have no competing interests to declare that are relevant to the content of this article.

References

1. Amel, O., Stassin, S., Mahmoudi, S.A., Siebert, X.: Multimodal approach for harmonized system code prediction. arXiv preprint arXiv:2406.04349 (2024)
2. Andrianandrianina Johanesa, T.V., Equeter, L., Mahmoudi, S.A.: Survey on AI applications for product quality control and predictive maintenance in industry 4.0. Electronics **13**(5), 976 (2024)
3. Bagavathiappan, S., Lahiri, B.B., Saravanan, T., Philip, J., Jayakumar, T.: Infrared thermography for condition monitoring-a review. Infrared Phys. Technol. **60**, 35–55 (2013)
4. Benkedadra, M., et al.: CIA: controllable image augmentation framework based on stable diffusion. In: 2024 IEEE 7th International Conference on Multimedia Information Processing and Retrieval (MIPR), pp. 600–606. IEEE (2024)
5. Cardoso, D., Ferreira, L.: Application of predictive maintenance concepts using artificial intelligence tools. Appl. Sci. **11**(1), 18 (2020)
6. Chen, J.C., Guo, G., Wang, W.N.: Artificial neural network-based online defect detection system with in-mold temperature and pressure sensors for high precision injection molding. Int. J. Adv. Manuf. Technol. **110**(7), 2023–2033 (2020)
7. Colantonio, L., Equeter, L., Dehombreux, P., Ducobu, F.: A systematic literature review of cutting tool wear monitoring in turning by using artificial intelligence techniques. Machines **9**(12), 351 (2021)
8. Dong, Z., Pan, Y., Yang, J., Xie, J., Fu, J., Zhao, P.: A multiphase dual attention-based lstm neural network for industrial product quality prediction. IEEE Trans. Industr. Inform. (2024)
9. Fauvel, K., Lin, T., Masson, V., Fromont, É., Termier, A.: XCM: an explainable convolutional neural network for multivariate time series classification. Mathematics **9**(23), 3137 (2021)
10. Gupta, L.K., Koundal, D., Mongia, S.: Explainable methods for image-based deep learning: a review. Arch. Comput. Methods Eng. **30**(4), 2651–2666 (2023)
11. Hancock, J.T., Khoshgoftaar, T.M.: Survey on categorical data for neural networks. J. Big Data **7**(1), 1–41 (2020). https://doi.org/10.1186/s40537-020-00305-w
12. Kalasampath, K., Spoorthi, K., Sajeev, S., Kuppa, S.S., Ajay, K., Angulakshmi, M.: A literature review on applications of explainable artificial intelligence (XAI). IEEE Access (2025)
13. Kotsiopoulos, T., Sarigiannidis, P., Ioannidis, D., Tzovaras, D.: Machine learning and deep learning in smart manufacturing: the smart grid paradigm. Comput. Sci. Rev. **40**, 100341 (2021)

14. Lee, W.J., Wu, H., Yun, H., Kim, H., Jun, M.B., Sutherland, J.W.: Predictive maintenance of machine tool systems using artificial intelligence techniques applied to machine condition data. Procedia CIRP **80**, 506–511 (2019)
15. Lerat, J.S., Mahmoudi, S.A.: Scalable deep learning for industry 4.0: speedup with distributed deep learning and environmental sustainability considerations. In: International Conference of Cloud Computing Technologies and Applications, pp. 182–204. Springer (2024)
16. Li, Z., Liu, F., Yang, W., Peng, S., Zhou, J.: A survey of convolutional neural networks: analysis, applications, and prospects. IEEE Trans. Neural Netw. Learn. Syst. **33**(12), 6999–7019 (2021)
17. Mahmoudi, S.A., Gloesener, M., Benkedadra, M., Lerat, J.S.: Edge AI system for real-time and explainable forest fire detection using compressed deep learning models. In: Proceedings Copyright, pp. 847–854 (2025)
18. Mahmoudi, S.A., Stassin, S., Daho, M.E.H., Lessage, X., Mahmoudi, S.: Explainable deep learning for Covid-19 detection using chest X-ray and CT-scan images. In: Garg, L., Chakraborty, C., Mahmoudi, S., Sohmen, V.S. (eds.) Healthcare Informatics for Fighting COVID-19 and Future Epidemics. EICC, pp. 311–336. Springer, Cham (2022). https://doi.org/10.1007/978-3-030-72752-9_16
19. Microsoft: Predictive maintenance modelling guide data sets (2024). https://www.kaggle.com/datasets/arnabbiswas1/microsoft-azure-predictive-maintenance. Accessed 23 Nov 2024
20. Nemer, M.A., Azar, J., Demerjian, J., Makhoul, A., Bourgeois, J.: A review of research on industrial time series classification for machinery based on deep learning. In: 2022 4th IEEE Middle East and North Africa COMMunications Conference (MENACOMM), pp. 89–94. IEEE (2022)
21. Oliveira, R.M.A., Sant'Anna, Â.M.O., da Silva, P.H.F.: Explainable machine learning models for defects detection in industrial processes. Comput. Ind. Eng. **192**, 110214 (2024)
22. Patro, S., Sahu, K.K.: Normalization: a preprocessing stage. arXiv preprint arXiv:1503.06462 (2015)
23. Polenta, A., Tomassini, S., Falcionelli, N., Contardo, P., Dragoni, A.F., Sernani, P.: A comparison of machine learning techniques for the quality classification of molded products. Information **13**(6), 272 (2022)
24. Sahakyan, M., Aung, Z., Rahwan, T.: Explainable artificial intelligence for tabular data: a survey. IEEE Access **9**, 135392–135422 (2021)
25. Selvaraju, R.R., Cogswell, M., Das, A., Vedantam, R., Parikh, D., Batra, D.: Grad-CAM: visual explanations from deep networks via gradient-based localization. In: Proceedings of the IEEE International Conference on Computer Vision, pp. 618–626 (2017)
26. Shi, Y., Chen, Y., Zhang, L.: Product quality time series prediction with attention-based convolutional recurrent neural network. Appl. Intell. **54**(21), 10763–10779 (2024)
27. Shorten, C., Khoshgoftaar, T.M.: A survey on image data augmentation for deep learning. J. Big Data **6**(1), 1–48 (2019)
28. Tang, D., Chen, J., Ren, L., Wang, X., Li, D., Zhang, H.: Reviewing cam-based deep explainable methods in healthcare. Appl. Sci. **14**(10), 4124 (2024)
29. Yadav, D.K., Kaushik, A., Yadav, N.: Predicting machine failures using machine learning and deep learning algorithms. Sustain. Manuf. Serv. Econ. **3**, 100029 (2024)

Question Answering in a Low-Resource Language: Dataset and Deep Learning Adaptations for Sinhala

Janani Ranasinghe(✉) ⓘ and Ruvan Weerasinghe ⓘ

Informatics Institute of Technology, Colombo, Sri Lanka
{janani.20210926,ruvan.w}@iit.ac.lk

Abstract. Significant advancements have been made in Natural Language Processing (NLP) in recent years, particularly in Question-Answering (QA). The availability of pre-trained Large Language Models (LLMs) and annotated datasets has driven these improvements. However, most resources are designed for high-resource languages like English, while low-resource languages face challenges due to data limitations. Sinhala, the most widely spoken language in Sri Lanka, with over 20 million speakers, still lacks sufficient annotated datasets and monolingual models for downstream tasks like QA. To address this gap, this study presents a Sinhala QA dataset, SiQuAD, translated from SQuAD v1.1, containing 16,000 unique question-answer pairs. Experiments covering monolingual, cross-lingual, and multilingual approaches are conducted, with the best-performing model achieving an F1 score of 73%, indicating promising capabilities while highlighting room for improvement and future research. The dataset will be made publicly available.

Keyword: Sinhala language processing · Multilingual NLP · Question answering · SQuAD · Span-based · Machine reading comprehension

1 Introduction

Natural Language Processing (NLP) has achieved remarkable success in recent years, particularly in Question Answering (QA) and Machine Reading Comprehension (MRC) tasks. As Yang et al. [39] highlight, the ability to reason and infer using natural language is a sign of intelligence. In simple terms, the ability to question and answer can be considered a metric for determining the intelligence of a model. While answering a question itself is a significant skill, recent studies have introduced MRC, which extends beyond just answering a question by incorporating comprehension skills given a context or scenario [31].

The sudden surge in research in the QA domain is largely driven by the availability of large-scale publicly annotated datasets. The introduction of datasets like QUAC [10], SQuAD v1.1 [30], SQuAD v2.0 [22], and HotPotQA [39] can be considered the stepping stone for most of these advancements. Another reason for this is the introduction of

© The Author(s), under exclusive license to Springer Nature Switzerland AG 2025
A. Hadjali et al. (Eds.): DeLTA 2025, CCIS 2627, pp. 336–352, 2025.
https://doi.org/10.1007/978-3-032-04339-9_22

transformer-based models [4]. Transformer architecture has changed the entire domain of NLP.

However, a visible gap exists in the development of QA systems for low-resource languages. Almost all the state-of-the-art models and benchmarking datasets are built focusing on high-resource languages. Languages like English have an excess of annotated datasets and models, but when it comes to languages like Sinhala, there is an extreme scarcity in datasets and models. This remains a major barrier for research in this domain.

Despite these challenges, researchers have achieved significant results in Sinhala NLP tasks. Most research has focused on foundational tasks like tokenization, embedding, and trying to understand the language. This can be due to the uniqueness of the Sinhala Language. Other than foundational tasks, text classification, sentiment analysis, machine translation, and several other areas in Sinhala NLP have had more interest from researchers [28]. While there have been breakthroughs in Sinhala NLP research, the specific domain of Question Answering remains largely unexplored.

To help bridge this gap, our study introduces a Sinhala Question Answering dataset, translated from the SQuAD v1.1. Even though SQuAD v1.1 was introduced in 2016, there has not been much effort to develop a similar dataset for Sinhala. To the best of the authors' knowledge, there aren't any publicly available datasets for span-based Sinhala Question Answering.

In this study, SQuAD v1.1 is translated into the Sinhala language to create the Sinhala SQuAD, SiQuAD. The dataset consists of 16,000 question-answer pairs. Further, we will evaluate the performance of existing transformer-based models on this dataset. XLM-RoBERTa [40], multilingual BERT [22], distilBert [34], along with sinBERT [15] will be fine-tuned and used for evaluations.

XLM-R_{LARGE} was the best-performing model, achieving an F1 score of 73% and an Exact Match score of 59%. This study aims to identify the best approaches for Sinhala QA using pre-trained models and to test the applicability of the translated dataset for these tasks.

Given the challenges and barriers in Sinhala NLP, we hope this study will serve as a foundation and benchmark for further research in Sinhala QA. The main contributions of this study are as follows,

1. We introduce the first publicly available Sinhala Question-Answering dataset, SiQuAD.
2. We evaluate and analyze the adaptability of existing transformer-based approaches for Sinhala

This work lays the groundwork for future research on Sinhala QA systems, offering valuable resources for both researchers and practitioners in the field.

2 Related Work

2.1 Question-Answering in English

QA systems can be divided into two types based on the information source used to derive the answers. They are textual QA and knowledge base (KB) QA. Textual QA uses unstructured sources, while KB QA uses structured KBs. Based on the answer extraction method, QA systems have two types. They are extractive and generative. Extractive QA extracts the answer from a given context. The answer must be present in the context provided. In generative QA, the answer is generated using natural language, usually sequence-to-sequence (seq-2-seq) methods [17]. If a QA system accepts questions only from a specific pre-defined domain, they are categorized as a closed-domain QA system. If the QA system accepts questions without any domain restriction, they are called an open-domain QA system [12].

Most of the initial work was based on rule-based systems or statistics. Traditional QA systems have an architectural structure consisting of three main steps. They are question analysis, context retrieval, and answer extraction. With the advent of DL into the QA domain, MRC was integrated into the answer retrieval process. This was a breakthrough because in earlier days, answer retrieval was considered an IR task. With MRC, answer retrieval involved understanding the context rather than just retrieving an answer based on the statistics [17].

The attention mechanism [4] and later, the Transformer architecture marked perhaps the most significant shift in the QA domain. Transformers can capture long-term dependencies, but the fixed window size is a known limitation. Even with this limitation, these models outperform models like LSTM [31].

For modern DL architecture to perform well, a large amount of training data is required. When it comes to QA datasets, depending on the answer retrieval type, there can be different types of datasets. The most common types are cloze-style, Multi-hop, span-based, and User log-based datasets [9]. In cloze-style datasets, the task is to fill in the blanks. Here, the model's ability to predict the missing words is tested. These are usually considered foundational for language comprehension and QA. The CNN/Daily Mail dataset is a cloze-style dataset [8]. Multi-hop datasets add more complexity to QA models by making the models reason through multiple passages or even several documents. In 2018, Yang et al. [39] introduced HotpotQA, which is one of the most popular datasets for this task. User log-based and conversational datasets have become increasingly relevant for modeling real-world interactions, reflecting genuine user inquiries. MS MARCO is a prime example, containing real queries from Microsoft's Bing search logs [6].

Span-based datasets are designed to extract an answer directly from a given context passage. It is mandatory for the answer to be present in the provided context. Stanford Question Answering Dataset (SQuAD) can be considered one of the most significant milestones in QA [30]. SQuAD was built using English Wikipedia passages as context. SQuAD 2.0 extended the original dataset by adding unanswerable questions, thereby testing models' abilities to recognize cases where an answer is absent [32]. Other notable datasets are TriviaQA, which focuses on trivia-style answer retrieval [23], Natural Questions (NQ) for the retrieval of answers of varying lengths from Wikipedia [24] and NewsQA for answer retrieval from news articles [35]. These datasets are widely considered standard QA benchmarks.

2.2 Question-Answering Beyond English

The primary limitation in developing QA systems for low-resource languages is the lack of annotated data. This data bottleneck hinders the use of most of the state-of-the-art QA models for low-resource languages because they require large amounts of training data. Even though there are challenges and limitations, researchers have made strides in QA for certain low-resource languages to bridge this gap.

Numerous models and approaches have been used in the domain of QA for low-resource languages. There are monolingual models where the training is done in a single language. BERT is one such example [22]. There are several language-specific BERT variations. HindBERT is available for Hindi. MahaBERT is available for Marathi [33]. ParsBERT is available for the Persian language [1]. Cross-lingual and Multilingual Models are another option when the target language has less data. Cross-lingual models can transfer to a different language without any training data for the target language. Techniques like zero-shot learning are used in these models. mBERT (Multilingual BERT) is a commonly used multilingual model. It's trained in 104 languages. XLM-RoBERTa is a Cross-lingual Language Model trained on 100 languages [11].

There are two main approaches to creating datasets for low-resource languages. They are creating seed datasets or translating datasets from a high-resource language. Creating a seed dataset can be costly, time-consuming, and challenging. And, in cases like TiQuAD [19] When there are not enough articles available digitally in the target language, data collection can be hard. Therefore, translating from a high-resource language to the target language is a more feasible approach. When using translated versions of datasets, the annotation could change. The translated answer can sometimes not be found in the text. Since having the answer in the context is a must for span-based datasets, the answers must be annotated separately for this issue. And even if the answer is present in the context, the starting and end points of the answer span could change, as shown in ParSQuAD [1]. Therefore, again, the answer span positions need to be annotated. Machine translation (MT) remains the popular approach among researchers, with tools like Google Translate API being widely used [7, 13, 20]. More advanced MT systems were used by certain studies. For example, the CUNI transformer model was used for the Czech SQuAD [27], Meta AI's NLLB-200 was used by SQuAD-sr [2]. The Hindi SQuAD dataset uses an IndicTrans machine translation tool to convert the SQuAD into Hindi and then separately do the answer-locating process [33]. In contrast, some studies used manual translation, focusing more on the linguistic fidelity [3]. Hybrid approaches were also used, where manual and machine translation were combined [2]. Using translated versions of high-resource language datasets is a possible way to mitigate the limitations of datasets for low-resource languages.

2.3 Sinhala Language and NLP

Sinhala is an Indo- Aryan language used by more than 20 million people in Sri Lanka [15]. Despite its widespread use, NLP resources and research focused on Sinhala are significantly under-resourced compared to English and other Indic languages. Most of the studies done for Sinhala NLP aim at understanding the unique morphological and linguistic features of the language. These studies have paved the way for more complex downstream tasks like sentiment analysis, text classification, text summarization, machine translation, optical character recognition tasks, phonological applications, spell and grammar checkers, sign language-related tasks, and chatbots [28]. However, the specific domain of QA systems remains largely unexplored. There have been several QA-related studies in Sinhala. None of them have made the datasets or implementations public.

In 2016, Mahoshadha was introduced, which is considered one of the first studies done in the QA domain for Sinhala [21]. Mahoshadha aimed at retrieving information from a given corpus. A QA model capable of answering arithmetic questions asked in the Sinhala language was introduced in 2018. This was done as an improvement to Mahoshadha [37]. A thematic relations-based QA generator was made for Sinhala in 2020 [16]. As the dataset, 56 sample sentences were used. The sentences had restricted structure and were only able to answer simple questions. Existing models are mostly rule-based methods. Recently, significant research has been conducted on the development of chatbots for Sinhala [5, 14, 18]. More recent work [36], presents a Visual Question Answering (VQA) approach using Knowledge bases.

However, while these systems contribute to conversation QA tasks and VQA, and rule-based QA tasks, they do not address the critical need for textual extractive QA and Reading Comprehension (RC) research. Furthermore, one of the key barriers to developing robust Sinhala QA systems is the lack of publicly available annotated datasets specifically tailored for Sinhala QA tasks. Without such datasets, it is difficult to train and evaluate models effectively for Sinhala text comprehension. Consequently, existing systems fall short in addressing the broader challenge of creating robust, domain-independent, text-based QA/RC systems for Sinhala, highlighting a clear gap in the current research landscape.

3 Dataset Preparation

3.1 Translating SQuAD to Sinhala

SQuAD v1.1 [30] was systematically translated to Sinhala using the Google Cloud Translation API[1]. Context paragraphs and their respective question-answer pairs were extracted and translated in batches to adhere to the API limitations. When the context and associated QA pairs were returned, the original data structure was preserved. A total of 20,000 context paragraphs were translated, resulting in 96,759 question-answer pairs before cleaning and removing noise.

3.2 Post-Processing and Cleaning

Several post-processing and cleaning steps were done to remove noise data and to format and structure the data. First, the noise data was cleaned out, such as zero-width joiners, redundant spaces, and unnecessary punctuation. English characters were left out during translation due to machine translation inaccuracies, therefore, these entries had to be

---
[1] https://cloud.google.com/translate/docs.

removed. Next, the answers were checked to find out-of-context answers. These question-answer pairs were removed. Some answers were in the context span, but the answer start positions had changed compared to the English dataset. Therefore, the answer start positions were calculated again.

Once the cleaning and post-processing were completed, the final dataset contained 16,000 unique QA pairs. There were 16,000 unique questions and 8188 unique context paragraphs. The training set contained 13,500 unique questions, and the test and development sets contained 1250 unique questions each. Erro! A origem da referência não foi encontrada. Shows a summary of the dataset.

Table 1. Summary of dataset statistics.

Split	Contexts	Questions	Answers
Train	5911	13500	13500
Development	1141	1250	1250
Test	1136	1250	1250
Total	8188	16000	16000

4 Dataset Analysis

Several analyses were carried out to assess the special features in the Sinhala dataset.

4.1 Question-Type Analysis

The questions in the test set were clustered into eight groups based on the manually created set of question phrases. What, which, who, and how accounted for about 85% of all questions. When and where followed next and why types of questions had the lowest count. Around 2.5% of the questions were categorized in the other category, which didn't fall into any other predefined categories. Figure 1 shows how the question type count varies in the training dataset. Notably, similar variations in question type distributions are observed in the SQuAD 1.1 dataset [30], reinforcing the common patterns seen across different QA datasets.

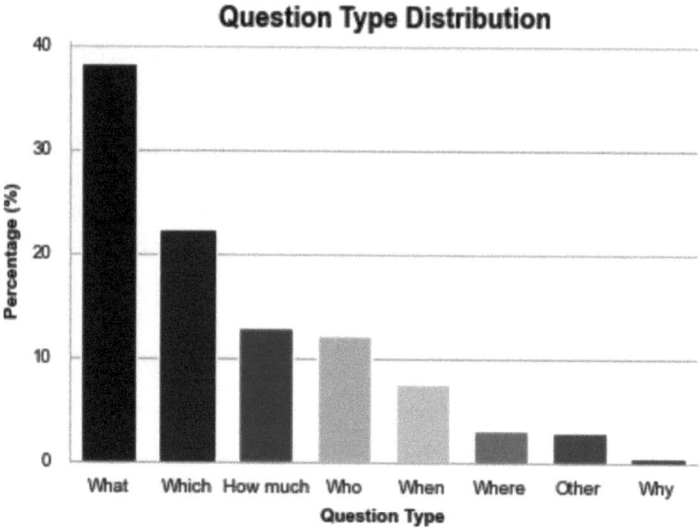

Fig. 1. Proportion of Question Types in the Training Dataset.

A notable characteristic in the Sinhala dataset is the diversity of question phrases. Sinhala exhibits a wide range of syntactic and lexical variations in how questions are phrased. Unlike in English, where the question phrases are mostly fixed like the WH-format (e.g., "what", "how", "who"). Table 2 contains the most common Sinhala question phrases against the English counterpart extracted from the translated dataset. This diversity can complicate QA tasks, as the model must learn to generalize across a broader variety of linguistic structures, compared to languages with more standardized question formats.

Table 2. Different Sinhala Question phrases present in the dataset.

English Question Phrases	Sinhala Question Phrases
What	කුමක්ද, මොනවාද,
When	කවදාද, කෙදිනද
Where	කොහේද, කොතැනින්ද
Why	ඇයි
How much	කොච්චර, කොපමණ, කෙතරම්
Which	කුමන, මොන
Who	කවුද, කාටද, කවුරුන්

4.2 Sequence Lengths

The context lengths in the Sinhala dataset range between 9 and 393 words. The question lengths range between 2 and 33 words, and the answer lengths range from 1 to 23 words.

Compared to the English translations, the sequence lengths had slight variations. Context lengths had a slight difference on average, while answer lengths and question lengths were mostly the same in both datasets. For further reference, Fig. 2 shows a comparison of lengths between the translated and original datasets.

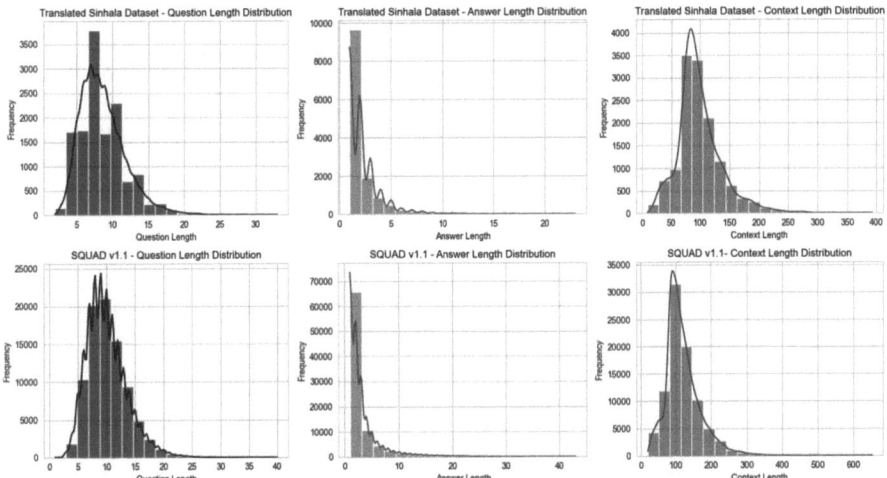

Fig. 2. Question, Context, Answer length distribution plots.

5 Experiments

5.1 Evaluation Metrics

To evaluate model performance on the Translated Sinhala Question Answering dataset, we employ BLEU, ROUGE, Exact Match (EM), and F1 score, ensuring a comprehensive assessment of answer quality.

BLEU (Bilingual Evaluation Understudy) [29] is a measure of similarity between the predicted answer and the ground truth answer by calculating the n-gram precision. Brevity Penalty is incorporated to address the issues with short answers. Since Sinhala is a morphologically rich language, we adapted BLEU to operate at the grapheme level instead of the usual word or subword level. This allowed for more granular evaluation by not penalizing changes in suffixes or prefixes that can occur frequently in languages like Sinhala.

ROUGE (Recall-Oriented Understudy for Gisting Evaluation) [25], assess the textual overlap between predicted and ground truth answers using n-grams, weighted matches, or longest common sequence (LCS). In this study, we used ROUGE-L since it allows flexibility in word order and morphological variations.

Exact Match (EM) strictly checks if the predicted answer is the same as the ground truth answer. Even though EM is useful in measuring the precision, due to the strictness in comparing the answers, this metric is not the best choice to use for Sinhala on its own. Since Sinhala has a diverse vocabulary and different variations of words.

The F1 score is the harmonic mean between precision and recall, which assesses the token-level overlap between the ground truth and the predicted answer. Since EM is highly restrictive, incorporating F1 alongside EM provides a balanced evaluation, capturing both exactness and partial matches effectively.

5.2 Experimental Setup

The Experimental setup consists of three experimental configurations. XLM-R$_{LARGE}$ [12], XLM-R$_{BASE}$ [12], mBERT [22], sinBERT [15], and DistilBERT [34] models were used for the experiments. The models had parameters ranging from 66 million to 550 million parameters (Table 3).

Table 3. Model details.

Model	Language/s	Attention Heads	Parameters	Layers
XLM-R$_{LARGE}$	100	16	550 M	24
XLM-R$_{BASE}$	100	12	270 M	12
DistilBERT	104	12	66 M	6
mBERT	104	12	172 M	12
sinBERT	Sinhala	12	125.9 M	12

XLM-R is a variant of the RoBERTa model [12]. It is trained in 100 languages, including Sinhala. It is among the best-performing models for Sinhala [15]. mBERT is the multilingual variant of the BERT model [22]. It contains 104 languages. However, it is not pre-trained on Sinhala data. DistilBERT[2] is the distilled version of BERT [34]. It is also not pre-trained on Sinhala data. SinBERT$_{LARGE}$ is a pre-trained language model specifically designed for Sinhala text classification. It is based on RoBERTa architecture and is trained on the "sin-cc-15M" corpus [15]. These models were chosen based on their strong performance in multilingual and cross-lingual tasks, their ability to handle low-resource languages like Sinhala. The variety in these models allows for a comprehensive evaluation of performance across multilingual, cross-lingual, and Sinhala-specific scenarios.

The experiments can be grouped into three main setups based on the training data: (1) Monolingual setup: The model is trained and evaluated on the translated Sinhala dataset. (2) Cross-lingual setup: The model is trained on a high-resource language dataset and evaluated on the translated Sinhala dataset. (3) Multilingual setup: The model is trained on a mix of the translated Sinhala dataset and a high-resource dataset and evaluated on the translated Sinhala. Figure 3 illustrates the experimental settings.

---

[2] https://huggingface.co/mrm8488/distilbert-multi-finetuned-for-xqua-on-tydiqa.

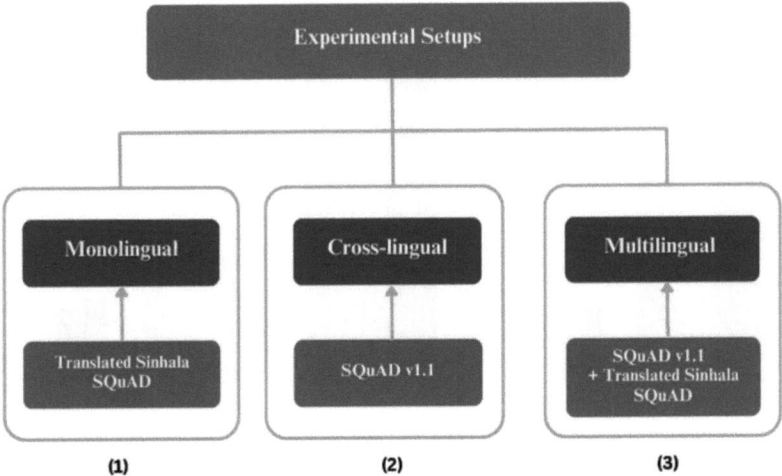

Fig. 3. Experimental Setups.

AdamW [26] is used as the optimizer for all experiments with a weight decay of 0.1, and a learning rate of 5e–6. The batch size was set at 8. All experiments were trained for 5 epochs. The experiments were conducted using the HuggingFace Transformers library [38]. A dedicated A100 GPU was employed for training.

6 Results and Discussion

The experimental results are presented and discussed in the following section.

6.1 Monolingual QA

In this setup, the models are trained on the Sinhala-translated dataset and then evaluated on the translated test and dev sets. During these experiments, the models are exposed only to the Sinhala language. The models used in this setup included XLM-R_{LARGE}, XLM-R_{BASE}, mBERT, sinBERT, and DistilBERT. XLM-R_{LARGE} achieved an F1 score of 73.29% and an EM of 60.08%, the best performance across all models in the monolingual data setup. mBERT had the second-best performance with an F1 score of 33.09%, and then sinBERT and DistilBERT followed. The relatively weak performance of DistilBERT can be attributed to its lack of pretraining on Sinhala data. Figure 4 shows the evaluation results in the monolingual setup. While the results demonstrate that larger multilingual models fine-tuned solely on Sinhala can effectively adapt to Sinhala, the large performance gap between XLM-R_{LARGE} and smaller models highlights the continued importance of model capacity and extensive pretraining for achieving strong results in low-resource languages.

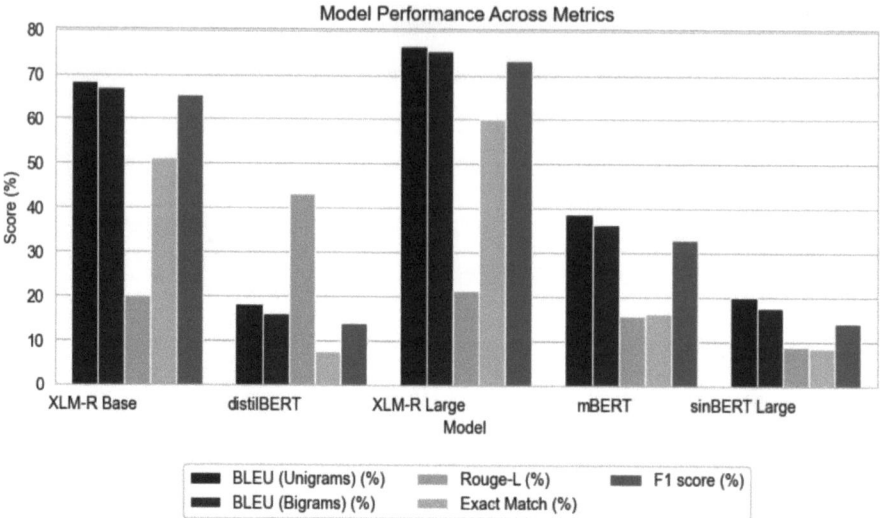

Fig. 4. Performance of models in the Monolingual setup.

6.2 Zero-Shot Cross-Lingual QA

In this setup, the transferability of QA to Sinhala using cross-lingual models was evaluated. Only the best-performing monolingual model was used in this setup. Accordingly, XLM-R$_{LARGE}$ model was trained on a high-resource language, English, and then tested and evaluated on the Sinhala translated dataset. This setup achieved 69% in the F1 score and 51.76% in EM. The results demonstrate the effectiveness of transfer learning in low-resource settings. However, the performance gap compared to the monolingual setup suggests that direct exposure to the target language during training still plays a crucial role, and purely cross-lingual transfer may not fully close the resource gap for complex reasoning tasks.

6.3 Multilingual QA

In this experimental setup, we employed a multilingual training approach by combining the Sinhala-translated SQuAD dataset with a subset of the original English SQuAD 1.1 dataset. To manage computational resources effectively, only 20% of the English SQuAD dataset, approximately 17,520 question-answer pairs, was utilized. For this setup, only the XLM-R$_{LARGE}$ model was used since it was the best-performing model in the monolingual setup. A consistent increase in performance was noticed in this setup. XLM-R$_{LARGE}$ model achieved a 73.52% F1 score and 59% exact match, improving over the monolingual models. This suggests that models trained with multiple languages can leverage shared linguistic features, improving generalizability and transferability. This approach benefits situations where the target language has data scarcity, so they can combine limited data with that of a high-resource dataset and still achieve better performance. Table 4 shows the evaluation results for all the experiments carried out. However, while the gains are notable, the improvements remain modest, indicating that multilingual transfer still faces challenges when large linguistic and structural differences exist between source and target languages.

Table 4. Evaluation Results.

Model	EM	F1	BLEU	ROGUE-L
Translated				
Distil-BERT	7.77	14.08	18.18	43.25
mBERT	16.48	33.09	38.86	15.96
sinBERT	8.72	14.20	20.12	15.96
XLM-R$_{BASE}$	51.28	65.44	68.18	19.99
XLM-R$_{LARGE}$	60.08	73.29	76.50	21.40
SQuAD				
XLM-R$_{LARGE}$	51.76	69.48	72.58	20.95
SQuAD + Translated				
XLM-R$_{LARGE}$	59.44	73.52	76.22	21.37

6.4 Performance Breakdown by Question Type

In this section, we examined how the question type affected the performance of the model. This experiment revealed that depending on the question type, the performance of the model differed. Figure 5 shows the evaluation results per question type. This can be due to the complexity of the reasoning needed for each type of question to be answered.

For the "When" question, the model performed the best, achieving an F1 score of 78%. This can be because the model identifies that the question is regarding temporal information and searches for the relevant information in the context. However, "Why" questions performed the least. This is mainly because these types of questions require additional reasoning.

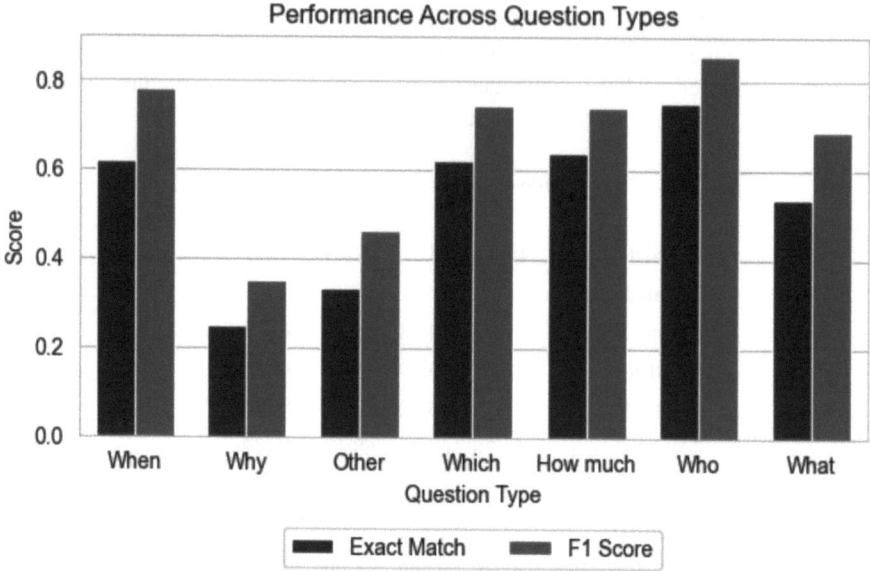

Fig. 5. Performance of Question Types.

"Other" category also performed less, mostly because these questions are comparatively ambiguous. "Which" questions performed well. These questions usually require selecting between options. "How much" questions also did well since these questions usually require numerical answers, which makes it easy for the model to predict the answer. "Who" questions performed the best since it's clear that a person or Named Entity is requested.

Lastly, "What" questions, asking for general explanations, show moderate results (EM: 0.54, F1: 0.69) due to the variety of potential answers. Overall, the model performs best on concrete questions like "Who" and "How much," but struggles with abstract or causal questions like "Why." These differences reflect the varying reasoning requirements for each question type.

7 Open Challenges and Future Directions

A key limitation in this study was the lack of pre-trained models specifically designed for Sinhala. Even though multilingual models like mBERT, and DistilBert have shown success across various languages, they were not pre-trained on Sinhala data. Consequently, the fine-grained details of Sinhala's complex morphology and diverse syntax may not be adequately represented, affecting model performance. As a result, the models used in this study struggled to reach performance compared to English.

Despite the rigorous translation and cleaning process, the dataset presented several challenges. One major issue lies in the inherent differences between Sinhala and English syntax and semantics, which may result in discrepancies during translation and affect model comprehension. During the translation and cleaning process, most of the data was lost due to inconsistencies in the answer spans and indices. This led to a smaller, noisier dataset, further complicating the task.

Additionally, the evaluation metrics used in this study were not specifically tailored for Sinhala QA tasks. Sinhala has various prefixes and suffixes; therefore, using metrics like exact match can be overly strict. Furthermore, common evaluation metrics such as F1 score, BLEU, and ROUGE do not effectively capture the morphological richness of Sinhala, which could lead to inaccurate performance assessments.

To address the limitations and challenges in this study, several avenues for future work can be explored. First, advanced data augmentation techniques for Sinhala, such as semi-supervised learning or synthetic data generation, could help resolve the issues with the dataset size. Building a native seed dataset would not only improve the quality of the dataset but also reduce the noise in the data, as well as improve the naturalness of the dataset.

Future research should also focus on developing pre-trained models specifically for Sinhala. Building robust, language-specific models, like sinBERT [15], would significantly improve the performance of Sinhala QA and other NLP tasks. These resources need to be made accessible to the public so that researchers and practitioners can make use of them for future work and build on top of them.

Additionally, building evaluation metrics focusing on the language nuances in Sinhala needs to be prioritized. Future research should explore extending the dataset to follow the SQuAD 2.0 [32] to have unanswerable questions. This addition would help the model learn to effectively differentiate between questions that have answers and those that do not, thereby improving its ability to handle more diverse and realistic QA scenarios.

8 Conclusion

In this paper, we presented a translated Sinhala Question Answering Dataset, SiQuAD, derived from the English SQuAD v1.1. After thorough cleaning and post-processing, we curated a total of 16,000 unique question-answer pairs. Our experiments showed promising results. The top-performing model XLM-R_{LARGE} achieved an F1 score of 73% and an EM of 59%, showing strong potential but leaving room for further improvement. Additionally, the cross-lingual and multilingual approaches proved to be effective, reinforcing the importance of leveraging multilingual pretraining for low-resource languages like Sinhala. However, there were several challenges and limitations in our study. The translation from English to Sinhala introduced some discrepancies and noise to the dataset due to the syntactic and semantic differences between the two languages. These issues, combined with the limitations of the evaluation metrics not being tailored to Sinhala's morphological complexity, slightly affected model performance. Although the results are promising, they highlight the need for further improvement in data quality, evaluation approaches, and model optimization. This work lays the groundwork for Sinhala QA, providing a dataset and a performance benchmark. We hope this dataset serves as a foundation and a benchmark when it comes to Sinhala, encouraging future research in Sinhala NLP.

Disclosure of Interests. None.

References

1. Abadani, N., et al.: ParSQuAD: machine translated SQuAD dataset for Persian question answering. In: 2021 7th International Conference on Web Research (ICWR), pp. 163–168. IEEE, Tehran (2021). https://doi.org/10.1109/ICWR51868.2021.9443126
2. Aleksa Cvetanović, P.: Tadić: synthetic dataset creation and fine-tuning of transformer models for question answering in Serbian. Telecommun. Forum. (2023). https://doi.org/10.1109/tel for59449.2023.10372792
3. Armengol-Estapé, J., et al.: Are multilingual models the best choice for moderately under-resourced languages? a comprehensive assessment for Catalan. In: Findings of the Association for Computational Linguistics: ACL-IJCNLP 2021, pp. 4933–4946. Association for Computational Linguistics, Online (2021). https://doi.org/10.18653/v1/2021.findings-acl.437
4. Vaswani, A., et al.: Attention is all you need. Neural Inform. Process. Syst. **30**, 5998–6008 (2017)
5. Avishka, W., et al.: A novel conceptual chatbot architecture for the Sinhala language – a case study on food ordering scenario. In: 2022 2nd International Conference on Advanced Research in Computing (ICARC), pp. 254–259. IEEE, Belihuloya (2022). https://doi.org/10.1109/ICARC54489.2022.9753725
6. Bajaj, P., et al.: MS MARCO: a human generated MAchine Reading COmprehension dataset (2018). http://arxiv.org/abs/1611.09268
7. Basaj, D., et al.: LAS: language agnostic system for question answering. In: 2018 Fifth International Conference on Social Networks Analysis, Management and Security (SNAMS), pp. 260–263. IEEE, Valencia (2018). https://doi.org/10.1109/SNAMS.2018.8554469
8. Chen, D., et al.: A Thorough Examination of the CNN/Daily Mail Reading Comprehension Task (2016). http://arxiv.org/abs/1606.02858
9. Shao, C.-C., et al.: DRCD: a Chinese Machine Reading Comprehension Dataset. arXiv: Computation and Language (2018)
10. Choi, E., et al.: QuAC : Question Answering in Context (2018). http://arxiv.org/abs/1808.07036
11. Conneau, A., et al.: Unsupervised cross-lingual representation learning at scale. In: Proceedings of the 58th Annual Meeting of the Association for Computational Linguistics, pp. 8440–8451. Association for Computational Linguistics, Online (2020). https://doi.org/10.18653/v1/2020.acl-main.747
12. Conneau, A., et al.: Unsupervised Cross-lingual Representation Learning at Scale (2020). http://arxiv.org/abs/1911.02116. https://doi.org/10.48550/arXiv.1911.02116
13. Croce, D., et al.: Neural learning for question answering in Italian. In: Ghidini, C., et al. (eds.) AI*IA 2018 – Advances in Artificial Intelligence, pp. 389–402. Springer, Cham (2018). https://doi.org/10.1007/978-3-030-03840-3_29
14. Dasanayaka, C., et al.: Multimodal AI and large language models for orthopantomography radiology report generation and Q&A. ASI **8**(2), 39 (2025). https://doi.org/10.3390/asi8020039
15. Dhananjaya, V., et al.: BERTifying Sinhala -- a comprehensive analysis of pre-trained language models for Sinhala text classification (2022). http://arxiv.org/abs/2208.07864. https://doi.org/10.48550/arXiv.2208.07864
16. Dissanayake, T., Hettige, B.: Thematic Relations Based QA Generator for Sinhala
17. Zhu, F., et al.: Retrieving and Reading: A Comprehensive Survey on Open-domain Question Answering. arXiv.org. (2021)
18. Fernando, W.S., et al.: LawKey — Law Constitution Chatbot. In: 2024 8th SLAAI International Conference on Artificial Intelligence (SLAAI-ICAI), pp. 1–6. IEEE, Ratmalana (2024). https://doi.org/10.1109/SLAAI-ICAI63667.2024.10844946

19. Gaim, F., et al.: Question-answering in a low-resourced language: benchmark dataset and models for Tigrinya. In: Proceedings of the 61st Annual Meeting of the Association for Computational Linguistics, vol. 1, Long Papers, pp. 11857–11870. Association for Computational Linguistics, Toronto (2023). https://doi.org/10.18653/v1/2023.acl-long.661
20. Gupta, D., et al.: A unified framework for multilingual and code-mixed visual question answering. In: Proceedings of the 1st Conference of the Asia-Pacific Chapter of the Association for Computational Linguistics and the 10th International Joint Conference on Natural Language Processing, pp. 900–913 (2020)
21. Jayakody, J.A.T.K., et al.: "Mahoshadha", the Sinhala Tagged Corpus Based Question Answering System, pp. 313–322 (2016). https://doi.org/10.1007/978-3-319-30933-0_32
22. Devlin, J., et al.: BERT: pre-training of deep bidirectional transformers for language understanding. North American Chapter of the Association for Computational Linguistics (2019). https://doi.org/10.18653/v1/n19-1423
23. Joshi, M., et al.: TriviaQA: A Large Scale Distantly Supervised Challenge Dataset for Reading Comprehension (2017). http://arxiv.org/abs/1705.03551
24. Kwiatkowski, T., et al.: Natural Questions: A Benchmark for Question Answering Research
25. Lin, C.-Y.: ROUGE: A Package for Automatic Evaluation of Summaries
26. Loshchilov, I., Hutter, F.: Decoupled Weight Decay Regularization (2019). http://arxiv.org/abs/1711.05101. https://doi.org/10.48550/arXiv.1711.05101
27. Macková, K., Straka, M.: Reading Comprehension in Czech via Machine Translation and Cross-lingual Transfer (2020). http://arxiv.org/abs/2007.01667. https://doi.org/10.48550/arXiv.2007.01667
28. de Silva, N., de Silva, N.: Survey on Publicly Available Sinhala Natural Language Processing Tools and Research. arXiv: Computation and Language (2019)
29. Papineni, K., et al.: BLEU: a method for automatic evaluation of machine translation. In: Proceedings of the 40th Annual Meeting on Association for Computational Linguistics - ACL 2002, p. 311. Association for Computational Linguistics, Philadelphia, Pennsylvania (2001). https://doi.org/10.3115/1073083.1073135
30. Rajpurkar, P., et al.: SQuAD: 100,000+ Questions for Machine Comprehension of Text, pp. 2383–2392 (2016). https://doi.org/10.18653/v1/d16-1264
31. Rejimoan, R., et al.: A comprehensive review on deep learning approaches for question answering and machine reading comprehension in NLP (2023). https://doi.org/10.1109/delcon57910.2023.10127327
32. Rajpurkar, P., et al.: Know What You Don't Know: Unanswerable Questions for SQuAD (2018). http://arxiv.org/abs/1806.03822
33. Sabane, M., et al.: Breaking Language Barriers: A Question Answering Dataset for Hindi and Marathi (2024). http://arxiv.org/abs/2308.09862
34. Sanh, V., et al.: DistilBERT, a distilled version of BERT: smaller, faster, cheaper and lighter (2020). http://arxiv.org/abs/1910.01108. https://doi.org/10.48550/arXiv.1910.01108
35. Trischler, A., et al.: NewsQA: A Machine Comprehension Dataset (2017). http://arxiv.org/abs/1611.09830
36. Wansekara, I., Jayasekara, A.G.B.P.: Intelligent Sinhala question and answer system by incorporating visual clues. In: 2024 4th International Conference on Electrical Engineering (EECon), pp. 95–100. IEEE, Colombo (2024). https://doi.org/10.1109/EECon64470.2024.10841868
37. Chathurika, W.M.T., et al.: Solving Sinhala Language Arithmetic Problems using Neural Networks. arXiv: Computation and Language (2018)
38. Wolf, T., et al.: Transformers: state-of-the-art natural language processing. In: Proceedings of the 2020 Conference on Empirical Methods in Natural Language Processing: System Demonstrations, pp. 38–45. Association for Computational Linguistics, Online (2020). https://doi.org/10.18653/v1/2020.emnlp-demos.6

39. Yang, Z., et al.: HotpotQA: A Dataset for Diverse, Explainable Multi-hop Question Answering (2018). http://arxiv.org/abs/1809.09600
40. Liu, Y., et al.: RoBERTa: A Robustly Optimized BERT Pretraining Approach. arXiv: Computation and Language (2019)

Author Index

A

Al Machot, Fadi   306
Alagha, Ahmed   263
Aldausari, Nuha   96
Alonso, Eduardo   14
Andrianandrianina Johanesa, Tojo Valisoa   325

B

Bagri, Ikram   163
Bansal, Anukriti   217
Bayeh, Ali   229
Benjelloun, Mohammed   240, 280
Betrouni, Nacim   240
Blaha, Martin   82

C

Cheung, Mark   132
Cooper, David   96

D

Dehombreux, Pierre   325
Deng, Yancong   1
Drábek, Jan   82

E

Equeter, Lucas   325
Eryilmaz, Selim Eren   151

F

Fang, Ke   1

G

Goel, Kartikay   217
Guneeth, Palakurthy   217
Gupta, Amish   70
Gupta, Saumilya   217

H

Hafeez, Abdul Basit   14
Hani, Moad   240
Heyden, William   306
Honnavalli, Prasad B.   70
Hosseini, Mahdi S.   263
Hraiba, Aziz   163

I

Izmailov, Rauf   132

K

Kanwal, Preet   70
Kaplun, Dmitrii   28, 51
Khellaf, Abdelhakim   263
Kyrkou, Christos   112

L

Laimiņa, Tamara   204
Lukjanovs, Jegors   204

M

Mahmoudi, Saïd   240
Mahmoudi, Sidi Ahmed   325
Maity, Gouranga   28
Majumder, Siddhant   51
Mandal, Diptarka   51
Mercadier, Mathieu   184
Mohammadi, Gelareh   96
Mouhoub, Malek   229
Mousrij, Ahmed   163
Muntean, Cristina Hava   184

N

Najem, Rabie   280
Noppanamas, Thanakron   294
Novák, Jiří   82

O

Ouardirhi, Fatima Zahra   240

P

Panagi, Andriani 112
Pani, Saptarshi 28
Phooomvuthisarn, Suronapee 294
Porwal, Shrusti 217

R

Ranasinghe, Janani 336
Riaz, Atif 14

S

Sadaoui, Samira 229
Sai Mohananshu, J. 70
Sarkar, Ram 28, 51
Sarkar, Sujan 51
Satish, Naresh Kumar 184
Sharma, Vasudev 263
Siddiqui, Muhammad Salman 306
Sidorina, Daria 51
Simiscuka, Anderson Augusto 184

Singh, Preety 217
Sisojevs, Aleksandrs 204

T

Tahiry, Karim 163
Tatarinov, Alexey 204
Touil, Achraf 163
Triepels, Ron 151
Trinh, Vincent Quoc-Huy 263

U

Ullah, Habib 306

V

Varecha, Jaroslav 82
Venkatesan, Sridhar 132
Voznesensky, Alexander 28

W

Weerasinghe, Ruvan 336

MIX
Papier aus verantwortungsvollen Quellen
Paper from responsible sources
FSC® C105338

If you have any concerns about our products,
you can contact us on
ProductSafety@springernature.com

In case Publisher is established outside the EU,
the EU authorized representative is:
**Springer Nature Customer Service Center GmbH
Europaplatz 3, 69115 Heidelberg, Germany**

Printed by Libri Plureos GmbH
in Hamburg, Germany